U.S. Department of the Interior
U.S. Geological Survey

MINERALS YEARBOOK

Area Reports: International 2012

Asia and the Pacific

Volume III

Vincennes University
Shake Learning Resources Center
Vincennes, In 47591-9986

U.S. DEPARTMENT OF THE INTERIOR
SALLY JEWELL, Secretary

U.S. GEOLOGICAL SURVEY
Suzette M. Kimball, Acting Director

UNITED STATES GOVERNMENT PRINTING OFFICE, WASHINGTON: 2015

Manuscript approved for publication on February 27, 2015.

For more information on the USGS—the Federal source for science about the Earth, its natural and living resources, natural hazards, and the environment:
World Wide Web: http://www.usgs.gov
Telephone: 1–888–ASK–USGS

Any use of trade, product, or firm names in this publication is for descriptive purposes only and does not imply endorsement by the U.S. Government.

Although this report is in the public domain, permission must be secured from the individual copyright owners to reproduce any copyrighted materials contained within this report.

Suggested citation:
U.S. Geological Survey, 2015, Area reports—International—Asia and the Pacific: U.S. Geological Survey Minerals Yearbook 2012, v. III, 224 p.

For sale by the Superintendent of Documents, U.S. Government Printing Office
Internet: bookstore.gpo.gov Phone: toll free (866) 512-1800; DC area (202) 512-1800
Fax: (202) 512-2104 Mail: Stop IDCC; Washington, DC 20402-0001

ISSN 0076-8952
ISBN 978 1 4113 3676 6

Foreword

This edition of the U.S. Geological Survey (USGS) Minerals Yearbook discusses the performance of the worldwide minerals and materials industries during 2012 and provides background information to assist in interpreting that performance. Content of the individual Minerals Yearbook volumes follows:

- Volume I, Metals and Minerals, contains chapters about virtually all metallic and industrial mineral commodities important to the U.S. economy. Chapters on survey methods, summary statistics for domestic nonfuel minerals, and trends in mining and quarrying in the metals and industrial mineral industries in the United States are also included.
- Volume II, Area Reports: Domestic, contains a chapter on the mineral industry of each of the 50 States and Puerto Rico and the Administered Islands. This volume also has chapters on survey methods and summary statistics of domestic nonfuel minerals.
- Volume III, Area Reports: International, is published as four separate reports. These regional reports contain the latest available minerals data on more than 190 foreign countries and discuss the importance of minerals to the economies of these nations and the United States. Each report begins with an overview of the region's mineral industries during the year. It continues with individual country chapters that examine the mining, refining, processing, and use of minerals in each country of the region and how each country's mineral industry relates to U.S. industry. Most chapters include production tables and industry structure tables, information about Government policies and programs that affect the country's mineral industry, and an outlook section.

The USGS continually strives to improve the value of its publications to users. Constructive comments and suggestions by readers of the Minerals Yearbook are welcomed.

Suzette M. Kimball, Acting Director

Contacts

Information about the U.S. Geological Survey, its programs, staff, and products may be accessed on the Internet at http://www.usgs.gov or by contacting the Earth Science Information Center at 1–888–ASK–USGS. For specific information about this publication, contact the Center secretary at (703) 648–4961. Additional minerals information may be accessed on the Internet at http://minerals.usgs.gov/minerals.

Acknowledgments

The Country Specialists in the National Minerals Information Center, U.S. Geological Survey, in preparing the International Review regional books of volume III of the Minerals Yearbook, extensively use statistics and data on mineral production, consumption, and trade provided by various foreign Government minerals and statistical agencies through various official publications. The cooperation and assistance of these organizations are gratefully acknowledged. Statistical and informational material was also obtained from reports of the U.S. Department of State, United Nations publications, and the domestic and foreign technical and trade press. Of particular assistance were reports submitted by the Resource Reporting Officers and other officers of the Department of State in U.S. Embassies worldwide. Their contributions are sincerely appreciated. Internal statistical support was provided by the staff of the Data Collection and Coordination Section of the National Minerals Information Center.

The regimes of some countries reviewed in this volume may not be recognized by the U.S. Government. The information contained herein is technical and statistical in nature and is not to be construed as conflicting with or being contradictory of U.S. foreign policy.

The staff of the National Minerals Information Center gratefully acknowledges the invaluable contributions that John H. DeYoung, Jr., has made to the quality of the Minerals Yearbook during his tenure as the Director of the National Minerals Information Center from January 1996 until June 2013.

                                                Steven M. Fortier
                                                Director, National Minerals Information Center

Contents

Foreword, by Suzette M. Kimball	iii
Contacts	iv
Acknowledgments, by Steven M. Fortier	v
Asia and the Pacific, by Yolanda Fong-Sam, Chin S. Kuo, Lin Shi, Pui-Kwan Tse, Susan Wacaster, and David R. Wilburn	1.1
Afghanistan, by Chin S. Kuo	2.1
Australia, by Pui-Kwan Tse	3.1
Bangladesh, by Yolanda Fong-Sam	4.1
Bhutan and Nepal, by Lin Shi	5.1
Burma, by Yolanda Fong-Sam	6.1
Cambodia, by Yolanda Fong-Sam	7.1
China, by Pui-Kwan Tse	8.1
Fiji, by Lin Shi	9.1
India, by Chin S. Kuo	10.1
Indonesia, by Chin S. Kuo	11.1
Japan, by Chin S. Kuo	12.1
Korea, North, by Lin Shi	13.1
Korea, Republic of, by Lin Shi	14.1
Laos, by Yolanda Fong-Sam	15.1
Malaysia, by Pui-Kwan Tse	16.1
Mongolia, by Susan Wacaster	17.1
New Caledonia, by Susan Wacaster	18.1
New Zealand, by Pui-Kwan Tse	19.1
Pakistan, by Chin S. Kuo	20.1
Papua New Guinea, by Susan Wacaster	21.1
Philippines, by Yolanda Fong-Sam	22.1
Sri Lanka, by Chin S. Kuo	23.1
Taiwan, by Pui-Kwan Tse	24.1
Thailand, by Lin Shi	25.1
Vietnam, by Yolanda Fong-Sam	26.1

The Mineral Industries of Asia and the Pacific

By Yolanda Fong-Sam, Chin S. Kuo, Lin Shi, Pui-Kwan Tse, Susan Wacaster, and David R. Wilburn

The Asia and the Pacific region, which includes 30 countries and territories, has a total area of about 30 million square kilometers, which accounts for about 20% of the world total. The total population of about 3.9 billion accounted for about 55% of the world total in 2012. China and India, which were the world's two most populous countries, accounted for about 67% of the region's total population and about 37% of the world's total population. The economies of Afghanistan, Bhutan, Cambodia, China, East Timor, Laos, Mongolia, and Papua New Guinea were the fastest growing in the region in 2012, with real gross domestic product (GDP) growth rates of more than 7% (tables 1, 2).

Australia and China were among the world's leading mineral producers. Australia has large resources of bauxite, coal, cobalt, copper, diamond, gold, iron ore, lead, lithium, manganese, mineral sands, tantalum, and uranium. China has large resources of antimony, arsenic, barite, bauxite, coal, fluorite, gold, graphite, iron ore, magnesium, rare earths, strontium, tin, tungsten, and zinc. India was also one of the world's significant mineral producers; it has large resources of barite, bauxite, chromium, iron ore, manganese, rare earths, and salt. Other significant mineral producers in the region were Indonesia, which has large resources of bauxite, coal, copper, gold, nickel, and tin; Mongolia, which has large resources of copper, fluorspar, and molybdenum; Papua New Guinea, which has large resources of copper and gold; the Philippines, which has large resources of copper, gold, and nickel; and Thailand, which has large resources of feldspar, gypsum, and potash (table 4).

Despite the large amount and wide variety of resources of nonfuel minerals and coal in Australia, China, India, Indonesia, Mongolia, Papua New Guinea, the Philippines, and Thailand, the regional supplies of numerous nonfuel minerals [including aluminum, bauxite, copper, diamond, gold, iron ore, lead, phosphate rock, platinum-group metals (PGMs), silver, and zinc] and such major mineral fuels as coal, natural gas, crude petroleum, and refined petroleum products, were insufficient to satisfy the demand in the region in recent years. The shortage was caused largely by a substantial increase in the consumption of nonfuel minerals and mineral fuels by China and India; by continued high levels of consumption by such resource-poor industrialized countries as Japan, the Republic of Korea, Singapore, and Taiwan; and by the growing economies of such middle-income developing countries as Indonesia, Malaysia, and Thailand. In 2012, the region of Africa and the Middle East supplied a large percentage of the Asia and the Pacific region's requirements for natural gas, crude petroleum, and refined petroleum products. Africa, North America, and South America supplied a substantial percentage of the region's raw material requirements for ferrous and nonferrous metals.

China and Japan were the two major regional markets for crude and processed minerals. Japan was the region's leading consumer of imported ferrous and nonferrous metals because of its large manufacturing sector and poor indigenous resources. China, however, remained the region's leader in terms of growth in consumption, especially for such mineral commodities as aluminum, cement, coal, copper, iron and steel, lead, natural gas, crude petroleum, phosphate rock, rare earths, tin, and zinc. India, Indonesia, Malaysia, the Republic of Korea, Singapore, Taiwan, Thailand, and Vietnam also were significant consumers of such mineral commodities as aluminum, cement, copper, gold, iron ore, lead, phosphate rock, silver, steel, and zinc.

Acknowledgments

The U.S. Geological Survey (USGS) acknowledges and thanks the following foreign Government agencies, international institutions, and private research organizations for providing mineral production statistics, basic economic data, and exploration and mineral-related information:

For mineral production statistics—
- Australia—Australian Bureau of Agricultural and Resource Economics (ABARE) and Western Australia Department of Minerals and Petroleum Resources;
- Bhutan—Ministry of Trade and Industry, Department of Geology and Mines;
- Brunei—Prime Minister's Department, Petroleum Unit;
- Cambodia—Ministry of Industry, Mines and Energy, Department of Mineral Resources Development;
- India—Indian Bureau of Mines;
- Japan—Ministry of Economy, Trade and Industry, Research and Statistics Department;
- Laos—Ministry of Industry and Handicraft, Department of Geology and Mines;
- Malaysia—Ministry of Natural Resources and Environment, Minerals and Geoscience Department;
- Mongolia—Mineral Resources and Petroleum Authority;
- Nepal—Ministry of Industry, Commerce and Supplies, Department of Mines and Geology;
- Republic of Korea—Korea Institute of Geoscience and Mineral Resources;
- Sri Lanka—Geological Survey and Mines Bureau;
- Thailand—Ministry of Industry, Department of Primary Industries and Mines; and
- Vietnam—Vietnam Institute of Geosciences and Mineral Resources.

For key economic data—
- Asian Development Bank in Manila, Philippines;
- International Monetary Fund in Washington, DC; and
- The World Bank in Washington, DC.

For exploration and other mineral-related information—
- Australian Bureau of Statistics in Canberra, Australia; and
- SNL Metals and Mining (formerly SNL Metals Economics Group) in Halifax, Nova Scotia, Canada.

General Economic Conditions

The global economy remained weak in 2012, but economic activity started to pick up at the end of the year. The economies of countries in the Asia and the Pacific region grew at a respectable rate during 2012, and the region was expected to maintain its momentum of gross domestic product (GDP) expansion during the next 2 years. The global economy showed a more promising outlook in 2013, and the countries of the euro area, Japan, and the United States reported positive growth quarter-on-quarter for 2013 (International Monetary Fund, 2014).

The Chinese Government continued to tighten the availability of credit gradually, and its monetary policy was focused on supporting structural adjustment and promoting reform and transformation of the economy's growth pattern. In 2008–09, the Government authorizeded a large stimulus package to support the development of the industrial sector. The Central Government's debt increased to about 50% of the GDP in 2012 from less than 21% in 2007. Local government debt also increased significantly during the same period. In 2012, the Chinese Government refrained from rolling out another large stimulus package but continued to enact moderate measures in the infrastructure, public housing, and renewable energy sectors to ensure that economic growth did not fall below the target of 7.5%. China's economic outlook remained bright, although the balance between investment-led and consumption-based growth was a significant challenge to the Government. The Chinese Government had plans to reduce state intervention in the economy but faces the challenge of how to implement reforms to restructure the economy toward domestic demand and inclusive growth. The reforms are likely to have a profound effect on the country's capital, labor, and land markets. These changes could transform the country's economy from a social market economy to a market economy in the future (World Bank, The, 2013, p. 28–37).

In India, the slowdown in domestic investment affected the trend towards sustained economic growth in the country. Industrial development in India faced many obstacles, such as delays in land acquisitions and environmental approvals, which hampered investment in mining, and inadequate supplies of coal and water, which affected power generation. Given the gradual economic recovery in Europe and the United States, exports of goods were expected to increase in the next several years, and India's economy is expected to improve. The passage of several bills, including the Land Acquisition Bill, the Companies Bill, the Food Security Bill, and the Pension Bill was expected to help improve the investment climate in India (Asian Development Bank, 2013, p. 191–196).

Japan's economy rebounded from having a negative growth rate in 2011 to a positive rate of growth in 2012 because of a large increase in private consumption. A new set of policies introduced by the Japanese Government included aggressive monetary policy, a short-term flexible fiscal policy with medium-term consolidation, and institutional and regulatory measures designed to strengthen the country's competitiveness with respect to other countries. The Bank of Japan announced that it would leave the asset purchase program unchanged in 2013 but would begin open-ended asset purchases in 2014. The Japanese Government announced a 23 trillion yen (equivalent to 4.8% of the GDP) economic stimulus package to ease the access to credit for farmers and small- and medium-sized enterprises. The Japanese Government planned to increase the consumption tax to 8% from 5% in April 2014, which would likely affect consumer spending, and introduced a blueprint for reducing the corporate tax. Japan had the oldest population in the region, and the decrease in the share of the population that was of working age was expected to affect the country's economic development. For example, the country's construction sector was facing labor shortages (Citigroup Global Markets Inc., 2013, p. 44).

Weak external and domestic demand continued to affect economic growth in the Republic of Korea. Exports were expected to remain the main driver of economic growth, and the improved overseas economic conditions were likely to increase the domestic demand. The Bank of Korea delayed tightening of its monetary policy to support the country's economic growth (Asian Development Bank, 2013, p. 165–168).

Economic growth among the member countries of the Association of Southeast Asian Nations (ASEAN) (Brunei, Burma, Cambodia, Indonesia, Laos, Malaysia, the Philippines, Singapore, Thailand, and Vietnam) plummeted during the worst of the global financial crisis in 2009 but rebounded quickly during the next several years. Robust domestic demand was a key factor in the region's ability to weather the global economic crisis. Before the crisis, these countries depended on exports of goods to developed countries in Europe and to the United States. During the past decade, however, the trade structure had changed. Total trade within ASEAN and with China and India increased by much more than did trade with the traditional partners. China was ASEAN's leading trading partner, and India had also become a significant trading partner with ASEAN. Consequently, the future economic development in these countries is likely to be affected by the economic activities in China and India. The current and projected slowdown of economic activity in China is expected to influence ASEAN's growth outlook. To capitalize on ASEAN's intensive production network, China and India were promoting the free flow of goods, investments, services, and skilled workers among China, India, and the ASEAN countries by 2020 (Asian Development Bank, 2013, p. 11–14).

Mongolia's economic growth rate was again greater than 10% in 2012. This continued growth was attributable mainly to growth in the mining sector—this sector accounted for more than 80% of export revenue and about 17% of Government revenue. The country was facing a shortage of skilled workers and high inflation. China was receiving more than 90% of Mongolia's mineral exports, and thus Mongolia's economic growth was expected to depend on stable global mineral commodity prices and continued strong economic growth in China. Mongolia's mining sector was expected to continue to grow slowly during the next several years (Asian Development Bank, 2013, p. 170–173).

Legislation

China's rapid, capital-intensive, export-oriented growth had been successful during the past three decades; however, the global markets it relied on were expected to be weaker in the future. The country's economic pattern of growth was energy- and natural-resource intensive and environmentally unsustainable. The constrained supply of major mineral commodities and environmental degradation were limiting the country's ability to maintain its past level of economic growth. The Government indicated that it intended to support and build a more energy-efficient and ecologically friendly society by upgrading the value chain in manufacturing while enhancing innovation and promoting the development of new strategic industries. Its plan to reduce carbon emissions was to be focused on the energy-intensive sectors, such as cement, chemicals, iron and steel, and nonferrous metals. The Government stated that the country's economic growth was expected to be less dependent on export markets, such as those of Europe and the United States, and that it was planning to transform the economy from one that is export focused to one that is consumer driven.

In Indonesia, the mining law enacted in 2009 prohibits the export of unprocessed minerals from Indonesia beginning on January 12, 2014. The purpose of the new provision is to increase the value of commodity exports and encourage development of the mineral processing and smelting sector in the country. Halting exports of unprocessed minerals could affect the global mineral sector in the next several years. The Indonesian Government planned to modify the guidelines by allowing exports of concentrates of copper, iron, lead, and zinc if companies commit to build smelters in the country by 2017. Minerals that have to be refined before export are bauxite, chromium, gold, nickel, silver, and tin. Mineral exports accounted for about 5% of the country's total exports.

The Indonesian Government required that foreign companies reduce their stakes in mines by the 10th year of production so that domestic ownership is at least 51%. Consequently, the Government was renegotiating existing contracts with Freeport McMoRan Copper & Gold Inc. and Newmont Mining Corp. (both of the United States) to adjust the ownership percentage. Freeport McMoRan owned 90.64% of the joint venture that operated the Grasberg copper and gold mine and the Government owned 9.36%; the company agreed to divest a 9.36% interest to a potential acquirer, such as the Province of Papua. Rio Tinto plc of the United Kingdom was expected to retain the rights to 40% of the mine production from the Grasberg Mine to 2021. Newmont Mining owned a minority stake in PT Newmont Nusa Tenggara in 2009. Newcrest Mining Ltd. of Australia, which owned an 82.5% stake in the Gosowong Mine (Antam owned the remaining 17.5%), would not be affected until its existing contract of work (COW) runs out in 2029. Kingsrose Mining Ltd. of Australia under its existing COW was supposed to start selling down its 85% stake in PT Natarang Mining to 49% beginning in 2012. Intrepid Mines Ltd., also of Australia, which had a mining concession for the Tujuh Bukit copper-gold-silver mine in East Java Province, might also be affected.

In 2012, the Government of Vietnam approved several decisions and decrees; these included ones that supported the implementation of the 2010 Mineral Law Directive 02/CT-TTg (Directive 02), Decree no. 15/2012/ND-CP (Decree 15), and Decree No. 22/2012/ND-CP (Decree 22). These directives were intended to help to attract international investor interest in the country's emergent mineral industry. Also, the Government released Circular No. 41/2012/TT-BCT (Circular 41), which provides the guidelines for the export of minerals, went into effect on February 4, 2013, . The circular includes a list of minerals allowed to be exported, the specific quality of the material (that is, the percentage of the contained mineral), and the conditions under which the minerals may be exported.

Exploration

Australia's mineral exploration spending (for minerals and mineral fuels) for fiscal year 2013 (July 2012 to June 2013) was A$7.8 billion (US$7.4 billion), which was an increase of 9.9% from that of fiscal year 2012. The increase was accounted for largely by expenditures on oil and gas exploration, which increased to A$4.79 billion (US$4.55 billion). Exploration spending for iron ore, on the other hand, decreased by 13% to A$1.01 billion (US$960 million); gold, by 14% to A$662 million (US$628 million); coal, by 35% to A$544 million (US$517 million); and uranium, by 55% to A$70 million (US$66 million). The State of Western Australia accounted for more than 50% of the total exploration spending of Australia followed by Queensland, South Australia, the Northern Territory, and New South Wales. Brownfields exploration spending accounted for A$1.91 billion (US$1.82 billion) of the total (Bureau of Resources and Energy Economics, 2014, p. 169).

In 2012, China's exploration spending increased to $21.0 billion from $17.7 billion in 2011, of which oil and gas accounted for $12.5 billion and the nonfuel sector accounted for $7.5 billion. The Ministry of Land and Resources (MLR) announced that 10 large mineral resource provinces had been discovered during the past several years. These mineral resource provinces included copper in Xizang; gold in Dachang; iron ore in Awulale; nonferrous metals in Nyainqentanghla, Qimantage, Tianshan, and Yunnan; potash in Lop Nor Lake; and uranium in North China. At yearend 2012, China's mineral resources increased for coal to 55.7 billion metric tons (Gt); iron ore, 3.7 Gt; oil, 1.5 Gt; bauxite, 210 million metric tons (Mt); zinc, 6.4 Mt; lead, 3.4 Mt; copper, 3.2 Mt; molybdenum, 1.7 Mt; and natural gas, 961 billion cubic meters. The MLR encouraged companies to explore for and develop bauxite, copper, and iron ore deposits because China depended on imports of these mineral commodities. The Chinese Government would provide special funding for these projects (Ministry of Land and Resources, 2013, p. 9).

Based on the data compiled by the USGS, the three countries in Southeast Asia with the largest number of exploration sites were, in order of the number of active exploration sites, Indonesia, Papua New Guinea, and the Philippines, which together accounted for 66% of the active exploration sites in the region. Other countries that were active in exploration in 2012 include Burma, Cambodia, Fiji, Indonesia, the Republic of Korea, Laos, Malaysia, New Caledonia, New Zealand, the Solomon Islands, and Vietnam. Exploration for gold and silver accounted for approximately 60% of all exploration investment in the Pacific region, and exploration for base metals accounted for about 40%; a minor amount of exploration activity was also conducted for iron ore, PGMs, and other minerals in 2012. Additionally, a number of sites were determined to have reached the feasibility stage and to have potential for development (table 3).

Commodity Overview

The estimates for the production of major mineral commodities for 2015 and beyond have been based upon a supply-side assumptions, such as announced plans for increased production, new capacity construction, and bankable feasibility studies. The outlook tables in this summary chapter (tables 5 through 20) show historic and projected production trends; therefore, no indication is made about whether the data are estimated or reported, and revisions are not identified. Data on individual mineral commodities in tables in the individual country chapters are labeled to indicate estimates and revisions. The outlook segments of the mineral commodity tables are based on projected trends that could affect current producing facilities and on planned new facilities that operating companies, consortia, or Governments have projected to come online within the indicated timeframes. Forward looking information, which includes estimates of future production, exploration and mine development, the cost of capital projects, and timing of the start of operations, is subject to a variety of risks and uncertainties that could cause actual events or results to differ significantly from expected outcomes. Projects listed in the following section are presented as an indication of industry plans and are not a USGS prediction of what will take place.

Metals

Aluminum and Bauxite and Alumina.—The region's production of bauxite accounted for about 67% of the world total in 2012. Australia, which was the world's leading producer of bauxite, accounted for about 30% of the world total; it was followed by China (18%) and Indonesia (11%). Production of aluminum accounted for about 52% of the world total in 2012. China, which was the world's leading producer of aluminum, accounted for about 44% of the world total; it was followed by Australia and India (3% each) (table 4).

Regional production of bauxite and aluminum was expected to continue to increase at an average annual rate of about 2.6% for bauxite and 3.2% for aluminum between 2012 and 2019 (tables 5, 6). The projected figure for bauxite is based on reported capacity expansions in Australia, China, and Vietnam, and that for aluminum is based on projected capacity expansions in China.

Owing to increased demand for alumina in the world in recent years, Australian alumina producers planned to expand their refineries' output capacities. Bauxite Resources Ltd.'s Darling Range North project was expected to start production in 2013. Rio Tinto Alcan was conducting a feasibility study and an environmental impact study to develop the bauxite resource in an area south of the Embley River and the existing Weipa Mine.

The new operation would progressively replace depleted resources at the Andoom and the East Weipa mining areas in Weipa. The new development would increase output capacity to 50 million metric tons per year (Mt/yr) from the current 23 Mt/yr in the region south of the Weipa Peninsula, and it could extend the mine life in the area by 40 years. Whether or not Rio Tinto Alcan moves forward with the project depends on the supply of bauxite in the region. China's dependence on bauxite imports was expected to continue, and although Indonesia had been a source of bauxite for China, the Indonesian Government introduced regulations in 2012 to restrict the export of raw materials. As a result, the supply of bauxite in the Asia and the Pacific region could be uncertain. Once all major Government approvals have been granted, Rio Tinto Alcan's decision about whether to proceed with the project south of the Embley River will likely depend on market conditions at that time.

The government of the State of Queensland reopened the bidding for the development rights of the Aurnkun bauxite deposit in 2012 and subsequently finalized a short list of five bidders. They included Australian Indigenous Resources Ltd., Cape Alumina Consortium, Aluminum Corporation of China Ltd. (Chalco) of China, Glencore International plc of Switzerland, and Rio Tinto. The shortlisted companies had until September 2013 to submit detailed proposals for the development of the deposit. The right to mine would not be tied to the requirement to build an alumina refinery in the area.

In China, several bauxite mines were under construction in the Provinces of Guizhou, Henan, and Shanxi. China's bauxite production capacity was expected to increase by 7 Mt by 2017. Geologists continued to explore for bauxite resources in these Provinces and discovered several significant deposits in the Provinces of Guizhou and Henan that had bauxite resources of more than 1 Gt combined.

Vietnam National Coal Mineral Industries Holding Co. Ltd. completed the construction of the Tan Rai alumina and bauxite complex in 2012. Bauxite output was expected to be sufficient for production of 600,000 metric tons per year (t/yr) of alumina, which would meet the demand in Vietnam.

Despite the Chinese Government's macroeconomic policy regarding investment in certain commodities, the output of aluminum metal continued to increase rapidly. To take advantage of investment incentives offered by local governments, many aluminum companies, including Chalco, China Power Investment Corp., Shandong Xinfa Group, Tiashan Aluminum-Power Co. Ltd., and Zhonghe Aluminum Co. Ltd., moved some of their operations to the northwestern part of the country. The government of Xinjiang Uygur Autonomous Region urged enterprises to develop an integrated coal-power-metallurgy industry in the region. Primary aluminum output capacity in Xinjiang increased to 2.3 Mt/yr in 2012 from 50,000 t/yr in 2007; Qinghai Province, to 2.2 Mt/yr from 1 Mt/yr; Gansu Province, to 2.1 Mt/yr from 96,000 t/yr; and Shandong Province, to 4.9 Mt/yr from 2.1 Mt/yr. Domestic analysts estimated that about 5.3 Mt/yr of output capacity would be installed in 2013, of which Xinjiang would add 2.1 Mt/yr and Shandong would add 1 Mt/yr. The aluminum output capacity in Xinjiang was expected to increase to 13 Mt/yr by yearend 2015.

China's aluminum output capacity was expected to increase to more than 35 Mt/yr by 2019.

Cobalt.—The region's mined cobalt output accounted for about 14% of the world total in 2012. Most of the region's cobalt is produced as either a byproduct or coproduct of nickel operations. Regional production of mined cobalt was expected to continue to increase at an average annual rate of about 3.5% between 2012 and 2019 (table 7). In Papua New Guinea, the Ramu nickel/cobalt project started production in 2012 and was expected to reach its full production capacity in 2016. The Coral Bay Nickel Corp. planned to expand its Rio Tuba Nickel Mine production capacity at Palawan in the Philippines.

China was the leading cobalt producing country in the region. China had limited cobalt resources and was required to import a large quantity of cobalt concentrates to support the development of the cobalt battery sector. In 2012, China imported 166,491 metric tons (t) of cobalt concentrates and 9,979 t of unwrought cobalt. The cobalt content in these imported concentrates was about 6%, which was equal to about 10,000 t of cobalt. Domestic analysts estimated that the supply of cobalt in the Chinese market was about 37,000 t in 2012. The consumption of cobalt in batteries increased to 67% of the total cobalt consumption in 2012 from 50% in 2008; the total estimated consumption of cobalt was about 31,000 t in China. The slow growth of the domestic and global economies contributed to the decrease of cobalt consumption in China. As the global economy recovers during the next several years, the demand for cobalt batteries is expected to increase sharply. China is a major world producer of batteries for electronic devices, and the country is expected to increase its imports of cobalt concentrates and metal to meet its demand in the future.

Copper.—The region's production of mined copper accounted for about 20% of the world total in 2012. China was the leading regional producer followed by Indonesia and Australia. Production of primary refined copper accounted for about 43% of the world's total output in 2012. China was the leading world and regional producer of primary and secondary refined copper. Australia, India, Japan, and the Republic of Korea were also significant producers of refined copper in the region (table 4).

Between 2012 and 2019, regional production of mined and refined copper was expected to continue to increase at an average annual rate of about 5.9% and 2.6%, respectively. This estimate is based on reported capacity expansions of mined copper in Australia, China, Mongolia, and Pakistan, and on reported capacity expansions of refined copper in China (tables 8, 9).

Australia's copper mine production was expected to increase at an annual rate of less than 5%. Operations that were expected to increase the country's copper mine output included CuDeco Ltd.'s Rocklands project, Golden Cross Resources' Copper Hill project, and MMG Ltd.'s Golden Grove project. Australia was expected to export more than 50% of its output of copper concentrates.

In Mongolia, Oyu Togoi LLC's Oyu Tolgoi mining complex planned to produce at least 544,000 t/yr of copper in concentrates for the first 10 years of operation starting in 2013. The Erdenet Oovo Mine, which was jointly owned by Erdenet Mining Corp. and Samsung Corp. of Japan, was scheduled to start production in 2012 and would reach full operational capacity in 2013. Mongolia was expected to export more than 90% of its output of copper concentrates, probably to China. Also, Tethyan Copper Co. in Pakistan planned to start producing 200,000 t/yr of copper in 2015.

In 2012, China's copper smelting and refining output capacities increased by 350,000 t and 930,000 t, respectively, and reached 4.4 Mt/yr and 7.9 Mt/yr, respectively. Several greenfield and brownfield copper projects were recently completed or were under construction, and the country's refining output capacity was expected to increase to about 9 Mt in 2017. The output of domestic mined copper was also expected to increase. Copper resources discovered in the western part of the country in Xinjiang Uygur Autonomous Region and Xizang Autonomous Region were expected to be put into development during the next several years. Copper resources in the Gangdise metallogenic belt in Xizang and the Tishan area in Xinjiang would be developed during the next several years. Significant copper resources were discovered recently in the southwestern part of the country and were expected to help replace depleted copper resources in the eastern part of the country.

Gold.—The region's production of mined gold accounted for about 31% of the world total in 2012 (table 4). China was the leading producer of mined gold in the region, followed by Australia and Indonesia (table 10). China and Australia ranked first and second, respectively, in the world production of gold. Indonesia and Papua New Guinea also were significant gold producers in the region. Between 2012 and 2019, regional production of mined gold was expected to continue to increase at an average annual rate of about 3.0%.

Australian gold production was expected to increase in the next few years. Citigold Ltd.'s Charters Town project and YTC Resources Ltd.'s Hera project were expected to be brought onstream during the next several years. In China, gold production was expected to increase by only about 2.5% per year, as the country was unlikely to have the strong production growth rate of the past several years. Many of China's gold mines had relatively high production costs and were small in scale. Gold production in Mongolia could increase because of access to exceptionally high-grade ore from the Oyu Tolgoi copper-gold mine, and output was expected to remain steady after 2013. The increase in gold production in Papua New Guinea would depend upon when the Frieda River Mine and the Yandera Mine are put into operation.

The Asia and the Pacific region was the world's major market for gold and accounted for about 54% of the world's total gold consumption in 2012. India was the world's leading consumer of gold, accounting for about 27% of the world total. Owing to continued strong economic growth and rising urban incomes (which led to increased demand for gold jewelry), China was the second-ranked gold consumer, and its share of the world total increased to about 26%. Jewelry accounted for about 90% of the world's gold consumption. The growth in consumption in the Asia and the Pacific region is attributable to increased gold jewelry demand in China and India. Rising incomes in the region were expected to increase the affordability of jewelry and other fabricated gold products. China and India were likely to continue to be the driving forces behind the increase in jewelry demand. China was likely to overtake India as the leading gold consuming country in 2013 because the Indian Government had issued regulations to limit demand, including establishing higher import duties and a quota on imports (World Gold Council, 2012, p. 21).

Iron and Steel.—The region's production of iron ore was estimated to account for, in terms of gross weight, about 70% of the world total in 2012. China ranked first in the world in the production of iron ore (in terms of iron content), and Australia and India ranked second and third in the region. The region's production of crude steel was estimated to account for about 66% of the world total. China, which was by far the world's leading producer of crude steel, accounted for about 46% of the world total, and Japan, India, and the Republic of Korea ranked among the top 10 producing countries in the world. China's crude steel output was more than the combined production total of, in order of output, Japan, the United States, Russia, India, and the Republic of Korea (table 4).

Regional production of iron ore and crude steel was expected to increase at an average annual rate of about 2.9% and 3.2%, respectively, between 2012 and 2019 (tables 11, 12). China was expected to lead in the expansion of crude steel production in the region.

East Asian countries, such as China, Japan, and the Republic of Korea, were the world's leading consumers and importers of iron ore. Australia was the region's and the world's leading supplier of iron ore. Australian iron ore producers had been expanding their iron ore production facilities during the past several years to meet expected increased demand from Australia's neighboring countries, and this increased investment in mine expansions and new mines was expected to support strong growth in iron ore exports from Australia. A number of greenfield and brownfield iron ore projects were at various stages of development. Rio Tinto Ltd.'s expansion of the capacity of its Pilbara iron ore operations to a total of 360 Mt/yr was scheduled to be completed in 2015. BHP Billiton Ltd. was expected to increase its production capacity to 220 Mt/yr in 2014. Fortescue Metals Group's Chichester Hub and Solomon Hub expansion projects were projected to increase the company's iron ore output capacity to 155 Mt/yr in 2014. In China, the Government granted mining licenses for the development of iron ore mines in the Provinces of Anhui, Liaoning, and Sichuan, and in Xinjiang Uygur Autonomous Region.

India's iron ore output increased sharply to meet domestic and regional demand. India was China's third-ranked iron ore supplier after Australia and Brazil and was expected to remain in that position for the next several years. Brazil and South Africa would also continue to be major iron ore suppliers to the region. Imports of iron ore by East Asian countries were expected to continue to increase, especially imports by China to meet its steel industry's demand.

The Chinese Government continued its effort to curb the fast growing crude steel output capacity in the country, including by ordering iron and steel producers to phase out obsolete facilities. Major iron and steel enterprises continued to expand their output capacity, however, either by replacing smaller furnaces with larger furnaces or by relocating to coastal areas and building steel complexes with larger output capacities. Anshan Iron and Steel (Group) Co. built greenfield iron and steel production facilities at Bayuquan in Yingkou and Chaoying, which increased the company's output capacity by more than 8.5 Mt/yr during the past 5 years. Shougang and Tangshan Iron and Steel Co. jointly built a 10-Mt/yr iron and steel complex in Hebei Province. The Government approved Baogang and Wuhan Iron and Steel Group's construction of two iron and steel complexes in the Provinces of Guangdong and Guangxi, respectively. The Government also approved the merger of 12 privately owned iron and steel producers in Hebei Province to form a large iron and steel enterprise, Tangshan Bohai Iron and Steel (Group) Co. Ltd., which proposed to build a 15-Mt/yr iron and steel complex in the coastal area of Hebei. The government of Shandong Province planned to build a 17.5-Mt/yr iron and steel complex at Dongjiakou.

Nickel.—The region's production of mined nickel and refined nickel accounted for about 34% and 41%, respectively, of the world total in 2012. The Philippines was the leading mined-nickel-producing country in the region, followed by Australia and Indonesia (table 13). Between 2012 and 2019, regional production of mined nickel was expected to continue to increase at an average annual rate of about 3.7%.

The increase in the production of mined nickel took place mainly in Indonesia, New Caledonia, and Papua New Guinea. The Koniambo nickel-cobalt mine in New Caledonia was scheduled to start production in 2013. The Ramu nickel-cobalt mine in Papua New Guinea started operations in 2012 and was expected to reach its full output capacity in 2014. Also, the Tagaung Taung nickel mine in Burma started production in 2012. Australia's nickel mine production was forecast to decrease because several unprofitable operations were scheduled to shut down. The Indonesian Government banned the export of raw materials from the country, which could affect the development of Indonesia's nickel industry in the country and could lead to the shutting down of some operations.

Nickel is an important raw material for stainless steel production. China was the leading stainless-steel-producing country in the world. In recent years, China's nickel pig iron production had increased significantly. Nickel pig iron is produced by smelting low-grade nickel ore (laterite, which contains 1% to 2% nickel) as a substitute for conventional refined ferronickel (which contains 20% to 40% nickel) for stainless steel production. Indonesia and the Philippines were major suppliers of laterite ore to China for production of nickel pig iron. India's nickel consumption is expected to increase during the next several years to support projected growth in India's iron and steel industry.

Tin.—The Asia and the Pacific region was the dominant producer of mined tin and tin metal in the world. Production of mined tin and refined tin accounted for 65% and 83%, respectively, of the world total in 2012. China ranked first in the world in the production of mined tin and refined tin, and Indonesia ranked second. Other significant refined tin producers in the region were Malaysia and Thailand (table 4). Regional production of refined tin was expected to continue to increase at an average annual rate of about 2.3% between 2012 and 2019 (table 17). Tin was used principally in the manufacturing of electronics, glass, iron and steel, and packaging.

China, which was the world's leading consumer of tin metal, accounted for about 40% of the world's total consumption. The region as a whole consumed about 50% of the world's total output of tin metal. Indonesia was the leading tin exporting country in the region. China's tin smelting capacity exceeded its mine output capacity; therefore, the country imported tin concentrates from such countries as, in order of volume (tonnage) of imports, Australia, Burma, Bolivia, Malaysia, Laos, and Vietnam to meet its needs. Owing to increased domestic tin consumption, the export volume of tin from China was expected to decrease in the future. Likewise, the Indonesian Government banned the export of raw material and, as a result, the volume of tin exports from Indonesia to the world market also decreased. Burma was expected to replace Indonesia as the leading mined tin exporting country in the region in the future.

Industrial Minerals

Lithium.—Lithium is the lightest metallic element, and it is widely used in the manufacture of batteries and electronics. Australia was the leading lithium producer in the Asia and the Pacific region. Talison Minerals Group's lithium operation is located at Greenbushes in Western Australia. Talison planned to increase the output capacity of its processing plant to 740,000 t/yr of lithium concentrates in 2012. Windfield Holding Pty Ltd. [a subsidiary of Chengdu Tianqi Industry (Group) Co. Ltd. of China] agreed to acquire the balance of the ordinary shares of Talison that it did not already own as well as stock options in Talison for C$7.50 (US$7.90) per share. In China, lithium was produced from brine and spodumene. Qinghai CITIC Guoan Technology Development Co. Ltd., Tibet Mineral Development Co. Ltd., and Xinjiang Haoxin Lithium Salt Development Co. Ltd. were the major lithium producers, and the Government approved Chengdu Tianqi to develop the Ya'an lithium mine in Sichuan Province. These companies all planned to expand their output capacities, although high production costs and low recovery rates forced the companies to source raw material from overseas. Australia accounted for about 80% of China's lithium consumption. Overall regional lithium production was expected to increase at an average annual rate of about 1% between 2012 and 2019 (table 19).

Mineral Fuels

Coal.—The Asia and the Pacific region's overall production of coal, which included anthracite, bituminous, and lignite, accounted for about 60% of the world total in 2012. Production of anthracite coal, however, accounted for about 94% of the world total, and production of bituminous coal accounted for about 68%. China, which was by far the world's leading producer of anthracite and bituminous coals, accounted for about 68% and 49%, respectively, of the world total. Australia, India, and Indonesia were the other significant coal producers in the region (table 4). China overtook Japan to become the world's leading importer and the region's leading consumer of coal. Japan and the Republic of Korea imported virtually all the coal required by their iron and steel and utility industries. Australia and Indonesia ranked as the world's leading coal exporters. The major regional coal exporters were Australia, Indonesia, and Vietnam.

Overall regional coal production was expected to increase at an average annual rate of about 2% between 2012 and 2019 (table 20). Australia's export of coal is expected to increase because of the continued strong demand from China and India. The additional output would be made possible by the completion of expansions at GlencoreXstrata's Rolleston Mine and Peabody Energy's North Goonyella and Middlemount operations.

The Indonesian Government had indicated that it would limit coal production in order to preserve resources, and that it planned to increase coal production by only 5% beginning in 2015. The Government intended to expand domestic coal consumption, however, and coal-fired powerplants were under construction. Thus, the increased production would supply mainly the domestic market, and the country's coal exports were expected to increase only slightly in the future.

Trade Review

During the past three decades, the main source of economic growth in the Asia and the Pacific region has shifted from the export of manufactured goods toward the export of machinery. This shift was initially led by Japan, followed by the newly industrialized economies of Hong Kong, the Republic of Korea, Singapore, and Taiwan, and, more recently, by Indonesia, Malaysia, the Philippines, and Thailand. Trade liberalization and investment policy reforms in developing countries in the region have reduced barriers to trade and investment. Both the cross-border transshipment of production components and assembly within the region increased during the past several years, and the composition of exports was shifted toward intermediate goods. The share of parts and components in manufactured imports also was trending upward in the region. By 2012, the volume of imports of parts and components had more than doubled in China, the Philippines, Thailand, and Vietnam. China had become one of the major export destinations for all economies in the region.

Outlook

The global economy is expected to improve during the next several years, and the Asia and the Pacific countries are expected to benefit from that recovery. The Chinese Government's long-term development policy is focused on higher quality economic growth that includes improving air and water quality and reducing the growth of greenhouse gas emissions. The Chinese Government plans to promote market-based reform by setting interest rates and foreign exchange rates and instituting tax reform. The Government plans to continue with the expansion of industrial production and infrastructure development in China. As a result, the demand for energy and minerals is expected to continue to increase at a moderate rate during the medium term. Australia, which is the leading exporter of mineral commodities, is likely to benefit the most from increased demand for minerals from China. The Indonesian Government's restrictions on the export of raw materials could affect the country's economic growth during the next several years and could cause mining companies to phase out their operations in Indonesia.

Economic growth in Mongolia will depend greatly on the development of its mineral industry. Foreign direct investment in both the mining and nonmining sectors in the country is expected to increase following the Parliament's enactment of a new investment law in 2013. The Parliament also plans to amend the Mineral Law and to issue guidelines for implementing the Mineral Law, the Petroleum Law, and the Law on Specially Protected Lands in 2014. The Government of Mongolia appears to be taking a pragmatic approach to promoting a more favorable environment for foreign investors to invest in the mining sector. The construction of railways between China and Mongolia is planned reduce the cost of shipping coal and copper ore to China from Mongolia.

References Cited

Asian Development Bank, 2013, Asian development outlook 2013: Manila, Philippines, Asian Development Bank, 306 p.

Bureau of Resources and Energy Economics, 2014, Resources and energy quarterly—Australian private mineral exploration expenditure: Canberra, Australian Capital Territory, Australia, Bureau of Resources and Energy Economics, 206 p.

Citigroup Global Markets Inc., 2013, Global economic outlook and strategy: New York, New York, Citigroup Global Markets Inc., December 2, 80 p.

International Monetary Fund, 2014, World economic outlook update: Washington, DC, International Monetary Fund press release, January 21, 3 p.

Ministry of Land and Resources, 2013, China land and mineral resources report 2012: Beijing, China, Ministry of Land and Resources, April, 46 p.

World Bank, The, 2013, World Bank East Asia and Pacific economic update—Rebuilding policy buffers, reinvigorating growth: Washington, DC, The World Bank, October, 133 p.

World Gold Council, 2014, Gold demand trends—Full year 2013: London, United Kingdom, World Gold Council, February, 30 p.

TABLE 1
ASIA AND THE PACIFIC: AREA AND POPULATION IN 2012

Country	Area[1] (square kilometers)	Estimated population[2] (thousands)
Afghanistan	652,230	29,824
Australia	7,741,220	22,683
Bangladesh	143,998	154,695
Bhutan	38,394	741
Brunei	5,765	412
Burma	676,578	52,797
Cambodia	181,035	14,864
China	9,596,961	1,350,695
Fiji	18,274	874
India	3,287,263	1,236,686
Indonesia	1,904,569	246,864
Japan	377,915	127,561
Korea, North	120,538	24,763
Korea, Republic of	99,720	50,004
Laos	236,800	6,645
Malaysia	329,847	29,239
Mongolia	1,564,116	2,796
Nepal	147,181	27,474
New Caledonia	18,575	258
New Zealand	267,710	4,433
Pakistan	796,095	179,160
Papua New Guinea	462,840	7,167
Philippines	300,000	96,706
Singapore	697	5,312
Solomon Islands	28,896	549
Sri Lanka	65,610	20,328
Taiwan	35,980	23,316[3]
Thailand	513,120	66,785
Timor-Leste	14,874	1,210
Vietnam	331,210	88,775
Total	29,958,011	3,873,616
World total (land only)[4]	148,940,000	7,046,368

[1]Source: U.S. Central Intelligence Agency, The World Factbook 2012.

[2]Source: The World Bank, 2012 World Development Indicators Database.

[3]Source: Statistics Monthly, Accounting and Statistics, Executive Yuan, Taiwan, June 2013.

[4]Source: The World Bank, 2013 World Development Indicators Database.

TABLE 2
ASIA AND THE PACIFIC: GROSS DOMESTIC PRODUCT IN 2012[1,2]

Country	Gross domestic product based on purchasing power parity		Real gross domestic product growth rate (percentage)		
	Gross value (million dollars)	Per capita (dollars)	2010	2011	2012
Afghanistan	33,790	1,055	8.2	6.1	12.5
Australia	961,014	41,954	2.7	2.4	3.7
Bangladesh	302,803	1,962	6.4	6.5	6.1
Bhutan	4,880	6,563	8.3	8.5	9.2
Brunei	21,635	54,114	2.6	3.4	0.9
Burma	102,622	1,611	5.5	5.9	6.3
Cambodia	36,540	2,395	6.0	7.1	7.3
China	12,261,270	9,055	10.3	9.3	7.7
Fiji	4261	4,740	0.3	1.9	2.2
India	4,715,600	3,842	10.1	6.3	3.2
Indonesia	1,203,630	4,923	6.1	6.5	6.2
Japan	4,575,530	35,855	4.0	-0.6	1.9
Korea, North[3]	40,000	1,800	4.0	0.8	NA
Korea, Republic of	1,597,620	31,949	6.2	3.6	2.0
Laos	18,918	2,846	7.9	8.0	7.9
Malaysia	494,686	16,889	7.2	5.1	5.6
Mongolia	15,028	5,313	6.4	17.5	12.3
Nepal	40,026	1,456	4.6	3.4	4.8
New Caledonia[4]	9,280	37,700	NA	NA	NA
New Zealand	130,882	29,481	1.7	1.4	2.6
Pakistan	546,691	3,055	3.8	3.6	4.3
Papua New Guinea	18,677	2,736	7.0	10.7	8.1
Philippines	419,572	4,379	7.6	3.6	6.8
Singapore	322,989	60,799	14.5	5.1	1.3
Solomon Islands	1,858	3,287	6.5	10.7	4.8
Sri Lanka	124,895	6,045	8.0	8.2	6.4
Taiwan	894,315	38,356	10.9	4.0	1.3
Thailand	645,172	9,502	7.8	0.1	4.5
Timor-Leste	23,194	20,112	6.1	12.0	8.2
Vietnam	336,214	3,787	6.8	6.2	5.2
Total	29,903,592	XX	XX	XX	XX
World total	83,193,418	XX	XX	XX	XX

NA Not available. XX Not applicable.

[1]Source: International Monetary Fund, World Economic Outlook Database, October 2013.

[2]Gross domestic product listed may differ from that reported in individual country chapters owing to differences in source date of reporting.

[3]Based on 2011 to 2012 estimates. Source: U.S. Central Intelligence Agency, The World Factbook 2013.

[4]Based on 2008 estimate. Source: U.S. Central Intelligence Agency, The World Factbook 2013.

TABLE 3
ASIA AND THE PACIFIC: SELECTED EXPLORATION SITES IN 2012[1]

Country	Type[2]	Site	Commodity	Company	Resources[3]
Australia	F	Central Eyre	Iron ore	Iron Road Ltd.	177 Mt Fe (ID)
Do.	E	Hillside	Cu, Au	Rex Minerals Ltd.	636,000 t Cu, 540,000 oz Au (PR)
Do.	P	Garden Well/Rosemont	Au	Regis Resources Ltd.	2.3 Moz Au (R)
Do.	F	Mt. Ida	Iron ore	Jupiter Mines Ltd.	321 Mt Fe (ID)
China	P	Jiama	Cu, Au, Pb, Ag, Mo	China Gold International Resources Corp.	2.8 Mt Cu, 2.6 Moz Au, 139 Moz Ag, 95,000 t Pb, 109,000 t Mo (R)
Do.	P	Ying	Ag, Pb, Zn, Au	Silvercorp Metals Inc.	79 Moz Ag, 396,000 t Pb, 137,000 t Zn, 19,000 oz Au (R)
Mongolia	E	Mandal Moly	Mo, W	Moly World Ltd.	254,000 t Mo, 54,000 t W (D)
Do.	E	Selenge	Iron ore	Haranga Resources Ltd.	44 Mt Fe (D)
Philippines	F	Bananghilig	Au	Medusa Mining Ltd.	608,000 oz Au (D)
Do.	E	Basay	Cu	Copper Development Corp.	629,000 t Cu (IF)

Do. Ditto

[1]Abbreviations used for commodities in this table include the following: Ag—silver; Au—gold; Cu—copper; Fe—iron; Mo—molybdenum; Pb—lead; W—tungsten; and Zn—zinc. Abbreviations used for units of measure include the following: Moz—million troy ounces; Mt—million metric tons; oz—troy ounces; t—metric tons.

[2]E—Active exploration; F—Feasibility work ongoing or completed; P—Exploration associated with producing site.

[3]Expressed in terms of contained metal in ore based on 2012 data reported from various sources; D—measured + indicated; ID—indicated; IF—inferred; PR—probable; R—proven + probable. Resource data have not been verified by the U.S. Geological Survey. In cases where resources data are not released, the site was considered noteworthy based on the level of exploration activity or regional significance.

TABLE 4
ASIA AND THE PACIFIC: PRODUCTION OF SELECTED COMMODITIES IN 2012[1]

(Thousand metric tons unless otherwise specified)

Country	Aluminum			Metals Copper			Gold, mine output, Au content (kilograms)	Iron and steel Iron Ore, gross weight	Pig iron	Steel, crude	Lead Mine output, Pb content	Refined, primary
	Alumina	Bauxite	Metal[2]	Mine output, Cu content	Refined, primary							
Afghanistan	--	--	--	--	--	NA[e]	--	--	--	--	--
Australia	20,914	76,282	1,864	914	461	250,000	521,000[e]	3,711	4,894	648	160
Bangladesh[e]	--	--	--	--	--	--	--	--	--	--	--
Bhutan[e]	--	--	--	--	--	--	--	--	--	--	--
Brunei	--	--	--	--	--	--	--	--	--	--	--
Burma	--	--	--	19	19	NA	--	NA	NA	NA	--
Cambodia[e]	--	--	--	--	--	--	--	--	--	--	--
China[e]	37,700	47,000	24,500	1,550	3,930	403,000	1,310,000	663,500[3]	723,880[3]	2,800	3,300
Fiji	--	1	--	--	--	1,653	--	--	--	--	--
India[3]	3,900	19,000	1,700	34	671	--	144,000	48,000[3]	77,600[3]	118	122[3]
Indonesia[e]	--	29,000	248[3]	360	272	58,800[3]	--	--	3,700	--	--
Japan	250[e]	--	168	--	1,271	7,233	--	81,405	107,232	--	91
Korea, North[e]	--	--	--	12	15	2,000	5,300	900	1,300	13	9
Korea, Republic of	--	--	--	--	591	336	593	NA	69,073	4	280
Laos	--	--	--	63	86	6,415	48	--	--	5	--
Malaysia	--	122	(4)	--	--	4,625	10,278	--	5,612	--	--
Mongolia	--	--	--	122	2	5,995	7,561	--	68	--	--
Nepal[e]	--	--	--	--	--	--	--	--	--	--	--
New Caledonia	--	--	--	--	--	--	--	--	--	--	--
New Zealand	--	--	327	--	--	10,164	--	669[e]	912[e]	27	--
Pakistan[e]	--	12	--	19	--	--	380	380	450	--	--
Papua New Guinea	--	--	--	125	--	52,100	--	--	--	--	--
Philippines	--	--	--	65	98	15,762	346	--	1,200[e]	--	--
Singapore	--	--	--	--	--	--	--	--	NA	--	--
Sri Lanka[e]	--	--	--	--	--	--	--	--	--	--	--
Taiwan	--	--	--	--	--	--	--	11,800	21,083	--	--
Thailand	--	--	--	--	--	4,158	103	--	4,000[e]	--	--
Vietnam[e]	--	100	--	11	8	3,500	2,874[3]	NA	2,992[3]	6	--
Total	62,800	172,000	28,800	3,300	7,430	826,000	2,000,000	810,000	1,020,000	3,620	3,960
Share of world total	66%	67%	52%	20%	43%	31%	70%	75%	66%	70%	80%
United States	4,390	NA	2,070	1,170	962	235,000	54,200	32,100	88,700	345	111
World total	95,000	256,000	55,000	16,800	17,100	2,700,000	2,980,000	1,080,000	1,560,000	5,200	4,980

See footnotes at end of table.

TABLE 4—Continued
ASIA AND THE PACIFIC: PRODUCTION OF SELECTED COMMODITIES IN 2012[1]

(Thousand metric tons unless otherwise specified)

Country	Manganese ore, mine output, Mn content	Mercury, mine output, Hg content (metric tons)	Nickel, metal content		Tin (metric tons)		Tungsten, mine output, W content (metric tons)	Zinc (metric tons)	
			Mine output	Refined metal	Mine output, Sn content	Metal, primary		Mine output, Zn content	Metal[2]
Afghanistan[e]	--	--	--	--	--	--	--	--	--
Australia	3,080	--	246	129	5,849	--	80	1,541,000	498,000
Bangladesh	--	--	--	--	--	--	--	--	--
Bhutan	--	--	--	--	--	--	--	--	--
Brunei	--	--	--	--	--	--	--	--	--
Burma	115	--	5	--	10,600	30	140 [e]	10,000	--
Cambodia[e]	--	--	--	--	--	--	--	--	--
China[e]	2,900	1,350	93	229	110,000	148,000	64,000	4,900,000	4,890,000
Fiji	--	--	--	--	--	--	--	--	--
India[e]	800	--	--	--	--	--	--	750,000	800,000
Indonesia	40	--	228	--	41,000	42,000	--	--	--
Japan	--	--	--	170 [e]	--	1,133	--	--	571,312
Korea, North[e]	--	--	--	--	--	--	100	70,000	75,000
Korea, Republic of	--	--	--	20 [e]	--	--	--	2,868	875,000
Laos	--	--	--	--	762	--	--	5,250	--
Malaysia	429	--	--	--	3,726	37,792	20	--	--
Mongolia	--	--	--	--	--	--	--	119,100	--
Nepal[e]	--	--	--	--	--	--	--	--	--
New Caledonia	--	--	132	--	--	--	--	--	--
New Zealand	--	--	--	--	--	--	--	--	--
Pakistan	--	--	--	--	--	--	--	12 [e]	--
Papua New Guinea	--	--	5	--	--	--	--	--	--
Philippines	2	--	424	--	--	--	--	19,559	--
Singapore	--	--	--	--	--	--	--	--	--
Sri Lanka	--	--	--	--	--	--	--	--	--
Taiwan	--	--	--	11 [e]	--	--	--	--	--
Thailand[e]	6	--	--	--	124	20,000	83	31,000 [3]	53,000 [3]
Vietnam[e]	7	--	--	--	5,400	4,000	--	25,000	18,000
Total	7,380	1,350	1,130	559	177,000	253,000	64,400	7,470,000	7,780,000
Share of world total	47%	89%	34%	41%	65%	83%	87%	56%	63%
United States	--	NA	--	--	--	--	--	738,000	261,000
World total	15,600	1,520	3,360	1,370	273,000	305,000	73,800	13,300,000	12,400,000

See footnotes at end of table.

TABLE 4—Continued
ASIA AND THE PACIFIC: PRODUCTION OF SELECTED COMMODITIES IN 2012[1]

(Thousand metric tons unless otherwise specified)

Country	Cement, hydraulic	Industrial minerals					Coal		Mineral fuels	Petroleum, crude (thousand 42-gallon barrels)
		Fluorspar (metric tons)	Graphite (metric tons)	Magnesite	Salt	Anthracite	Bituminous	Natural gas, dry, marketable/marketable (million cubic meters)		
Afghanistan[e]	37	--	--	--	180	--	780	145	(4)	
Australia	8,600 [e]	--	--	300 [e]	10,821	--	365,000	55,970	119,200	
Bangladesh[e]	NA	--	--	--	1,400	--	820	21,000	--	
Bhutan	521	--	--	--	--	--	99	--	--	
Brunei	NA	--	--	--	--	--	--	NA	NA	
Burma[e]	540	--	--	--	100	--	--	12,500	6,500	
Cambodia	980 [e]	--	--	--	NA	--	--	--	--	
China[e]	2,210,000 [3]	4,600,000	820,000	16,000	69,120 [3]	500,000	2,830,000	95,000	1,510,000	
Fiji	110 [e]	--	--	--	--	--	--	--	--	
India[e]	270,000	13,600	160,000	355	17,000	--	550,000	35,000	270,000	
Indonesia[e]	51,000	--	--	--	700	100,000	140,000	74,000	342,000	
Japan	54,737	--	--	--	925	--	700 [e]	3,500 [e]	4,995	
Korea, North[e]	6,400	12,500	30,000	150	500	41,492 [3]	--	--	--	
Korea, Republic of	48,000	--	--	--	309	2,000 [e]	--	--	--	
Laos	NA	--	--	--	12	134	--	--	--	
Malaysia	21,726	--	--	--	--	--	2,951	62,000	212,979	
Mongolia	350	346,000	--	--	2	--	28,561	--	3,636	
Nepal	3,900 [e]	--	--	--	--	--	11	--	--	
New Caledonia	124	--	--	--	--	--	--	--	--	
New Zealand	1,200 [e]	--	--	--	95 [e]	--	4,926	4,559	14,149	
Pakistan[e]	33,000	1,700	--	8	2,080	--	4,000	41,000	65,000	
Papua New Guinea[e]	--	--	--	--	--	--	--	100	10,000	
Philippines[e]	18,907 [3]	--	--	--	720	--	7,000	4,000	2,500	
Singapore	--	--	--	--	--	--	--	--	--	
Sri Lanka[e]	2,400	--	5,000	--	12	--	--	--	--	
Taiwan	15,806	--	--	--	--	--	--	390 [e]	72	
Thailand	41,047	9,602	--	--	--	--	--	21,766 [5]	37,164	
Vietnam	55,531	--	NA	--	1,178	42,383	--	9,403 [5]	122,747	
Total	2,840,000	4,980,000	1,020,000	16,800	105,000	686,000	3,930,000	440,000	2,720,000	
Share of world total	75%	69%	87%	73%	39%	94%	68%	13%	10%	
United States	74,900	--	--	W	37,200	2,150	848,000	717,000	2,370,000	
World total	3,810,000	7,180,000	1,170,000	23,000	270,000	731,000	5,820,000	3,420,000	28,600,000	

See footnotes at end of table.

TABLE 4—Continued
ASIA AND THE PACIFIC: PRODUCTION OF SELECTED COMMODITIES IN 2012[1]

eEstimated; estimated data, U.S. data, and world totals are rounded to no more than three significant digits. NA Not available. W Withheld to avoid disclosing company proprietary data; not included in world total. -- Zero or zero percent.
[1]Totals may not add due to independent rounding. Percentages are calculated on unrounded data. Table includes data available as of May 7, 2014.
[2]Primary and secondary production.
[3]Reported figure.
[4]Less than ½ unit.
[5]Natural gas, gross production.

TABLE 5

ASIA AND THE PACIFIC: HISTORIC AND PROJECTED BAUXITE MINE PRODUCTION, 2005–2019[1]

(Thousand metric tons, gross weight)

Country	2005	2010	2012	2015[e]	2017[e]	2019[e]
Australia	59,960	68,414	76,282	81,000	96,000	98,000
China	22,000	44,000	47,000	49,000	51,000	52,000
India	12,385	18,000	19,000	19,000	20,000	20,000
Indonesia	1,442	27,000	29,000	30,000	31,000	32,000
Malaysia	5	124	122	150	150	150
Other	33	112	110	1,500	3,500	3,800
Total	95,800	158,000	172,000	181,000	202,000	206,000

[e]Estimated.

[1]Estimated data and totals are rounded to no more than three significant digits; may not add to totals shown.

TABLE 6

ASIA AND THE PACIFIC: HISTORIC AND PROJECTED PRIMARY AND SECONDARY ALUMINUM METAL PRODUCTION, 2005–2019[1]

(Thousand metric tons)

Country	2005	2010	2012	2015[e]	2017[e]	2019[e]
Australia	2,030	2,060	1,864	1,650	1,600	1,600
China	9,740	20,200	24,500	27,000	29,000	30,000
India	942	1,607	1,700	1,800	1,900	2,000
Indonesia	252	253	248	250	260	270
Japan	240	180	168	170	180	200
New Zealand	373	343	327	330	330	330
Other	--	--	100	800	1,000	1,000
Total	13,600	24,600	28,900	32,000	34,000	35,000

[e]Estimated. -- Negligible or no production.

[1]Estimated data and totals are rounded to no more than three significant digits; may not add to totals shown.

TABLE 7

ASIA AND THE PACIFIC: HISTORIC AND PROJECTED COBALT MINE PRODUCTION, 2005–2019[1]

(Metal content in metric tons)

Country	2005	2010	2012	2015[e]	2017[e]	2019[e]
Australia	5,600	3,850	5,882	6,000	6,000	6,000
China	2,100	6,000	6,800	7,000	7,000	7,000
Indonesia	1,600	1,600	1,300	1,300	1,400	1,400
New Caledonia	--	--	3,500	3,500	4,000	4,000
Papua New Guinea	--	--	469	2,200	2,800	2,800
Philippines	300	2,200	2,600	4,000	5,000	5,000
Total	9,600	13,700	20,600	24,000	26,200	26,200

[e]Estimated. -- Negligible or no production.

[1]Estimated data and totals are rounded to no more than three significant digits; may not add to totals shown.

TABLE 8

ASIA AND THE PACIFIC: HISTORIC AND PROJECTED COPPER MINE PRODUCTION, 2005–2019[1]

(Metal content in thousand metric tons)

Country	2005	2010	2012	2015[e]	2017[e]	2019[e]
Australia	930	870	914	1,100	1,200	1,250
China	762	1,160	1,550	1,700	1,800	1,900
India	27	36	34	36	38	40
Indonesia	1,064	878	360	400	500	600
Mongolia	127	125	122	320	570	550
Papua New Guinea	193	160	121	70	50	150
Philippines	16	58	64	70	78	100
Other	73	55	56	98	150	240
Total	3,190	3,350	3,220	3,800	4,400	4,800

[e]Estimated.

[1]Estimated data and totals are rounded to no more than three significant digits; may not add to totals shown.

TABLE 9

ASIA AND THE PACIFIC: HISTORIC AND PROJECTED REFINED COPPER METAL PRODUCTION, 2005–2019[1]

(Thousand metric tons)

Country	2005	2010	2012	2015[e]	2017[e]	2019[e]
Australia	461	417	460	500	310	310
China	2,600	4,650	5,880	7,000	7,300	7,500
India	497	664	671	700	720	740
Indonesia	263	278	272	360	380	400
Japan	1,395	1,549	1,516	1,500	1,600	1,600
Korea, Republic of	519	565	591	600	600	600
Other	270	266	220	300	310	330
Total	6,010	8,390	9,610	11,000	11,200	11,500

[e]Estimated.

[1]Estimated data and totals are rounded to no more than three significant digits; may not add to totals shown.

TABLE 10

ASIA AND THE PACIFIC: HISTORIC AND PROJECTED GOLD MINE PRODUCTION, 2005–2019[1]

(Metal content in kilograms)

Country	2005	2010	2012	2015[e]	2017[e]	2019[e]
Australia	263,000	261,000	250,000	255,000	270,000	265,000
China	225,000	345,000	403,000	450,000	470,000	480,000
Indonesia	130,620	106,316	58,800	60,000	80,000	100,000
Japan	8,318	8,544	7,233	7,000	6,500	6,000
Laos	6,232	5,061	6,415	7,000	8,000	8,000
Mongolia	24,120	6,000	5,995	27,000	20,000	20,000
New Zealand	10,583	13,494	10,164	12,000	12,000	12,000
Papua New Guinea	68,483	62,900	52,100	57,000	60,000	73,000
Philippines	37,490	40,847	15,762	25,000	30,000	30,000
Other	19,000	19,800	19,600	23,000	26,000	27,000
Total	793,000	868,000	829,000	923,000	983,000	1,020,000

[e]Estimated.

[1]Estimated data and totals are rounded to no more than three significant digits; may not add to totals shown.

TABLE 11

ASIA AND THE PACIFIC: HISTORIC AND PROJECTED BENEFICIATED IRON ORE PRODUCTION, 2005–2019[1]

(Metal content in thousand metric tons)

Country	Average ore grade (% Fe)	2005	2010	2012	2015[e]	2017[e]	2019[e]
Australia	62	163,000	271,000	315,000	350,000	360,000	370,000
China	64	134,000	350,000	406,000	460,000	480,000	500,000
India	64	97,500	134,000	92,000	100,000	130,000	170,000
Other		3,000	5,000	8,200	8,300	8,600	8,900
Total		398,000	760,000	820,000	920,000	980,000	1,000,000

[e]Estimated.

[1]Estimated data and totals are rounded to no more than three significant digits; may not add to totals shown.

TABLE 12

ASIA AND THE PACIFIC: HISTORIC AND PROJECTED CRUDE STEEL PRODUCTION, 2005–2019[1]

(Thousand metric tons)

Country	2005	2010	2012	2015[e]	2017[e]	2019[e]
Australia	7,790	7,408	4,894	5,000	5,000	5,000
China	353,240	637,230	723,880	820,000	860,000	890,000
India	45,800	68,300	77,600	90,000	95,000	100,000
Japan	112,470	110,000	107,200	105,000	112,000	120,000
Korea, Republic of	47,820	58,914	69,073	69,000	69,000	69,000
Malaysia	5,296	5,693	5,612	6,500	7,000	7,000
Taiwan	18,567	20,498	21,083	23,000	23,000	23,000
Thailand	5,161	4,145	4,000	4,000	4,000	5,000
Other	7,800	12,400	11,000	15,000	20,000	25,000
Total	604,000	925,000	1,020,000	1,140,000	1,200,000	1,250,000

[e]Estimated.

[1]Estimated data and totals are rounded to no more than three significant digits; may not add to totals shown.

TABLE 13

ASIA AND THE PACIFIC: HISTORIC AND PROJECTED NICKEL MINE PRODUCTION, 2005–2019[1]

(Metal content in metric tons)

Country	2005	2010	2012	2015[e]	2017[e]	2019[e]
Australia	189,000	170,000	246,000	210,000	200,000	200,000
Burma	10	--	5,000	18,000	21,000	24,000
China	72,700	80,000	93,300	100,000	105,000	110,000
Indonesia	135,000	235,800	228,000	450,000	460,000	480,000
New Caledonia	111,939	129,800	131,700	164,000	185,000	185,000
Papua New Guinea	--	--	5,283	25,000	31,000	31,000
Philippines	26,636	207,000	424,000	450,000	450,000	450,000
Total	535,000	823,000	1,130,000	1,400,000	1,450,000	1,480,000

[e]Estimated. -- Negligible or no production.

[1]Estimated data and totals are rounded to no more than three significant digits; may not add to totals shown.

TABLE 14

ASIA AND THE PACIFIC: HISTORIC AND PROJECTED PALLADIUM MINE PRODUCTION, 2005–2019[1]

(Metal content in kilograms)

Country	2005	2010	2012	2015[e]	2017[e]	2019[e]
Australia	550	650	300	300	300	300
China	450	650	650	650	700	700
Total	1,000	1,300	950	950	1,000	1,000

[e]Estimated.

[1]Estimated data and totals are rounded to no more than three significant digits; may not add to totals shown.

TABLE 15

ASIA AND THE PACIFIC: HISTORIC AND PROJECTED PLATINUM MINE PRODUCTION, 2005–2019[1]

(Metal content in kilograms)

Country	2005	2010	2012	2015[e]	2017[e]	2019[e]
Australia	111	130	90	100	100	100
China	700	750	700	700	700	750
Total	811	880	790	800	800	850

[e]Estimated.

[1]Estimated data and totals are rounded to no more than three significant digits; may not add to totals shown.

TABLE 16

ASIA AND THE PACIFIC: HISTORIC AND PROJECTED TIN MINE PRODUCTION, 2005–2019[1]

(Metric tons)

Country	2005	2010	2012	2015[e]	2017[e]	2019[e]
Australia	2,819	6,600	5,849	6,000	6,000	6,000
Burma	708	4,030	10,600	11,000	12,000	12,000
China	126,000	115,000	110,000	120,000	110,000	110,000
Indonesia	78,404	43,258	41,000	41,000	42,000	40,000
Malaysia	2,857	2,668	3,726	2,500	2,500	2,500
Thailand	158	291	124	150	200	300
Vietnam	5,400	5,400	5,400	5,400	5,400	5,400
Other	800	900	760	800	800	800
Total	217,000	178,000	177,000	187,000	179,000	177,000

[e]Estimated.

[1]Estimated data and totals are rounded to no more than three significant digits; may not add to totals shown.

TABLE 17

ASIA AND THE PACIFIC: HISTORIC AND PROJECTED TIN METAL PRODUCTION, 2005–2019[1]

(Metric tons)

Country	2005	2010	2012	2015[e]	2017[e]	2019[e]
Australia	994	400	400	500	500	500
China	122,000	149,000	148,000	165,000	170,000	180,000
Indonesia	65,300	43,832	42,000	42,000	43,000	41,000
Japan	754	841	1,133	1,000	1,100	1,200
Malaysia	36,924	38,737	37,792	40,000	40,000	40,000
Thailand	31,600	19,423	20,000	20,000	23,500	30,000
Other	1,800	3,100	4,000	4,000	4,000	4,000
Total	259,000	255,000	253,000	273,000	282,000	297,000

[e]Estimated.

[1]Estimated data and totals are rounded to no more than three significant digits; may not add to totals shown.

TABLE 18

ASIA AND THE PACIFIC: HISTORIC AND PROJECTED DIAMOND PRODUCTION, 2005–2019[1]

(Thousand carats)

Country	2005	2010	2012	2015[e]	2017[e]	2019[e]
Australia	34,307	10,000	11,960	11,000	11,000	11,000
China	100	100	100	100	100	120
India	58	50	62	70	70	70
Indonesia	30	37	38	40	40	40
Total	34,500	10,200	12,200	11,000	11,000	11,000

[e]Estimated.

[1]Estimated data and totals are rounded to no more than three significant digits; may not add to totals shown.

TABLE 19

ASIA AND THE PACIFIC: HISTORIC AND PROJECTED LITHIUM PRODUCTION, 2005–2019[1]

(Metal content in metric tons)

Country	2005	2010	2012	2015[e]	2017[e]	2019[e]
Australia	4,800	8,200	12,700	13,000	13,500	14,000
China	3,600	6,000	9,500	10,000	10,000	10,000
Total	8,400	14,200	22,200	23,000	23,500	24,000

[e]Estimated.

[1]Estimated data and totals are rounded to no more than three significant digits; may not add to totals shown.

TABLE 20

ASIA AND THE PACIFIC: HISTORIC AND PROJECTED SALABLE COAL PRODUCTION, 2005–2019[1]

(Thousand metric tons)

Country	2005	2010	2012	2015[e]	2017[e]	2019[e]
Australia	370,000	499,000	430,000	530,000	540,000	560,000
China	2,260,000	3,240,000	3,660,000	3,850,000	4,000,000	4,100,000
India	360,000	507,000	580,000	600,000	620,000	640,000
Indonesia	192,920	256,789	240,000	260,000	280,000	300,000
Korea, North	23,500	41,000	41,492	42,000	42,000	42,000
Korea, Republic of	2,832	2,500	2,000	2,200	2,500	2,800
Mongolia	8,256	25,246	28,561	45,000	58,000	55,000
New Zealand	5,267	5,335	4,926	6,000	6,000	6,000
Pakistan	3,367	3,429	4,000	4,100	4,200	4,300
Philippines	3,165	6,650	7,000	7,000	9,000	9,000
Thailand	21,429	17,907	12,072	15,000	18,000	21,000
Vietnam	34,093	44,835	42,383	50,000	50,000	50,000
Other	1,450	3,950	5,400	5,400	5,300	5,000
Total	3,290,000	4,650,000	5,060,000	5,420,000	5,600,000	5,800,000

[e]Estimated.

[1]Estimated data and totals are rounded to no more than three significant digits; may not add to totals shown.

THE MINERAL INDUSTRY OF AFGHANISTAN

By Chin S. Kuo

Afghanistan is a land-locked country. Mineral and energy resources represent a potential source of new wealth for Afghanistan. The mineral sector was a significant contributor to its gross domestic product (GDP). The major metal and industrial mineral resources include chromium, copper, gold, iron ore, lead and zinc, lithium, marble, precious and semiprecious stones, sulfur, and talc (Peters and others, 2007). The mineral fuel resources consist of natural gas and petroleum. In 2012, a copper mine was being developed at Aynak, iron ore mines were being developed at Hajigak, and natural gas was produced at Sheberghan. Additional development of mineral resources that were not currently being produced was hindered by issues with improvements in infrastructure and security in the country. Investment in infrastructure and transportation projects for mining was a critical aspect of developing Afghanistan's mineral industry.

Minerals in the National Economy

With international assistance (foreign aid accounted for 90% of the country's revenue), Afghanistan's economy was recovering from a decade of conflict. Foreign aid was expected to begin to decrease in the near future, however, although foreign investment was expected to increase as mineral licensing rounds that had been underway since 2011 were completed. Gradual development of the country's mineral resources was expected to spur future economic growth. In 2012, the mineral sector accounted for 20% of Afghanistan's GDP. The country exported a small amount of precious and semiprecious gemstones and imported most of its energy, including petroleum products and electricity (Webb and Edel, 2012).

Government Policies and Programs

Corporate entities in Afghanistan are generally liable to pay corporate income tax at a flat rate of 20% applied to income as well as a business receipt tax of 2% applied to the corporation's gross revenue. Under the Income Tax Law of 2009, any contractor (a mining company) under an exploration and production-sharing contract will be treated as a qualifying extractive industries taxpayer and thus has the option to stabilize income tax at a rate of 30% for the entire term of the contract (8 years). Also, it is exempt from the business receipt tax, as it has paid royalties, and is eligible for accelerated depreciation and full carry-forward of losses. All exploration and development expenditures are deductible against operating revenues (Devine, 2012).

The Government of Afghanistan sought significant foreign direct investment in the mineral sector and had issued exploration tenders for four mineral prospects—Badakhshan, Zarkashan, Balkhab, and Shaida. The Badakhshan area in the Province of Badakhshan was a known gold-bearing quartz vein system. Based on trench sampling, Russian geologists had estimated reserves for the Veka Dur deposit and other quartz veins in the area to contain a total of 1,200 kilograms (kg) (39,000 troy ounces) of gold at a grade of 4.8 grams per metric ton (g/t). The Zarkashan project in the Province of Ghazni was a skarn mineralization with a core that has relatively high gold grades and a halo that has low gold grades. The Balkhab copper prospect in the Provinces of Balkh and Sar-e Pul showed evidence of copper mining activity in old surface and underground workings. A reconnaissance sampling was carried out in 2008. The Shaida mineralization in the Adraskan District of Herat Province was a copper porphyry deposit with copper grades of between 0.1% and 0.8%. The rock assemblage showed layered quartz plagioclase porphyry, quartz keratophyre, and aleuropelite interbedded with volcanic layers. Exploration licenses for three of the prospects—Badakhshan, Balkhab, and Shaida—had been awarded to three different companies (Chadwick, 2012).

The Ministry of Mines was running its second hydrocarbon licensing round for six blocks in the Afghan-Tajik Basin. Bidders would compete on the basis of their proposed royalty rate and exploration program. In case of a tie between two bidders, the Hydrocarbons Law, which was promulgated in 2009, provides that the contract would be awarded to the highest bidder with an Afghan partner. There is no distinction between royalty rates for oil and natural gas. The Ministry's profit share ranges between 50% and 70%, and the Ministry does not have a participating interest or any back-in rights (Devine, 2012).

Afghanistan's infrastructure development in 2012 included the opening run of the 75-kilometer railway that links Hairatan (near the border with Uzbekistan) and Mazar-i-Sharif. The line had been nominally operational and test-run since mid-2011 by Uzbekistan's state railway Uzbekistan Temir Yollari (UTY) under a 3-year concession agreement. UTY built the line with contributions from the Asian Development Bank ($165 million), the Government of Afghanistan ($20 million), and the local municipalities. The line was built to transport commercial cargoes and mineral ores (Railway Gazette, 2012).

Production

Owing to the lack of reported mineral production data, information about Afghanistan's mining activities was not readily available, but the activities in general appeared to be limited in scope except for planned operations by foreign companies. The country produced cement, coal, natural gas, and some industrial minerals for domestic consumption. The Government provided only partial output data for 2008 through 2010. For 2012, production of chromite was estimated by the U.S. Geological Survey (USGS) to be about 6,000 metric tons (t); marketed natural gas, 145 million cubic meters; and talc, 200,000 t. Production of rock salt and cement was estimated to have decreased by 5.3% and 2.6%, respectively, compared with that of 2011. Output of petroleum condensate, however, was estimated to have increased significantly by 14.3%, as China National Petroleum Corp. (CNPC) began trial oil production in October 2012 (table 1).

Structure of the Mineral Industry

Afghanistan's mineral industry was characterized by small-scale operations; output was supplied mainly to local and regional markets. Privatization of Afghanistan's state-owned companies, which controlled many of the country's mineral resources, was ongoing but not completed. The Government encouraged investment in the mineral sector by private domestic companies and foreign investors. Foreign companies from Canada, China, and India had begun to participate in the country's resource development (table 2).

Commodity Review

Metals

Copper and Gold.—The Aynak copper deposit, which was among the world's large copper deposits, is located southeast of Kabul in the Province of Logar. Metallurgical Group Corp. of China acquired the property in 2007 and continued with development of the $3 billion Aynak copper mining project, which had been attacked numerous times by insurgents. The development stage employed 500 Afghan citizens and a special security force of 1,500. Construction work had been delayed, and some Chinese workers had left the mine site because of the security situation. The Government planned to improve security and convince the Chinese workers to restart work. The discovery of ancient monastery relics and the clearing of land mines in the area were additional causes for the delay in construction. Copper production was expected to begin in 2016. The mine was projected to be able to generate $300 million in annual royalties for the Government. Social and environmental concerns linked to the development of the mine included displacement of villagers and disruptions to the water supply (Donati and Harooni, 2012).

Afghan Gold & Minerals Co., which was owned by a consortium led by JPMorgan Chase & Co. of the United States, was working on exploration of a gold deposit in northern Afghanistan and reviewing data from soil samples and drilling. The company planned to invest $50 million to develop a mine. The Government had also awarded Afghan Gold & Minerals the Balkhab copper exploration license in northwestern Afghanistan. Another company, Afghan Minerals Group, was granted a license to explore the Shaida copper deposit in the Province of Herat. Turkish-Afghan Mining Co. was picked to explore and develop the Badakhshan gold-copper project in the Province of Badakhshan (Khaama Press, 2012).

Iron Ore.—A consortium of seven Indian steel companies led by the Steel Authority of India Ltd. won the rights to develop the Hajigak iron ore deposit in 2011, and development contracts were expected to be finalized by the end of July 2012. When production from the $11 billion project begins, India could overtake China as the leading overseas investor in Afghanistan. The Hajigak iron ore deposit was considered to be one of the larger iron ore deposits in the world and had reserves of 1,800 million metric tons (Mt) (Webb and Edel, 2012).

Industrial Minerals

In 2011, the USGS (Kokaly and others, 2011) discovered some industrial minerals, such as dolomite, gypsum, and kaolin, in Afghanistan by using advanced remote sensing hyperspectral imaging. High concentrations of dolomite were identified in the central Provinces of Ghazni, Ghor, and Urozgan; gypsum, in the Province of Paewan; and kaolin, in the Province of Daykundi (Germain, 2012).

Rare Earths.—Rare earths had been identified since at least the mid-1970s in the Khanneshin deposit in Helmand Province. The geology of the deposit indicated Type 1 mineralization (semiconcordant bands and veins in alvikite) estimated to be 218 Mt of ore grading 2.77% light rare-earth elements (LREEs), and Type 2 mineralization (discordant dikes and sheets enriched in fluorine or phosphorus) estimated to be 15 Mt of ore grading 3.28% LREEs. The ores included khanneshite, monazite, and synchysite (Nicoletopoulos, 2012).

Talc.—Talc deposits have been identified in many districts in the Province of Nangarhar, and talc mining was being conducted on a small scale. The color of the talc in these deposits was very white, and the talc was in great demand. In the Khogyani District talc deposit (with mines at Kudikel and Markikhel), talc was found in bands within dolomite; in the Shinwari District talc deposit (with mines at Kot and Shinwari), talc was associated with magnesite. Both deposits showed low iron and calcium content, which made them suitable for making polymers and ceramics. The country's talc production was estimated to be about 200,000 metric tons per year (Wilson, 2012).

Mineral Fuels

Petroleum.—Exxon Mobil Corp. of the United States expressed interest in the 2012 licensing round of six blocks in the western portion of the Afghan-Tajik Basin. Seven companies were expected to participate and included Dragon Oil plc of Dubai, Kuwait Energy, ONGC Videsh Ltd. of India, Pakistan Petroleum Ltd., Petra Energia S/A of Brazil, PTT Exploration and Production Public Co. Ltd. (PTTEP) of Thailand, and Turkiye Petrolleri A.O. (TPAO) of Turkey. The deadline for bids was October 2012, and the contracts would be awarded in early 2013. The blocks were estimated to hold several hundred million barrels of oil equivalent. More than 60 geologic structures that may contain hydrocarbons had been discovered in the six blocks by gravity and magnetic surveys. CNPC won the development of three hydrocarbon blocks in the Amu Darya basin in 2011. These blocks held an estimated 82 million barrels of proved and probable oil reserves. CNPC offered to pay a 15% royalty on each barrel of oil produced and 30% corporate tax on its profits, as well as committing to build a $300 million oil refinery (Petroleum Economist, 2012).

Outlook

The development of Afghanistan's rich mineral resources could provide substantial impetus to economic growth in the country. The Government forecasted that mining would represent 25% of the Afghanistan's GDP in 2016 after copper production comes onstream. Foreign investment in infrastructure and transportation for mining is expected to be a key factor in the development of its mineral industry. Some gold and copper projects are in the development stages and are expected to start production in 2013 and 2016, respectively. The first gold mine has provided royalties to the Government. Contracts for iron ore and hydrocarbon projects have been awarded. The country is expected to offer more tenders for mineral and energy resource development in the near future.

References Cited

Chadwick, John, 2012, Exploration continues amid the turmoil: International Mining, v. 8, no. 4, April, p. 3.

Devine, Richard, 2012, Beyond the conflict: Petroleum Economist, October 23. (Accessed October 25, 2012, at http://www.petroleum-economist.com/article/3107094/beyond-the-conflict.html?LS=EMS735554.)

Donati, Jessica, and Harooni, Mirwais, 2012, Chinese halt of flagship mine works imperils Afghan future: Mineweb.com, September 28. (Accessed October 1, 2012, at http://www.mineweb.com/mineweb/view/mineweb/en/page/72068?oid=159334&sn=detail.)

Germain, Leah, 2012, USGS detects industrial minerals in Afghanistan: Industrial Minerals, no. 540, September, p. 16.

Khaama Press, 2012, Afghan Gold & Minerals Co. wins mining permits: Khaama Press, December 11. (Accessed March 4, 2013, at http://www.khaama.com/afghan-gold-minerals-company-wins-mining-permits-2045/print/.)

Kokaly, R.F., King, T.V.V., Hoefen, T.M., Dudek, K.B., and Livo, K.E., 2011, Surface materials map of Afghanistan—Carbonates, phyllosilicates, sulfates, altered minerals, and other materials: U.S. Geological Survey Scientific Investigations Map 3152–A, one sheet, scale 1:1,100,000. Also available at http://pubs.usgs.gov/sim/3152/A/.

Nicoletopoulos, Vasili, 2012, Afghanistan's mineral plan: Industrial Minerals, no. 533, February, p. 41.

Peters, S.G., Ludington, S.D., Orris, G.J., Sutphin, D.M., Bliss, J.D., and Rytuba, J.J., eds., 2007, Preliminary non-fuel mineral resources assessment of Afghanistan: U.S. Geological Survey Open-File Report 2007–1214, 822 p. (Also available at http://pubs.usgs.gov/of/2007/1214/.)

Petroleum Economist, 2012, ExxonMobil sniffs out Afghan opportunity: Petroleum Economist, v. 79, no. 7, September, p. 47.

Railway Gazette, 2012, Afghan railway starts commercial traffic: Railway Gazette, February 3. (Accessed February 10, 2012, at http://www.railwaygazette.com/nc/news/single-view/view/afghan-railway-starts-commercial-traffic.html.)

Webb, Stephen, and Edel, Robert, 2012, Mining in the Asia Pacific—A legal overview (Afghanistan): Mondaq.com. July 31. (Accessed August 2, 2012, at http://www.mondaq.com/article.asp?articleid=188776&print=1.)

Wilson, Ian, 2012, The emergence of Afghanistan as a significant talc supplier: Industrial Minerals, no. 540, September, p. 60.

TABLE 1
AFGHANISTAN: ESTIMATED PRODUCTION OF MINERAL COMMODITIES[1, 2]

(Metric tons unless otherwise specified)

Commodity[3]		2008	2009	2010	2011	2012
Cement, hydraulic		37,300 [4]	31,500 [4]	35,600 [4]	38,000	37,000
Chromite		6,500 [r]	6,700 [r]	5,727 [r, 4]	6,204 [r, 4]	6,000
Coal, bituminous		346,900 [4]	500,100 [4]	724,900 [4]	750,000	780,000
Gas, natural:						
Gross	million cubic meters	155 [4]	142 [4]	142 [4]	145	150
Marketed	do.	145	140	140	142	145
Gold	kilograms	--	--	NA	NA	NA
Gypsum		48,700 [4]	46,400 [4]	63,100 [4]	62,000	65,000
Marble		36,900 [4]	26,600 [4]	28,900 [4]	30,000	32,000
Nitrogen, N content of ammonia		18 [4]	22 [4]	27 [4]	28	30
Petroleum, condensate	42-gallon barrels	156 [4]	104 [4]	64 [4]	70	80
Salt, rock		158,200 [4]	180,300 [4]	186,100 [4]	190,000	180,000
Talc		200,000	200,000	200,000	200,000	200,000

[r]Revised. do. Ditto. NA Not available. -- Zero.

[1]Estimated data are rounded to no more than three significant digits.

[2]Table includes data available through June 12, 2013.

[3]Barite, natural gas liquids, precious and semiprecious stones, and lapis-lazuli were being produced, but sufficient data were not available to make reliable estimates of output.

[4]Reported figure.

TABLE 2
AFGHANISTAN: STRUCTURE OF THE MINERAL INDUSTRY IN 2012

(Metric tons unless otherwise specified)

Commodity		Major operating companies and major equity owners	Location of main facilities	Annual capacity[e]
Aluminum:				
Extrusion and powder coating		Qader Najib Ltd.	Kabul	NA
Manufacture		Salam Bilal Ltd.	Kandahar	360
Cement	metric tons per day	Afghan Cement LLC (subsidiary of Government-owned Afghan Investment Co.)	Ghori I, Pol-e-Khomri, Baghlan	400
Do.	do.	do.	Ghori II, Pol-e-Khomri, Baghlan	500
Do.	do.	do.	Ghori III, Pol-e-Khomri, Baghlan[1]	4,000
Coal		Afghan Coal LLC (subsidiary of Government-owned Afghan Investment Co.)	Ahandara, Dudkash, Karkar, and Khurdara near Pol-e-Khomri	NA
Copper, in concentrate		Aynak Minerals Co. Ltd. (China Metallurgical Group Corp., 75%, and Jiangxi Copper Co. Ltd., 25%)	Aynak, Logar[2]	180,000
Fertilizer, urea		Kud Bergh Fertilizer Ltd.	Qala Jangi near Mazar-i-Sharif	105,000
Gas, natural	cubic meters per day	Afghan Gas Ltd. (Government owned)	Jawzjan	70,000
Do.	do.	do.	Sheberghan	140,000
Gold	kilograms	Westland General Trading LLC	Nor Aaba, Takhar	NA
Lapis-lazuli	do.	Government owned	Sary-Sang, Badakhshan	9,000
Steel, manufacture		Khalil Najeeb Steel Mills Ltd.	Jalalabad, Kabul, and Mazar-i-Sharif	36,000
Talc		Amin Karimzai Ltd.	Khogyani and Shinwari Districts, Nangrahar	200,000

[e]Estimated; estimated data are rounded to no more than three significant digits. Do., do. Ditto. NA Not available.
[1]The Ghori III plant is expected to operate in 2013.
[2]The Aynak Mine is expected to start production in 2016.

THE MINERAL INDUSTRY OF AUSTRALIA

By Pui-Kwan Tse

Australia was subject to volatile weather in recent years that included heavy rains and droughts. The inclement weather conditions affected companies' abilities to expand their activities, such as port, rail, and road construction and repair, as well as to mine, process, manufacture, and transport their materials. Slow growth in the economies of the Western developed countries in 2012 affected economic growth negatively in many counties of the Asia and the Pacific region. China, which was a destination point for many Australian mineral exports, continued to grow its economy in 2012, although the rate of growth was slower than in previous years. As a result, Australia's gross domestic product (GDP) increased at a rate of 3.1% during 2012, which was higher than the 2.3% rate of growth recorded in 2011. The economic growth of Australia was owing mainly to the mining sector, which increased in value by 8.8% in 2012 compared with the value in 2011. Strong export growth, especially by the mineral and mineral fuels sectors, contributed to the economic growth. Increased demand from China supported exports of coking coal and iron ore (Australian Bureau of Statistics, 2013a, p. 4–5; Reserve Bank of Australia, 2013, p. 37).

Australia's total mineral exploration spending was estimated to be A$3.9 billion (US$4.1 billion) in fiscal year 2012 (the Australian fiscal year ran from July 1, 2011, to June 30, 2012), which was an increase of 34% from that of fiscal year 2011. The increase in exploration spending was the result of an increase in exploration for base metals, coal, gold, and iron ore. About 65% of the country's total exploration expenditure was spent on known deposits, and the remaining 35% was spent on new exploration projects. The State of Western Australia accounted for 53% of the total exploration spending followed by Queensland, 25%; South Australia, 8%; and others, 14%. Iron ore exploration spending accounted for 29% of the exploration spending followed by coal, 21%; base metals, 20%; gold, 19%; and other commodities, 11%. As a result of the spending on exploration, significant mineral resources were discovered. These included the Nova copper-nickel deposit and the Dampier heavy-mineral sand deposit in Western Australia and the Mallee Bull copper-gold-silver deposit in New South Wales (Geoscience Australia, 2013, p. 1–2).

Minerals in the National Economy

Australia's mineral sector contributed more than $142 billion, or about 10%, to the country's GDP in fiscal year 2012. The mineral sector employed 249,000 people. Expectations of sustained levels of global demand for minerals led to increased production of minerals and metals in Australia, and the mineral industry was expected to continue to be a major contributor to the Australian economy during the next several years (Australian Bureau of Resources and Energy Economics, 2013, p. 12–14).

Government Policies and Programs

The powers of Australia's Commonwealth Government are defined in the Australian Constitution; powers not defined in the Constitution belong to the States and Territories. Except for the Australian Capital Territory (that is, the capital city of Canberra and its environs), all Australian States and Territories have identified mineral resources and established mineral industries. Each State has a mining act and mining regulations that regulate the ownership of minerals and the operation of mining activities in that State. The States have other laws that deal with occupational health and safety, environment, and planning. All minerals in the land are reserved to the Crown; however, a very small percentage of minerals in Australia are owned by those who were granted titles to the land before the enactment of relevant State legislation that excludes mineral ownership. Companies or miners may obtain rights to conduct mining activities on unreserved Crown land where the permission of the landowner has been granted. Royalties on minerals are charged by State and Territorial governments. In most cases, royalties are payable on a percentage of value or a flat-rate per-unit basis. Each State sets its own rate. The Northern Territory's royalties are based on profit where the net value of a mine's production is used to calculate the applicable royalty. The royalty paid by a company is allowed to be deducted from reported income for income tax purposes. The amount of royalty paid can be reduced by deducting the costs incurred in the transportation of the mineral ore, concentrate, or metal.

The Australian Parliament passed the minerals resource rent tax (MRRT) bill in November 2011. A uniform national MRRT took effect on July 1, 2012. The MRRT, which applies only to coal and iron ore mining, is intended to target project profits rather than project production and to shift the tax burden from low-profitability projects to more profitable projects. The MRRT is set at an internationally competitive rate of 22.5%, and companies are charged the MRRT when the net mining profits are equal to or less than A$75 million (US$77.5 million). Companies are entitled to have an MRRT offset year if the company's group mining profit for the year is less than A$125 million (US$129.2 million). All Federal and State resource taxes would be credited towards tax payment. Fortescue Metals Group Ltd. (the third-ranked iron ore producer in Australia) filed a challenge to the tax in the Australian High Court, asserting that the MRRT discriminates among the States and curtails State sovereignty. The governments of the States of Queensland and Western Australia joined with Fortescue Metals Group in challenging the tax, arguing that the tax is unconstitutional. The Federal Government estimated that the MRRT would collect about A$2 billion (US$2.08 billion) in revenue during the first fiscal year in which it is in effect (fiscal year 2013). During the first 6 months of fiscal year 2013, the actual MRRT tax revenue was A$126 million (US$129 million) (Mining Weekly, 2012; Wilson, 2013).

The Department of Mines and Petroleum of the State of Western Australia introduced an A$80 million (US$84 million) exploration incentive scheme (EIS) fund to stimulate exploration in greenfield areas in Western Australia. The EIS program was to support companies generating new precompetitive geologic and geophysical information. Under the EIS, online systems for the management of administrative processes surrounding tenement applications, as well as other tenement-related processes (such as environmental databases), were being developed. The Government of Western Australia released guidelines to ensure a consistent planning process on mine closure in the State (Ellis, 2013; Risbey, 2013).

Production

Australia continued to be one of the world's leading producers of such mineral commodities as bauxite, coal, cobalt, copper, gem and near-gem diamond, gold, iron ore, lithium, manganese, tantalum, and uranium. The country's refined metal production capacity was moderate in the Asia and the Pacific region compared with that of China and Japan. Because of its large mineral resources, Australia was virtually self-sufficient in most mineral commodities. Petroleum production, however, supported only about 70% of the country's consumption. Australia was one of the world's leading exporting countries for alumina, coal, iron ore, and uranium. In general, the level of mineral and metal production was about the same in 2012 as it was in 2011. Some of the commodities for which production decreased in 2012 were iron and steel, refined lead, refined silver, and zircon. Mineral commodities for which reported production increased included mined antimony, cobalt, iron ore, ilmenite, and mined and refined nickel. The increase in iron ore output was from record production at mines operated by BHP Billiton Ltd., Fortescue Metals Group, and Rio Tinto Ltd. BHP Billiton's Olympic Dam returned to full production in 2011. An increase in mined nickel production reflected increased output from BHP Billiton's Nickel West and Western Areas NL's Spotted Quoll and Forrestania operations (table 1).

Structure of the Mineral Industry

The Australian mineral industry is characterized by free enterprise in which private companies are involved in exploration, mine development, mineral production, mineral processing, and marketing. A number of Australian mineral companies were affiliates or subsidiaries of European and U.S. companies, which controlled a large part of the mining, smelting, and refining sectors and a significant portion of the mineral fuels sector (table 2).

Each State and Territorial government administers the mineral industries within its own borders, which includes registering land titles; issuing exploration and development permits; conducting inspections and assuring compliance with health, safety, and environmental regulations; and levying royalties and taxes. Because the Commonwealth Government may restrict mineral exports for the good of the country, it effectively has control over most mineral production.

Mineral Trade

Australia continued to rely heavily on exports of the majority of its mineral production to sustain the country's mineral industry development. In 2012, the value of Australia's total foreign trade of goods was A$617.9 billion (US$642.6 billion), of which the value of exports was A$301.0 billion (US$313.0 billion) and the value of imports was A$316.9 billion (US$329.6 billion). As a result of moderated energy and mineral commodity prices, Australia's export revenue decreased to A$120 billion (US$126 billion) in 2012 from A$180 billion (US$190 billion) in 2011. Mineral and metal exports accounted for about 40.2% of the total value of exports. Mineral commodities for which the export volume was higher than in 2011 included bauxite, thermal coal, copper, iron ore, lead, manganese ore, nickel, uranium, and zinc. Australia's mineral and metal exports went mostly to Asian countries, such as China, Japan, the Republic of Korea, India, and Thailand (in descending order by volume of exports). Australia remained one of the world's leading exporters of alumina, coal, iron ore, mined lead, rutile, and zircon. Crude petroleum and refined petroleum products remained Australia's leading imported fuel and mineral commodity category, followed by gold, iron and steel, potassium fertilizer, and silver (Australian Bureau of Statistics, 2013b, p. 29–31).

Commodity Review

Metals

Aluminum.—Australia was the leading bauxite-producing country in the world. Bauxite was mined at the Gove Mine in Northern Territory; the Weipa Mine in the northern part of Queensland; and the Huntly, the Willowdale, and the Worsley Mines in Western Australia. Australia was also the leading alumina-producing country in the world. All Australia's alumina refineries were located in close proximity to their bauxite mines and shipping facilities. Western Australia remained the leading bauxite-producing State and accounted for about 59.1% of the country's total output of bauxite followed by Queensland, 30.5%, and Northern Territory, 10.4%. Australia exported 10.4 million metric tons (Mt) of bauxite compared with 11.3 Mt in 2011. Western Australia accounted for about 60% of the country's alumina output. The country exported 18.3 Mt of alumina in 2012, which was about 13% more than in 2011. China retook its place as the leading destination for exported Australian alumina; it received about 26% of the total exported volume, followed by the United Arab Emirates, 16%; South Africa, 12%; and other countries, less than 10% each. The consumption of domestic aluminum smelters was less than 20% of the country's total alumina output, and the remainder was exported. In 2012, Australia exported 1.65 Mt of aluminum. Japan was the leading destination for Australian aluminum exports and accounted for 33.8% of the total, followed by the Republic of Korea, 16.4%; Taiwan, 11.1%; Thailand, 9.0%; and Indonesia, 8.0%; the remainder went to other countries (Australian Bureau of Resources and Energy Economics, 2013, p. 162; Department of Mines and Petroleum, 2013b, p. 9).

The government of Western Australia granted a 5-year extension to Alcoa of Australia Ltd.'s (a subsidiary of Alcoa Inc. of the United States) expansion of the Wagerup alumina refinery's output capacity to 4.7 million metric tons per year (Mt/yr) from 2.6 Mt/yr in 2012. Alcoa's expansion project remained on hold because of unfavorable economic conditions and the need to obtain a competitive price for alumina in the world market and additional energy supplies in Western Australia. Alcoa had two operating mines in the Darling Range of Western Australia, and the mineral leases for the mines had been extended to 2045 and could be renewed beyond 2045. Bauxite output from these mines was supplied to Alcoa alumina refineries. The available alumina content in the Darling Range mines was about 32.9% aluminum oxide, and for each ton of alumina produced, between 2.5 metric tons (t) and 3.8 t of bauxite was consumed (Department of Mines and Petroleum, 2013a, p. 35; Alcoa Inc., 2013, p. 6–20).

Rio Tinto Alcan was conducting a feasibility study and an environmental impact study to develop the bauxite resource in an area south of Embley River and the existing Weipa Mine. The new operation would progressively replace depleted resources at the Andoom and the East Weipa mining areas in Weipa. It could extend the mine life in the area by 40 years. The new development would increase output capacity to 50 Mt/yr from the current 23 Mt/yr in the region south of the Weipa Peninsula and would enable the continuity of supply to the company's two Gladstone alumina refineries. The Weipa area had indicated bauxite resources of 1.35 billion metric tons (Gt) containing an average of 51.2% aluminum oxide. The Queensland Coordinator-General provided the required conditions for approval for the South of Embley project in 2012. Rio Tinto Alcan prepared a final environmental impact study that incorporated its response to public submissions and submitted the study to the Queensland and Commonwealth Governments for approval. Whether or not Alcan moves forward with the project could depend on the supply of bauxite in the region. China's dependence on bauxite imports was expected to continue, and although Indonesia had been a source of bauxite for China, the Indonesian Government introduced regulations in 2012 to restrict the export of raw materials. As a result, supply of bauxite in the Asia and the Pacific region could be uncertain. Once all major Government approvals have been granted, Rio Tinto Alcan's decision about whether to proceed with the project will likely depend on market conditions at that time (Rio Tinto plc, 2013a; 2013b, p. 55).

In 2008, the Board of BHP Billiton Ltd. approved an investment of $3 billion to expand the output capacity of the Worsley alumina refinery to 4.6 Mt/yr from 3.5 Mt/yr and to increase the capacity of the bauxite mining operation at Boddington to 19 Mt/yr of bauxite. The Worsley expansion project was fully completed and put into operation in 2012. The Boddington bauxite mining area had indicated resources of 587 Mt containing 32.3% aluminum oxide and 2.5% silicon oxide and had proven ore reserves of 263 Mt containing 31.1% aluminum oxide and 1.8% silicon oxide in 2012 (BHP Billiton Ltd., 2013, p. 63).

In 2004, the government of Queensland invalidated the permit for the bauxite deposit near Aurnkun that had been awarded to Pechiney S.A., and Aluminum Corp. of China Ltd. (Chalco) was subsequently awarded a permit to mine the bauxite deposit and to build an alumina refinery at the site. Chalco concluded that under economic conditions in the world in 2010, the company would have difficulty implementing the project in accordance with the development agreement between the company and the government of Queensland. The Queensland government re-opened the bidding for the development rights of the Aurnkun bauxite deposit in 2012 and subsequently finalized a short list of five bidders, which included Australian Indigenous Resources Ltd., Cape Alumina Consortium, Chalco, Glencore International plc of Switzerland, and Rio Tinto. The shortlisted companies had until September 2013 to submit detailed proposals for the development of the deposit. The right to mine would not be tied to the requirement to build an alumina refinery in the area (Foley, 2012).

Australia's primary aluminum production ranked Australia with Canada, China, and Russia as the world's leading aluminum-producing countries. Aluminum output was produced mainly from Alcoa of Australia's Point Henry and Portland smelters in Victoria, Hydro Aluminium Kurri Kurri Pty. Ltd.'s Kurri Kurri smelter in New South Wales, and Pacific Aluminum's Bell Bay smelter in Tasmania, as well as the Boyne Island smelter in Queensland and the Tomago smelter in New South Wales. Norsk Hydro ASA of Norway decided to close its aluminum operation at its Kurri Kurri smelter in October 2012 because of low aluminum prices in the world market and increased production costs. Alcoa considered shutting down its Point Henry smelter because the global aluminum price was at or below the break-even point of production costs (Australian Aluminium Council Ltd., 2012; Fitzgerald, 2013).

Antimony.—Compared with China, Australia was a relatively minor antimony producer in the world. Australia's antimony was produced from Mandalay Resources Ltd.'s Costerfield Mine in Victoria and Straits Resources Ltd.'s Hillgrove Mine in New South Wales. Straits Resources placed the Hillgrove Mine on care-and-maintenance status and planned to sell the mine to Bracken Resources Pty Ltd. (Straits Resources Ltd., 2012).

Mandalay Resources acquired the Augusta Mine in December 2009 after its operation was suspended in 2008 because of low antimony prices. Mandalay restarted the exploration in 2010 and discovered new reserves deeper on the Augusta E and Augusta W lodes. The company discovered additional resources in the Cuffley lode and found new veins in the district and subsequently renamed the Augusta Mine as the Costerfield Mine. At yearend 2012, the mine had ore reserves of 534,000 t at an average grade of 9.4 grams per metric ton (g/t) gold and 4.0% antimony. The mine produced 2,481 t of antimony, 90.5 t of silver, and 1.1 t of gold in 2012, which was higher than the company's output target. Mandalay planned to increase antimony production to about 2,800 t in 2013 (Mandalay Resources Ltd., 2013, p. 3).

In Nullagene, eastern Pilbara, Western Australia, the Blue Spec deposit was first discovered in 1906, and the Gold Spec deposit was discovered in 1956. Anglo American plc of the United Kingdom shut down the Blue Spec Mine in 1978 because of poor metal recoveries, and Chase Minerals Ltd. shut down the Gold Spec operation in 1992. Northwest Resources Ltd.

acquired these properties in the 2000s. The Blue Spec Shear (also known as Nullagene), which included the Blue Spec and the Gold Spec Mines, had total mineral reserves of 646,000 t grading 15.8 g/t gold and 1.2% antimony. Northwest Resources planned to develop an underground mine that would produce 1,900 metric tons per year (t/yr) of antimony and 2.0 t/yr of gold for a 5-year mine life starting in 2013 (Northwest Resources Ltd., 2013, p. 6–7).

Copper.—Australia's copper resources occur largely at Olympic Dam in South Australia and at Mount Isa in Queensland. Other significant copper resources are located at the CSA and the Northparkes deposits in New South Wales; the Ernest Henry, the Mammoth, and the Osborne deposits in Queensland; and the Golden Grove and the Nifty deposits in Western Australia. Australia's mined copper output ranked the country among the top five producers in the world. In 2012, South Australia accounted for 31% of the country's mined copper output, followed by Queensland, 27%; Western Australia, 21%; and New South Wales, 18%. Tasmania's mined copper output was mainly from Mount Lyell, which accounted for 3% of total mined copper output.

Australia's copper mine production for the year was slightly lower than that of 2011. The lower mined copper production was the result of a planned decrease in production at the Ernest Henry Mine as the mine transitioned from an open pit operation to an underground operation. Also, Kagara Ltd.'s mining operations in Queensland were placed on care-and-maintenance status because the company went into voluntary administration in early 2012. Several new mines were expected to start up, including Sandfire Resources NL's Degrussa and MMG Ltd.'s Golden Grove operations, and, as a result, mined copper output was expected to increase during the next 2 years. The decreases in refined copper production reflected the power outages that had been disrupting output at the Olympic Dam and the Port Pirie operations and lower concentrate production at the Ernest Henry Mine.

Australia exported a total of 2.0 Mt of copper concentrates compared with 1.8 Mt in 2011. China was the leading destination for exports of Australian copper concentrates and received 33% of the total exported; India, 30%; Japan, 23%; the Republic of Korea, 7%; and others, 7%. Australia decreased its refined copper exports to 370,000 t in 2012 from 379,000 t in 2011. China was the leading destination and received 31% of the total exported; Malaysia, 24%; Thailand, 16%; Taiwan, 12%; and Indonesia, 9% (Australian Bureau of Resources and Energy Economics, 2013, p. 165).

Sandfire Resources NL discovered the high-grade DeGrussa volcanogenic copper-gold deposit in the northeastern part of its Doolgunna tenement area, which is located 900 kilometers (km) northeast of Perth, in 2009. Exploration work continued in 2010. As of March 2012, the mine contained indicated and inferred mineral resources of 11.91 Mt of ore at average grades of 5.3% copper and 1.6 g/t gold. Construction of the DeGrussa project started in 2011 and was completed in the fourth quarter of 2012. Mining began in February 2012, and the 1.5-Mt/yr concentrator was fully operational in October. The project was being developed as an open pit and underground mine. The company planned to have two-stage open pit mining in operation within 2 years to mine 143,000 t/yr of high-grade direct-shipment ore at average grades of 25.6% copper and 2.5 g/t gold and 298,000 t/yr of sulfide material at average grades of 6.0% copper and 2.4 g/t gold. The company planned for the underground operation to extract a total of 10.72 Mt of ore grading 5% copper and 1.7 g/t gold during a mine life of about 7 years. Sandfire Resources had secured the sale of 100% of its direct-shipment ore to MRI Trading AG and Yunnan Copper Corporation Ltd. of China in 2011. The nondirect shipment ore would be processed together with underground ore to produce copper concentrate at an average grade of 27% copper for exporting to international customers. The company also planned to recover copper from copper oxide ore through a heap-leaching and solvent extraction-electrowinning process (Sandfire Resources NL, 2013, p. 11).

Newcrest Mining Ltd.'s Cadia Valley mines were located in the central part of western New South Wales. After 14 years of operation, Newcrest placed the Cadia Hill open pit copper and gold mine on care-and-maintenance status in June 2012 after mining of the Cutback 3 area was completed. The company's production plan did not include plans to mine the Cutback 4 area. The Ridgeway underground copper-gold mine is located 3 km from the Cadia Hill Mine. The company used sublevel cave extraction and block caving technology to mine Ridgeway. The mine contained about 180 Mt of mineral resources grading 0.35% copper, 0.69 g/t silver, and 0.35 g/t gold. The Cadia East Mine is adjacent to the Cadia Hill Mine and had mineral resources of 2,800 Mt grading 0.26% copper, 0.57 g/t silver, and 0.41 g/t gold. The construction of the Cadia East Mine started in 2010 and was completed in 2012. Newcrest also expanded the capacity of the existing Cadia Valley processing plant to 26 Mt/yr from 24 Mt/yr. The company expected that the output of the Cadia Valley operations would increase to 90,000 t of copper and 25 t of gold in 2016 (Newcrest Mining Ltd., 2013a, p. 4–5; 2013b, p. 7–9).

China Minmetals Corp. (CMC) through its subsidiary China Minmetal Nonferrous Metals Co. Ltd. established the Mineral and Metal Group Australia Ltd. (MMG) in 2009 to acquire the majority of OZ Minerals's assets in Australia, Indonesia, and Thailand. In 2010, MMG was acquired by Minmetals Resources Ltd., which was a subsidiary of CMC and was listed on the Hong Kong Stock Exchange. In 2012, Minmetals Resources Ltd. changed the registered name of the company to MMG Ltd. MMG Ltd. operated the Century, the Golden Grove, the Rosebery, and the Sepon Mines in Australia. The Golden Grove operation included the Gossan Hill Mine and the Scuddles Mine. In 2010, MMG completed a feasibility study evaluating the development of an open pit operation to mine the oxide resource above the Gossan Hill underground mine. The open pit was expected to extend the copper mining operation to 2016. In 2011, MMG approved $22 million for the development of an open pit at Gossan Hill as part of the Golden Grove operation. The open pit operation began in January 2012 and was expected to produce a total of 235,000 t/yr of copper concentrates containing 25% copper during its mine life (Resource Information Unit, 2012, p. 273–274).

Gold.—Gold mine output in Australia ranked the country among the world's top three producers, together with China and the United States. In 2012, Australia's mined gold output decreased by about 4% from that of 2011, and output of refined gold decreased by about 3%. The decrease in production was attributed to a number of mines taking advantage of high gold prices to target lower ore grades that would have been uneconomic to extract at lower prices. Western Australia remained the leading gold-producing State, with a 71.6% share, followed by New South Wales, 10.8%; Queensland, 6.4%; and the Northern Territory, South Australia, Tasmania, and Victoria accounting for the remaining 11.2% share. The country's gold resources occur and are mined in all States, as well as in the Northern Territory, and much of the gold was produced from large open pit mines. Owing to higher prices of gold in the world markets, gold operators could afford to reduce the grade of ore fed into their processing plants in order to extend mine life. In 2012, Australia exported 282 t (compared with 308 t in 2011) of refined gold produced from domestic mines or from imports of gold dore and scrap that were shipped from overseas, refined into gold bullion, and then reexported. Weaker global demand for gold bullion coins and bars had contributed to the decrease in refined gold exports. The United Kingdom replaced India as the leading destination for Australian refined gold. The United Kingdom, India, Singapore, and Thailand accounted for 56.4% of Australia's total gold exports (Australian Bureau of Resources and Energy Economics, 2013, p. 167).

Regis Resources Ltd. had operations at Dukeston in the northeastern goldfield in Western Australia and McPhillamys gold project in the central part of western New South Wales. The Dukeston gold project was located 130 km north of Laverton. The company completed the construction of the Moolart Well Mine in 2010, which produced about 3.1 t/yr (100,000 troy ounces per year) of gold for 5 years. The company also completed the construction of the Garden Well Mine at Dukeston in 2012. The mine life of the Garden Well Mine was about 9 years at an average production rate of 5.6 t/yr (180,000 troy ounces per year) of gold. The Garden Well Mine had mineral resources of 61.9 Mt grading 1.29 g/t gold in 2011. The company started the construction of the Rosemont Mine, which is located 9 km northwest of the Garden Well Mine, in 2012. The mine had reserves of 12.0 Mt grading 1.72 g/t gold and 33.2 Mt of indicated and inferred resources grading 1.62 g/t gold in 2013. Regis planned to mine 1.5 Mt/yr of ore to produce about 2.5 t (80,000 troy ounces) of gold in 2013. The Petra gold deposit is located 15 km east-southeast of the Moolart Well Mine. The company planned to continue exploring for gold resources in Dukeston during the next several years. Regis completed the acquisition of the McPhillamys gold project, which was located in the Bathurst region, from Alkane Resources Ltd. and Newmont Exploration Pty Ltd. in November 2012. The McPhillamys gold project had a total mineral resource of 57.4 Mt of ore grading 1.36 g/t gold. The company planned to continue exploring in the region in 2013 (Regis Resources Ltd., 2012, p. 5–9; 2013, p. 4–8).

The Paddington goldfield, which is located about 35 km north of Kalgoorlie, Western Australia, included the Enterprise, the Havana, the Homestead, the Janet Ivy, the Navajo Chief, the Nemesis, and the Robinson deposits. Norton Gold Fields Ltd. acquired the Paddington goldfield in 2007. Construction of the Homestead underground mine was started in 2009 and the mine was put into operation in 2010. The company started the construction of the Navajo Chief open pit mine in 2010, and the mine started production in late 2010. Ores from these mines were shipped to the Paddington Mill for processing. In 2012, Zijin Mining Group Co. Ltd. of China through its Hong Kong-based wholly owned subsidiary Jinyu International Mining Co. Ltd. offered an off-market takeover to become a majority shareholder in Norton. The Board of Norton Gold Fields Ltd. approved US$40 million for the development of the Enterprise open pit mine in 2012. The Paddington goldfield had proven and probable mineral reserves of 22.8 Mt of ore grading 1.53 g/t gold. The company continued to explore for mineral resources in the area (Norton Gold Fields Ltd., 2013, p. 26).

Evolution Mining Ltd. was formed through the merger of Catalpa Resources Ltd. and Conquest Mining Ltd. in 2011. Newcrest Mining was the major shareholder in Evolution Mining. Evolution Mining operated three gold mines in Queensland—the Cracow, the Mount Rawdon, and the Pajingo Mines—and the Edna May gold mine in Western Australia. In 2011, Evolution Mining invested US$180 million to develop the Mount Carlton open pit gold-silver-copper mine, which is located 150 km south of Townsville, Queensland. Construction of an 800,000-t/yr ore processing plant started in December and was scheduled to be completed at yearend 2012. In 2012, the mine had mineral resources of 22 Mt grading 19 g/t silver, 1.7 g/t gold, and 0.24% copper. The mine life was about 12 years. The production of concentrate was expected to begin in March 2013 (Evolution Mining Ltd., 2012, p. 28; 2013, p. 7).

Reed Resources Ltd. acquired the Meekatharra gold project from previous owners in 2011. The project included a tenement holding of about 1,000 square kilometers within the Murchison District of Western Australia. The company committed a US$40 million to upgrade and refurbish the 3-Mt/yr Bluebird processing plant in the Yaloginda region. In Stage 1 of the project, the plant would recover about 4.2 t (134,000 troy ounces) of gold during the first 19 months of the operation from the Yaloginda region. In Stage 2, the company planned to expand the open pit operation and exploit the underground operation at Paddy's Flat to the north and Reedys to the south of the ongoing operations at Yaloginda. The company would continue to explore for mineral resources in its tenement area. As of June 2012, the Meekatharra gold project had mineral resources of 63.9 Mt grading 1.8 g/t gold to sustain a 10-year mine life with an annual production of between 3.1 t (100,000 troy ounces) and 4.7 t (150,000 troy ounces) of gold at a total cost of less than US$1,000 per troy ounce (Reed Resources Ltd., 2012, p. 8; 2013, p. 4).

Iron Ore.—Australia was among the top three iron ore producers (in terms of iron content) in the world, along with Brazil and China. Australia's most significant iron ore mines were located in the Pilbara region of Western Australia, which accounted for 97.1% of the country's total iron ore production, followed by South Australia, 2.1%, and the Northern Territory and Tasmania, 0.4% each. Owing to its limited domestic demand and production capacities for iron and steel, Australia exported more than 90% of its iron ore output to such Asian countries as China (the world's leading importer of iron ore), Japan, the

Republic of Korea, and Taiwan. In 2012, Australia's iron ore and pellet exports increased to 493 Mt from 439 Mt in 2011. Faced with declining iron ore grades of domestic iron ore mines during the past two decades, Chinese iron and steel producers relied on imported iron ore to meet their demand, and this trend was expected to continue during the next 5 years. Australia's iron ore exports to China increased to 358 Mt in 2012 from 306 Mt in 2011. Australia's iron ore exports to the Republic of Korea increased to 46 Mt from 45 Mt and those to Japan and Taiwan remained the same at 75 Mt and 12 Mt, respectively (Australian Bureau of Resources and Energy Economics, 2013, p. 168).

As a result of an increase in investment during the past several years, expansions and new mines in Australia were expected to support strong growth in iron ore exports from Australia. Australian iron ore producers were expanding their iron ore production facilities to meet expected increased demand from Australia's neighboring countries. A number of greenfield and brownfield iron ore projects were at various stages of development. Rio Tinto expanded the capacity of its Pilbara iron ore operations to a total of 360 Mt/yr in 2015. BHP Billiton was expected to increase production capacity to 220 Mt/yr in 2014. Fortescue Metals Group's Chichester Hub and Solomon Hub expansion projects were projected to increase the company's iron ore output capacity to 155 Mt/yr in 2014 (BHP Billiton Ltd., 2013, p. 33; Rio Tinto plc, 2013b, p. 31).

Australia-based CITIC Pacific Mining Management Pty Ltd. (a subsidiary of Hong Kong-based CITIC Pacific Ltd., which was, in turn, a member of China's state-owned CITIC Group) had invested about $5 billion to develop its Sino iron ore project at Cape Preston, which is located 100 km southwest of Karratha in Western Australia. The company had planned to produce about 21 Mt/yr of 67% iron in concentrates and 6 Mt/yr of pellets in 2011. Concentrates would be moved by conveyor belt to barges, loaded into offshore vessels at Cape Preston, and then shipped to China. Owing to a shortage of skilled laborers, however, the first production line was completed only in November 2012. The second production line was scheduled to be completed in May 2013, and the remaining four production lines were planned to be put into operation in 2014. CITIC Pacific signed an agreement with Mineralogy Pty Ltd. to mine 2 Gt of magnetite ore between 2006 and 2008. CITIC Pacific had four options to acquire an additional 4 Gt of magnetite ore (1 Gt per option) at the same location. In April 2012, CITIC Pacific exercised the first option. Mineralogy alleged that CITIC Pacific was liable for royalties on the magnetite ore mined; however, CITIC Pacific argued that the royalty was due when the ore was ready for processing and not when it was mined. The dispute was to be ruled on by the Australian court in 2013. CITIC Pacific and Mineralogy were also in discussions about other issues related to access rights to the port that CITIC Pacific had built at Cape Preston (CITIC Pacific Ltd., 2013, p. 43–49).

The first phase of the development of the Karara iron ore mine, which was a joint venture of Gindalbie Metals Ltd. (50%) and Angang Group Investment (Australia) Pty Ltd. (50%) (a subsidiary of Anshan Iron and Steel Group Corp. of China), continued in 2012. The Karara deposit, which is located 220 km east of Geraldton, Western Australia, had magnetite iron ore resources of 1.4 Gt at average grades of 27.2% iron, 46.0% silicon oxide, 5.5% aluminum oxide, and 0.05% phosphorus. The area also had hematite iron ore resources of 6.1 Mt at average grades of 59.8% iron, 7.8% silicon oxide, 1.71% aluminum oxide, and 0.08% phosphorus. The company started mining the hematite iron ore in 2011 and planned to complete mining activities at the Blue Hill North area in early 2013. The concentrator had a design capacity to produce 8 Mt/yr of concentrates containing 68% iron. Iron ore would be transported by railway from Karara to Geraldton Port for shipping. The company performed a feasibility study on the proposed expansion of iron ore operations at Karara to a total of 16 Mt/yr of hematite and magnetite (Gindalbie Metals Ltd., 2012, p. 12–13; 2013).

The West Pilbara iron ore project, which was a joint venture project of Aquila Resources Ltd. (50%), AMCI Group (25.5%), and Pohang Iron and Steel Co. Ltd. (24.5%) of the Republic of Korea, is located 70 km south of Pannawonica, Western Australia. The feasibility study of Stage 1 development of the project was completed in 2010. The Mount Stuart and Red Hill deposits accounted for the majority of iron ore resources to be mined during Stage 1. The company planned to mine a total of 70 Mt of iron ore during a 14-year mine life at the Mount Stuart operation and a total of 289 Mt of iron ore in a 16-year mine life at the Red Hill operation. The Western Pilbara area had total iron ore resources (measured, indicated, and inferred) of 2.2 Gt that ranged from 54.4% to 60.8% iron, 3.9% to 11.9% silicon oxide, 2.4% to 3.8% aluminum oxide, and 0.04% to 0.16% phosphorus. When iron ore prices reached a 3-year low in September, the partners decided to suspend the project until June 2013. China's Boshan Iron and Steel Co. had a 14% share in Aquila Resources (Aquila Resources Ltd., 2012, p. 16–18; 2013; Sydney Morning Herald, The, 2013).

Lead, Silver, and Zinc.—Australia's lead, silver, and zinc mines were predominantly based on ore bodies with zinc as the major component and lead and silver as byproducts. An exception was BHP Billiton's Cannington underground mine in the State of Queensland, where lead and silver were major components and zinc was a minor component. In 2012, Australian zinc mine production increased slightly. The output of zinc was expected to increase during the next 2 years because Xstrata plc planned to expand the Black Star Open Cut Deeps at the Mount Isa Mine and the Handle Bar Hill Mine and also to develop the Lady Loretta deposit. Queensland remained the leading lead- and zinc-producing State in Australia. In 2012, Australia exported 469,000 t of lead concentrates compared with 428,000 t in 2011. China remained the leading destination for Australian lead concentrate exports and accounted for 36.0% of the total, which was an increase from 35.0% in 2011, followed by the Republic of Korea, 26.0%; Japan, 14.9%; and others, 23.1%. Australia exported 2.38 Mt of zinc concentrates in 2012. China replaced the Republic of Korea to become the leading destination for Australia's zinc exports, accounting for 31.3% of the total, followed by the Republic of Korea, 19.2%; the Netherlands, 14.4%; Japan, 12.2%; and other countries in the world, the remaining 22.9%. Australia also exported 201,000 t of refined lead, for which the Republic of Korea replaced Malaysia as the leading destination, followed by India, Malaysia, Vietnam, and Thailand. In 2012, zinc metal exports increased to 455,000 t and went to such destinations as, in descending order of volume exported, China, the United States, Taiwan, Hong Kong, and Malaysia. Australia's zinc production was expected to increase during the next 2 years (Australian Bureau of Resources and Energy Economics, 2013, p. 169, 181).

Xstrata's subsidiary, Xstrata Zinc, operated several lead and zinc mines and a processing plant in Mount Isa, Queensland. The company planned to expand the output capacities of its Black Star open pit mine and George Fisher underground mine. The executive committee of Xstrata approved $246 million to increase the output capacity of the George Fisher Mine. Zinc reserves in the mine had increased to 70 Mt in 2010 from 33 Mt in 2003 when Xstrata acquired the operation. Xstrata completed the expansion of the George Fisher underground mine in October, 6 months ahead of schedule. The output capacity was increased to 4.5 Mt/yr from 3.5 Mt/yr. The company planned to mine ore at a depth of 400 meters (m) below the surface, or 100 m below the current design of the Black Star open pit mine, and the life of the mine at the current production rate of 4.6 Mt/yr would be extended to 2016.

Mine construction at Xstrata's Lady Loretta lead-silver-zinc deposit in northwestern Queensland was also completed in 2012. The deposit, which is located 140 km northwest of the Mount Isa operation, had reserves of 13 Mt grading 15% zinc, 5.3% lead, and 89 g/t silver at yearend 2012. Lady Loretta was designed to produce 1 Mt/yr of ore; however, the company decided to expand the operating capacity to 1.6 Mt/yr by 2016. Ore from these mines would be processed at the Mount Isa concentrator. Xstrata submitted an environmental impact assessment for the phase 3 expansion of the McArthur River operation to the government of the Northern Territory for approval. The company planned to invest US$360 million to increase production to 300,000 t/yr from 200,000 t/yr of zinc in concentrate in 2014. The McArthur River Mine had reserves of 110 Mt grading 10% zinc, 4.7% lead, and 47 g/t silver. In the area of Mount Isa, Xstrata had reserves of 235 Mt grading 8.7% zinc, 4.2% lead, and 55 g/t silver (Xstrata plc, 2013a, p. 26; 2013b, p. 44).

Nickel.—Australia's main nickel ores were primary sulfides of nickel, which occur as lodes within mafic and ultramafic (iron- and magnesium-rich) igneous rocks that have a volcanic and subvolcanic origin. Western Australia was the leading State for mined nickel output and accounted for more than 90% of the country's total output. The top five nickel producers accounted for 80% of the total sales. BHP Billiton's Nickel West project was Australia's leading nickel operation. Nickel West included the Leinster and the Mount Keith Mines. A number of smaller sulfide nickel operations were operated by Mincor Resources NL and Xstrata Nickel Australia Pty Ltd. [a subsidiary of Xstrata plc (Xstrata)]. The increase in mined nickel output was a result of the redevelopment and restart of OJSC MMC Norilsk Nickel (Nornickel) of Russia's nickel operations in Australia and First Quantum Mineral Ltd.'s Ravensthorpe Mine in late 2011.

Nornickel shut down its nickel operations in 2009 and 2010, and most of its nickel operations remained closed in 2011. The company tried to enrich its nickel at the Lake Johnston operation, which was located about 500 km east of Perth in Western Australia. The concentrator was started and reached design capacity during the second half of 2011. The ore for the concentrator was sourced from the Maggie Hays Mine. Nornickel planned to use its hydrometallurgical technology (Activox® process) at its processing facility at Cawse to process nickel sulfide ore from the company's deposits in Australia. Nornickel planned to produce a nickel hydroxide solution that would contain about 50% nickel and then refine it into the metal product. In 2012, the Lake Johnston operation produced 8,975 t of nickel in concentrates. In early 2013, owing to the low world nickel price, the company placed the Lake Johnston operation on care-and-maintenance status. Nornickel also planned to sell its other Australian assets, including the Waterloo nickel operation and the Honeymoon Well nickel project, which the company had planned to develop by 2017. Australian nickel output was expected to decrease in 2013 as a result of Xstrata placing its Cosmos Mine on care-and-maintenance status and BHP reducing the output of its Nickel West operation by 30% (Heber, 2013; OJSC MMC Norilsk Nickel, 2013).

Tin.—Compared with other tin-producing countries in the Asia and the Pacific region, Australia was not a significant tin producer. Australia's tin was mined mainly in Tasmania, and to a lesser extent, in Western Australia. In Western Australia, tin production was mainly from Iluka Resources Ltd.'s heavy-mineral sand operation, but the company had not released any tin preconcentrate information. In Tasmania, tin was produced from Metal X Ltd.'s tin operations. Tin concentrates were smelted at Global Advanced Metals Pty Ltd.'s Greenbushes smelter. No primary refined tin production was reported in 2012. In 2012, Australia imported 506 t of refined tin and exported 13,399 t of tin concentrates (Australian Bureau of Resources and Energy Economics, 2013, p. 178).

In 2010, Metals X sold 50% of its interest in its Tasmanian tin assets to YT Parksong Australia Holding Pty Ltd. (a joint venture between L'sea Resources International Holdings Ltd. and Yunnan Tin Group of China). The former name of L'sea was Goodtop Tin International Holdings Ltd., which was incorporated in the Cayman Islands. The two parties established a joint-venture company, Bluestone Mines Tasmania Joint Venture Pty Ltd., to manage the assets. The joint venture completed the mine development at the North Renison decline in 2012 and started mining from both the North Renison and the South Renison declines at a rate of about 60,000 metric tons per month (t/mo) to produce about 7,000 to 8,000 t/yr of tin in concentrates. The joint venture estimated that the Renison Mine had mineral resources of 2.97 Mt grading 1.38% tin and 0.27% copper in 2012. The Mount Bischoff Mine ceased operations at yearend 2010 and was placed on care-and-maintenance status in 2012; significant tin resources remained at depth under the mine pit, and numerous historically mined areas remained unexplored (L'sea Resources International Holdings Ltd., 2013, p. 14; Metals X Ltd., 2013, p. 13–20).

Consolidated Tin Mines Ltd.'s major shareholder, Hong Kong-based Snow Peak Mining Pty Ltd. (SPM), completed the acquisition of Kagara Ltd.'s Central Region project for $40 million at Mount Garnet, near Cairns in northern Queensland. The Central Region project included the Baal Gammon open pit copper mine and the Mount Garnet processing plant, which had a designed capacity of 1 Mt/yr.

The processing plant had both copper and polymetallic circuits, and each circuit had a capacity to process 500,000 t/yr. SPM contracted Consolidated Tin to manage the processing plant and to process ore from the Baal Gammon Mine. The Baal Gammon Mine was owned by Monto Mineral Ltd. but was mined under a royalty agreement by Kagara. Consolidated Tin was expected to complete the feasibility study on the Mount Garnet tin project, and, if the prospecting feasibility study result is positive, Consolidation Tin and SPM would form a 50-50 joint venture to develop the Mount Garnet tin project. Tin production at Mount Garnet could start in 2014 and had the potential to produce 5,000 t/yr of tin. The processing plant could recover tin byproducts from Baal Gammon. The four deposits—Deadmans Gully, Gillian, Pinacles, and Windermere—in the Mount Garnet area had total resources of 10.57 Mt grading 0.44% tin (Consolidated Tin Mines Ltd., 2013a; 2013b, p. 4).

Titanium and Zirconium.—Australia's titanium and zircon were produced mainly from mineral sands. Iluka Resources Ltd. was the leading heavy-mineral producer in Australia, and its operations were located in the Eucla basin in South Australia, the Murray basin on the border of New South Wales and Victoria, and the Perth basin in Western Australia. Jacinth-Ambrosia in the Eucla basin was the major zircon production site in Australia. Rutile was produced from the Murray basin, where ilmenite and zircon were in the production stream. The Perth basin was the main supply source of ilmenite for synthetic rutile. The company operated two mineral separation facilities—Hamilton in Victoria and Narngulu in Western Australia. The Narngulu mineral separation plant was upgraded to process an additional 300,000 t/yr of heavy-mineral concentrate. Owing to weak demand for mineral sands in the global market, only two of its four synthetic rutile kilns were operated in 2012. These kilns used ilmenite to produce various synthetic rutile products containing a titanium oxide content of between 85% and 95%. The Chinese and United States construction and housing sectors were significant sources of demand for titanium dioxide and zircon. Because economic growth in China was expected to slow down and economic recovery in the United States remained weak, Iluka planned to reduce its production of rutile, synthetic rutile, and zircon and to idle some operations in Australia in 2013 (Iluka Resources Ltd., 2013a; 2013b, p. 12–16).

Tungsten.—Australia's tungsten was produced from three mines—Wolfram Camp and Mount Carbine in Queensland and Kara in Tasmania. The Wolfram Camp Mine, which is located 90 km west of Cairns, was discovered in 1894. In 2011, Metallic Minerals sold its 85% interest in Wolfram Camp to Deutsche Rohstoff AG of Germany. Deutsche Rohstoff acquired Tropical Metals Pty Ltd., which held a 15% interest in the Wolfram Camp Mine and 100% of the Bamford Hill deposit, which was located 25 km south of Wolfram Camp. The Wolfram Camp Mine was reopened in July 2012; the mine had resources of 1.42 Mt grading 0.6% tungsten trioxide and 0.12% molybdenum. The company planned to produce about 7,000 t of tungsten concentrates and 800 t of molybdenum concentrate during the next 4 years (Deutsche Rohstoff AG, 2012).

Carbine Tungsten Ltd.'s Mount Carbine Mine was closed in 1987 because of the low price of tungsten. In 2010, the company commissioned a feasibility study to recover tailings and mineralized wastes. The report indicated that the tailings could be treated to produce a salable mixed concentrate with a grade of 52% tungsten trioxide. The company decided to proceed to extract tungsten from the tailings. The tailings retreatment plant was completed in 2012. The company signed an offtake agreement with Mitsubishi Corp. of Japan to supply all concentrates produced from the retreatment plant at a price based on the monthly London Metal Bulletin price. Carbine also commissioned a hard-rock feasibility study within the existing mine lease area. The mine had resources of 47 Mt grading 0.13% tungsten trioxide. The company expected that the hard-rock project and tailings retreatment plant would produce about 21,800 metric ton units of tungsten trioxide per month (Carbine Tungsten Ltd., 2013, p. 5).

Industrial Minerals

Cement.—Australia had three major integrated cement companies (Adelaide Brighton Cement Pty Ltd., Blue Circle Southern Cement Ltd., and Cement Australia Pty Ltd.) and a number of small independent companies. The three major cement companies accounted for all integrated production of clinker and cement in Australia. Domestic clinker capacity was about 8 Mt/yr and cement capacity was about 10 Mt/yr. The highly efficient dry precalciner technology accounted for 87% of Australia's cement production in 2012. During the past several years, the three integrated cement producers produced about 9 Mt/yr for the domestic market. Small independent producers used imported clinker from Asian countries to produce cement and accounted for about 15% of the domestic supply of cement.

The Government implemented a carbon tax in 2011 that affected the cement sector in Australia. This is because carbon dioxide is emitted as a product of the chemical reaction during clinker production. To reduce carbon dioxide emission, some Australian cement companies were required to technically upgrade their production plants or relocate their operations overseas. Byproducts used in blending included fly ash from coal-fired powerplants and ground-granulated blast furnace slag from steel plants. The Government also introduced a Coastal Trading Bill in 2012 that would increase transshipping costs for dry bulk commodities, such as cement. Owing to weak demand for cement in the construction sector, Cement Australia Pty Ltd. shut down part of its operations in Queensland in late 2012 (Cement Industry Federation, 2013, p. 5).

Lithium.—Australia's lithium was produced by Talison Lithium Ltd.'s Greenbushes Mines and Galaxy Resources Ltd.'s Mount Cattlin Mine in Western Australia. The increase of Australia's economic demonstrated resources of lithium in 2011 from those of 2010 was a result of a large increase of the identified resources in the Greenbushes spodumene deposit. The lithium resource at the Mount Cattlin Mine was 17.2 Mt at an average grade of 1.09% lithium oxide. The mined pegmatite ore was processed onsite to produce a spodumene concentrate and a tantalum byproduct. The processing plant was designed to process 1 Mt/yr of ore to produce about 137,000 t/yr of spodumene concentrate grading 6% lithium oxide and 25 t/yr (56,000 pounds per year) of contained tantalum oxide for

18 years. In 2012, Galaxy Resources mined 454,912 t of ore at an average grade of 1.22% lithium oxide to produce 54,047 t of spodumene. Galaxy Resources exported its spodumene concentrate to its lithium carbonate plant in China. In July, the company decided to halt production at Mount Cattlin because an accident took place at Galaxy Resources' Jiangsu lithium carbonate plant in China. As a result, the Jiangsu plant was shut down for the second half of 2012. Spodumene concentrate was stockpiled at the Mount Cattlin site. The Board of Galaxy Resources decided to stop mining and instead signed a 3-year purchase agreement with Talison Lithium to supply spodumene concentrate to its Jiangsu plant (Galaxy Resources Ltd., 2013, p. 9).

As of September 2012, Talison Lithium's lithium resource at Greenbushes was 61.5 Mt at an average grade of 2.8% lithium oxide, and the estimated life of the Greenbushes Mine had been increased to 24 years. Talison Lithium invested $65 million to double the output capacity to 1.5 Mt/yr of ore feed to produce about 740,000 t/yr of lithium concentrate (about 100,000 t/yr of lithium carbonate equivalent). The construction of the Stage 2 expansion started in 2011 and was completed in the second quarter of 2012. In 2012, Talison Lithium and Windfield Holding Pty Ltd. [a subsidiary of Chengdu Tianqi Industry (Group) Co. Ltd. of China] agreed to acquire the balance of the ordinary shares that it did not already own and options in Talison for C$7.50 (US$7.90) per share. Tianqi held a 19.99% interest in Talison Lithium before the acquisition. The Australian Foreign Investment Board had no objections to Tianqi's acquisition in November 2012 (Talison Lithium Ltd., 2012, p. 13; 2013).

Magnesium Compounds.—All Australian magnesite deposits were mined by the open pit method. The Queensland Magnesia Pty Ltd.'s Kunwarara Mine, which is located 70 km northwest of Queenstown in Queensland, was the leading operating magnesite mine in the country. About 3 Mt/yr of ore was mined and processed at Kunwarara. The beneficiated magnesite was transported to the company's Parkhurst plant for calcination to produce the required magnesia products, such as high-grade deadburned, electrofused, and calcined magnesite. The Parkhurst plant had a designed capacity of 320,000 t/yr. In recent years, the Parkhurst plant operated at about 30% of its designed capacity. Sibelco Group of Belgium acquired Queensland Magnesia in 2012 (Resource Information Unit, 2012, p. 201).

There were two active magnesite mines—Thuddungra in New South Wales and Salt Creek in South Australia. The processing plant at the Thuddungra Mine had the capacity to produce 80,000 t/yr of high-purity magnesium carbonate that contains low contents of iron. During the past several years, production was between 35,000 t/yr and 40,000 t/yr. The Salt Creek Mine produced products that had magnesium carbonate content that ranged from 50% to 74% (Resource Information Unit, 2012, p. 110, 234).

Rare Earths.—China dominated global production of rare earths and accounted for more than 90% of the world total in 2011. China's share of rare-earth output was expected to decrease during the next several years. China was also a leading rare-earth consumer. During the past decade, the Chinese Government restricted rare-earth production and exports. As a result, the availability of rare earths in the international market became tighter during the past several years. Small-scale production of rare earths had been reported in Australia in the 20th century but records on these activities were incomplete. Lynas Corp. Ltd. started construction of an open pit mine and a concentration plant at the Mount Weld deposit in 2007; the deposit was located 35 km south of Laverton, Western Australia, and mining started at the Central Lanthanide pit in 2010. The construction of the concentration plant started in 2010, and the plant was put into operation in 2011 to produce at a target grade of 36% rare-earth oxide (REO) in concentrates; the recovery rate was expected to be 68.7%. The plant was designed to process 121,000 t/yr of ore and to produce 33,000 t/yr of rare-earth concentrate. The company reported a stockpile of 15,200 t of concentrates containing 5,410 t of REO at the end of December 2012. Some of the rare-earth concentrates were planned to be shipped to Lynas' advanced materials plant in Kuantan, Malaysia, in 2012. Owing to legal challenges from local residents in Kuantan, the Malaysian Government delayed issuing the temporary operating license to the plant.

Lynas also planned to develop the Duncan deposit, which is located southeast of the Central Lanthanide deposit. The mineral resource at Duncan was estimated to be 8.9 Mt grading 4.8% REO. The Duncan deposit could be exploited using the opencut mining method. The cost of developing the Duncan deposit was estimated to be $600 million. Lynas and Sojitz Corp. of Japan formed a strategic alliance and signed an offtake, distribution, and financing agreement to enable Lynas to accelerate the development of the phase 2 operation. Under the agreement, Sojitz was allocated a minimum of 8,500 t/yr of rare-earth products for the Japanese market for 10 years (Lynas Corp. Ltd., 2012, p. 51; 2013, p. 24).

Mineral Fuels and Related Materials

Coal.—Australia ranked behind China and India in the Asia and the Pacific region in coal output; the country, however, was the world's leading exporter of coal. Queensland and New South Wales were Australia's leading coal-producing States and accounted for more than 95% of the country's total output. In 2012, Australia mined 477 Mt of raw black (bituminous and anthracite) coal, of which 365 Mt was salable coal. Open pit coal mines accounted for about 79% of the total output. Coal from Queensland was mainly mined from the Bowen basin, which extends south from Collinsville to Blackwater and Moura, and from mines at Blair Athol, Newlands, and near Brisbane. Coal from New South Wales was mined near the eastern and western edges of the large Sydney Gunnedah basin. Australia exported more than 315.5 Mt of coal (which included 144.6 Mt of metallurgical coal and 170.9 Mt of thermal coal) compared with 280.6 Mt in 2011. Japan received 28.1% of Australia's metallurgical coal exports followed by India, 19.2%; China, 15.9%; the Republic of Korea, 8.2%; and others, 28.6%. Japan was also the leading destination for Australian thermal coal exports, receiving 44.0% of those exports followed by the Republic of Korea, 17.6%; China, 9.8%; Taiwan, 9.6%; and others, 19.0%. Domestic coal consumption was about 70 Mt, of which the power sector accounted for about 85% of total domestic consumption, followed by steel, 6.7%; cement,

1.3%; and others, 7%. Owing to increased demand from other countries in the Asia and the Pacific region, such as China and India, Australia's metallurgical coal exports were expected to increase during the next several years (Australian Bureau of Resources and Energy Economics, 2013, p. 163).

BHP Billiton approved funding for the development of the Caval Ridge project and the expansion of the Peak Downs Mine in the Bowen basin in Queensland. The total investment was $4.2 billion, of which BHP Billiton's share was $2.1 billion. BHP Billiton's partner, Mitsubishi Development Pty Ltd. of Japan provided the remaining funds. The Caval Ridge Mine would have the capacity to produce 5.5 Mt/yr of metallurgical coal, and the capacity of the Peak Downs Mine would increase by 2.5 Mt/yr and have a mine life of more than 60 years. BHP Billiton decided to delay the development of the Peak Downs Mine, but the construction of the Caval Ridge Mine remained on schedule to be completed in 2014. The Caval Ridge project was one of the four components of BHP Billiton Mitsubishi Alliance Coal Operations Pty Ltd.'s coal growth project in the Bowen basin. The Daunia Mine, which was a new open pit coal mine and coal handling preparation plant, was scheduled to be completed in 2013; the plant would have the capacity to produce 4 Mt/yr of coal for 21 years (BHP Billiton Ltd., 2013, p. 37).

Uranium.—Australia was the third-ranked uranium producer in the world after Kazakhstan and Canada. Australia's uranium production was mainly from three mines—the Beverley, the Olympic Dam, and the Ranger Mines. A number of undeveloped deposits also occur in the Northern Territory, and in Queensland, South Australia, and Western Australia.

The Australian Government permits uranium mining provided that all the relevant environmental safeguards and health requirements are met. Regulation of Australia's uranium mines is mainly a State and Territorial government responsibility. In October 2012, the government of Queensland decided to overturn the State's ban on uranium mining. Australia exported all its uranium output under long-term contracts. Australia's uranium production was expected to decrease during the next 2 years because of the shutdown of the Ranger Mine in December 2012. The Honeymoon project, which was a joint venture of Uranium One Inc. of Canada (51%) and Mitsui & Co. Ltd. of Japan (49%), is located 75 km northwest of Broken Hill, South Australia. The Honeymoon deposit had indicated resources of 4.2 Mt at an average grade of 0.129% uranium oxide. The company planned to produce 400 t/yr (880,000 pounds per year) of uranium oxide for 6 years. The mine produced 100.2 t (220,800 pounds) of uranium in 2012. Uranium One installed only 51 production wells instead of following the original plan of adding 96 production wells in 2012. Australian Government issued its approval for Mitsui to withdraw from the joint venture. Other new projects that were under feasibility study included Mega Uranium Ltd.'s Lake Maitland project in Western Australia and Marathon Resources Ltd.'s Mount Gee project in South Australia (Uranium One Inc., 2013, p. 20).

Outlook

Australia is a natural-resource-rich country with significant resources of metallic, nonmetallic, and fuel minerals. Mineral and energy commodity production and exports are an important part of the country's economy. As a result of strong world demand for mineral commodities, especially in the Asia and the Pacific region, the Australian economy is expected to continue to benefit from higher commodity export earnings. Expenditures on mineral and energy exploration in Australia are expected to increase owing to higher costs of labor and equipment and increased global demand for mineral resources in the near future. Mineral production, such as production of bauxite, copper, iron ore, natural gas, nickel, and zinc, is expected to increase during the next several years; however, the rate of increase is expected to be slower than in the previous several years. Major projects, such as the Yarwun alumina refinery project; BHP Billiton's RGP iron ore project; Hamersley Iron's Yandicoogina iron ore expansion; Fortescue Metals Group's iron ore project; Rio Tinto's Brockman 4, Hope Downs, and Mesa A iron ore projects and Clermont and Kestrel coal projects; and Xstrata's Mangoola coal project, are expected to come onstream within this decade. If the slow economic recovery in the United States and the European Union continues, the volume of imports of manufactured goods from China and other Asia countries to the United States and the European Union is expected to continue to decline. China plans to slow down its economic growth to between 7% and 8% in the next several years from 10% during the past 10 years; as a result, China's demand for most mineral commodities from Australia is expected to decrease, and companies in Australia could, therefore, delay their investment in these projects. Western Australia is Australia's leading State for metallic mineral exports, and New South Wales and Queensland are its major coal exporting States; however, to sustain export growth, the country's infrastructure would require significant expansion and upgrading so that minerals for export could be transported from inland to port terminals. A carbon tax and mineral resource rent tax would not affect Australian mineral investment significantly. Australia is expected to remain a major mineral and fuel exporting country.

References Cited

Alcoa Inc., 2013, 2012 annual report: Pittsburgh, Pennsylvania, Alcoa Inc., March, 196 p.

Aquila Resources Ltd., 2012, Annual report 2012: Perth, Western Australia, Australia, Aquila Resources Ltd., 92 p.

Aquila Resources Ltd., 2013, West Pilbara iron ore project resource statement increase: Perth, Western Australia, Australia, Aquila Resources Ltd., April 18, 6 p.

Australian Aluminium Council Ltd., 2012, Hydro Kurri Kurri Pty Ltd. (Accessed May 9, 2013, at http://aluminium.org.au/australian-aluminium/aluminium-hydro-kurri-kurri.)

Australian Bureau of Resources and Energy Economics, 2013, Resources and energy quarterly—March quarter 2013: Canberra, Australian Capital Territory, Australia, Australian Bureau of Resources and Energy Economics, March, 184 p.

Australian Bureau of Statistics, 2013a, Australian national accounts—National income, expenditure and product: Canberra, Australian Capital Territory, Australia, Australian Bureau of Statistics, 80 p.

Australian Bureau of Statistics, 2013b, Balance of payments and international investment position: Canberra, Australian Capital Territory, Australia, Australian Bureau of Statistics, 87 p.

BHP Billiton Ltd., 2013, Annual report 2012: Melbourne, Victoria, Australia, BHP Billiton Ltd., 276 p.

Carbine Tungsten Ltd., 2013, Half year financial report 31 December 2012: Abbotsford, Victoria, Australia, Carbine Tungsten Ltd., March 15, 21 p.

Cement Industry Federation, 2013, Australian cement industry statistics 2012: Forrest, Australian Capital Territory, Australia, Cement Industry Federation, 15 p.

CITIC Pacific Ltd., 2013, Annual report 2012: Hong Kong, China, CITIC Pacific Ltd., 230 p.

Consolidated Tin Mines Ltd., 2013a, Completion of acquisition of Kagara Central Region project: Cairns North, Queensland, Australia, Consolidated Tin Mines Ltd., January 30, 7 p.

Consolidated Tin Mines Ltd., 2013b, Consolidated Tin upgrades resource to 10.57 Mt at Mt Garnet tin project: Cairns North, Queensland, Australia, Consolidated Tin Mines Ltd., June 11, 17 p.

Department of Mines and Petroleum [State of Western Australia], 2013a, Western Australian mineral and petroleum statistics digest 2011–12: East Perth, Western Australia, Australia, Department of Mines and Petroleum, 84 p.

Department of Mines and Petroleum [State of Western Australia], 2013b, Western Australian mineral and petroleum statistics digest 2012: East Perth, Western Australia, Australia, Department of Mines and Petroleum, 45 p.

Deutsche Rohstoff AG, 2012, Wolfram Camp Mining ships first concentrate: Heidelberg, Germany, Deutsche Rohstoff AG, March, 2 p.

Ellis, Margaret, 2013, Western Australia's exploration incentive scheme: Journal of the Australasian Institute of Mining and Metallurgy, April, p. 24.

Evolution Mining Ltd., 2012, Annual report 2012: Sydney, New South Wales, Australia, Evolution Mining Ltd., 116 p.

Evolution Mining Ltd., 2013, Quarterly report for the period ending 31 December 2012: Sydney, New South Wales, Australia, Evolution Mining Ltd., 18 p.

Fitzgerald, Barry, 2013, Alcoa cut threatens Geelong smelter: The Australian [Sydney, New South Wales, Australia], May 3. (Accessed May 4, 2013, at http://www.theaustralian.com.au/business/mining-energy/alcoa-cut-threatens-geelong-smelter/story-e6frg9df-1226634237908.)

Foley, Mike, 2012, No refinery required—Gov. seeks Aurukun bauxite miner: Australian Journal of Mining, November 29. (Accessed November 29, 2012, at http://www.theajmonline.com.au/mining_news/news/2012/november/november-29-2012/no-refinery-required-gov.-seeks-aurukun-bauxite-miners.)

Galaxy Resources Ltd., 2013, 2012 annual report: West Perth, Western Australia, Australia, Galaxy Resources Ltd., 100 p.

Geoscience Australia, 2013, Australian mineral exploration—A review of exploration for the year 2012: Canberra, Australian Capital Territory, Australia, Geoscience Australia, February, 20 p.

Gindalbie Metals Ltd., 2012, Annual report 2012: Perth, Western Australia, Australia, Gindalbie Metals Ltd., 98 p.

Gindalbie Metals Ltd., 2013, March 2013 quarterly: Perth, Western Australia, Australia, Gindalbie Metals Ltd., 7 p.

Heber, Alex, 2013, WA nickel mine to close this month: Australian Mining, April 9. (Accessed June 25, 2013, at http://miningaustralia.com.au/news/wa-nickel-mine-to-close-this-month.)

Iluka Resources Ltd., 2013a, Idling of mining production at Eneabba: Perth, Western Australia, Australia, Iluka Resources Ltd. press release, January 17, 2 p.

Iluka Resources Ltd., 2013b, Iluka review 2012: Perth, Western Australia, Australia, Iluka Resources Ltd., 53 p.

L'sea Resources International Holdings Ltd., 2013, Announcement of annual results for the year ended 31 December 2012: Hong Kong, China, L'sea Resources International Holdings Ltd., March 28, 26 p.

Lynas Corp. Ltd., 2012, Annual report 2012: Sydney, New South Wales, Australia, Lynas Corp. Ltd., 110 p.

Lynas Corp. Ltd., 2013, Interim unaudited condensed consolidated financial report for the first year ended December 31, 2012: Sydney, New South Wales, Australia, Lynas Corp. Ltd., 24 p.

Mandalay Resources Ltd., 2013, Annual report 2012: Toronto, Ontario, Canada, Mandalay Resources Ltd., 25 p.

Metals X Ltd., 2013, Annual report 2012: East Perth, Western Australia, Australia, Metals X Ltd., 129 p.

Mining Weekly, 2012, Too taxing?: Mining Weekly, v. 18, no. 34, September 7–13, p. 8–9.

Newcrest Mining Ltd., 2013a, Mineral resources and ore reserve explanatory notes 2012: Melbourne, Victoria, Australia, Newcrest Mining Ltd., 34 p.

Newcrest Mining Ltd., 2013b, Quarterly report for the three months ending 31 December: Melbourne, Victoria, Australia, Newcrest Mining Ltd., 10 p.

Northwest Resources Ltd., 2013, Annual report 2012: Sydney, New South Wales, Australia, Northwest Resources Ltd., 57 p.

Norton Gold Fields Ltd., 2013, Annual report 2012—31 December: Perth, Western Australia, Australia, Norton Gold Fields Ltd., 121 p.

OJSC MMC Norilsk Nickel, 2013, MMC Norilsk Nickel announces preliminary consolidated production results for 4th quarter and 12 months 2012 and production outlook for 2013: Moscow, Russia, OJSC MMC Norilsk Nickel press release, January 30, 4 p.

Reed Resources Ltd., 2012, Annual report 2012: West Perth, Western Australia, Australia, Reed Resources Ltd., 104 p.

Reed Resources Ltd., 2013, Reed Resources Ltd. quarterly activities report for the quarter ended 31 March 2013: West Perth, Western Australia, Australia, Reed Resources Ltd., 14 p.

Regis Resources Ltd., 2012, 2012 annual report: Subiaco, Western Australia, Australia, Regis Resources Ltd., 94 p.

Regis Resources Ltd., 2013, Quarterly report to 31 December: Subiaco, Western Australia, Australia, Regis Resources Ltd., 13 p.

Reserve Bank of Australia, 2013, Statement on monetary policy: Canberra, Australian Capital Territory, Australia, Reserve Bank of Australia, February, 68 p.

Resource Information Unit, 2012, Register of Australian Mining: West Perth, Western Australia, Australia, Resource Information Unit, 973 p.

Rio Tinto plc, 2013a, Rio Tinto Alcan South of Embley project: London, United Kingdom, Rio Tinto plc, newsletter, February, 8 p.

Rio Tinto plc, 2013b, 2012 annual report: London, United Kingdom, Rio Tinto plc, 235 p.

Risbey, Danielle, 2013, Planning for mine closure in Western Australia: Journal of the Australian Institute of Mining and Metallurgy, April, p. 60.

Sandfire Resources NL, 2013, March 2013 quarterly report highlight: West Perth, Western Australia, Australia, Sandfire Resources NL, 14 p.

Straits Resources Ltd., 2012, Hillgrove sale process: West Perth, Western Australia, Australia, Straits Resources Ltd. press release, November 9, 1 p.

Sydney Morning Herald, The, 2013, Aquila puts $7.4b iron ore project on ice: The Sydney Morning Herald, February 4. (Accessed June 6, 2013, at http://www.smh.com.au/action/printArticle?id=4003974.)

Talison Lithium Ltd., 2012, Greenbushes lithium operations: Perth, Western Australia, Australia, Talison Lithium Ltd., December, 104 p.

Talison Lithium Ltd., 2013, Talison Lithium update on Tainqi transaction: Perth, Western Australia, Australia, Talison Lithium Ltd. press release, February 25, 3 p.

Uranium One Inc., 2013, 2012 annual report: Toronto, Ontario, Canada, Uranium One Inc., 83 p.

Wilson, Lauren, 2013, FMG won't pay MRRT for five years: The Australian [Sydney, New South Wales, Australia], April 9. (Accessed April 22, 2013, at http://www.theaustralian.com.au/national-affairs/fmg-wont-pay-mrrt-for-five-years/story-fin59niix-1226615323908.)

Xstrata plc, 2013a, Annual report 2012: Zug, Switzerland, Xstrata plc, 147 p.

Xstrata plc, 2013b, Mineral resources and ore reserves: Zug, Switzerland, Xstrata plc, 51 p.

TABLE 1
AUSTRALIA: PRODUCTION OF MINERAL COMMODITIES[1]

(Metric tons unless otherwise specified)

Commodity		2008	2009	2010	2011	2012
METALS						
Aluminum:						
Bauxite, gross weight	thousand metric tons	64,038	65,231	68,414	69,976	76,282
Alumina	do.	19,446	19,948	19,956	19,399	20,914
Metal, primary	do.	1,974	1,943	1,928	1,945	1,864
Antimony, Sb content of ores and concentrates[e]		1,500	1,000	1,106 [2]	1,577 [2]	2,481 [2]
Cadmium:[e]						
Mine output, Cd content		700	460	-- [r]	--	--
Metal, smelter, refined		350 [r]	370 [r]	350 [r]	390	380
Chromium, chromite, gross weight		224,809	119,314	180,000 [r]	323,800 [r]	452,300
Of which, chromite content[e]		90,000	45,000	35 [r]	45 [r]	--
Cobalt:						
Co content in laterite ore, Ni concentrate, and Zn concentrate		4,780	4,345	3,852 [r]	3,848 [r]	5,882
Metal, refined		3,620	4,050	4,120	4,720 [r]	4,860
Copper:						
Mine output, Cu content	thousand metric tons	885	859	870	958	914
Metal:						
Smelter, primary and secondary	do.	447	422	410	441 [r]	421
Refined, primary	do.	503	446	417	477	461
Gold:						
Mine output, Au content		215	224	261	260	250
Metal, refined:						
Primary		244	256	280	271	264
Secondary		117	123	71	48	44
Iron and steel:						
Iron ore:[e]						
Gross weight	thousand metric tons	342,000	394,000	433,000	488,000	521,000
Fe content	do.	208,000	228,000	271,000	277,000	315,000
Metal:						
Pig iron	do.	6,409	4,370	6,259	5,396 [r]	3,711
Ferroalloys:[e]						
Ferromanganese		147,000	87,000	138,000	146,000	102,000
Silicomanganese		125,000	74,000	131,000	130,000	96,000
Total		272,000	161,000	269,000	276,000	198,000
Steel, crude	thousand metric tons	7,724	5,135	7,408	6,538	4,894
Semimanufactured products[e]		10,200	7,530	9,100 [r]	9,750 [r]	8,000
Lead:						
Mine output, Pb content	thousand metric tons	645	566	625	621	648
Metal:						
Bullion	do.	167	150	142	139	147
Refined:						
Primary	do.	220	204	178	187	160
Secondary, excluding remelt	do.	24	25	26	26	24
Manganese ore, metallurgical:						
Gross weight	do.	4,812	4,451	6,474	6,963	7,531
Mn content	do.	2,310	2,140	2,650	2,860	3,080
Nickel:						
Mine output, Ni content	do.	188	165	170	212	246
Matte	do.	31	28	54	57	66
Metal, smelter, refined Ni and Ni content of oxide	do.	103	131	108	110	129

See footnotes at end of table.

TABLE 1—Continued
AUSTRALIA: PRODUCTION OF MINERAL COMMODITIES[1]

(Metric tons unless otherwise specified)

Commodity		2008	2009	2010	2011	2012
METALS—Continued						
Platinum-group metals:[e]						
Palladium, Pd content	kilograms	580	800	650	350 [r]	300
Platinum, Pt content	do.	120	230	130	95 [r]	90
Total	do.	700	1,030	780	445 [r]	390
Silver:						
Mine output, Ag content		1,926	1,633	1,864	1,725	1,728
Metal, refined		644	664	735	898	781
Tantalum, tantalite, Ta_2O_5 equivalent		680	105	--	--	--
Tin:						
Mine output, Sn content[3]		1,783	5,630	6,600	5,012 [r]	5,849
Metal, refined:						
Primary		170	--	--	--	--
Secondary[e]		400	400	400	400	400
Titanium concentrates, gross weight:						
Ilmenite	thousand metric tons	2,082	1,449	1,492	1,277	1,344
Leucoxene[e]		148,000	162,000	159,000	224,000	228,000
Rutile		325,000	285,000	429,000	474,000	439,000
Tungsten, mine output, W content		28	33	16	15	80
Zinc:						
Mine output, Zn content	thousand metric tons	1,519	1,290	1,479	1,515	1,541
Metal, smelter, primary	do.	499	525	499	507	498
Zirconium concentrates, gross weight	do.	514	400	549	762	605
INDUSTRIAL MINERALS						
Abrasives, natural, garnet		298,290	275,560	196,839	200,000	200,000
Barite[e]		17,000	12,000	12,000	12,000	12,000
Cement, hydraulic[e]	thousand metric tons	9,400	9,200	8,300	8,600	8,600
Diamond:						
Gem	thousand carats	273	220	100	86	65
Industrial	do.	15,397	10,575	9,900	7,500	11,895
Total	do.	15,670	10,795	10,000	7,586	11,960
Diatomite[e]		20,000	20,000	20,000	20,000	20,000
Feldspar, including nepheline syenite[e]		50,000	50,000	50,000	50,000	50,000
Gemstones, opal[e]	value, $million	41	33	40	40	41
Gypsum	thousand metric tons	3,734	3,436	3,000 [e]	3,000 [e]	2,500 [e]
Kyanite[e]		1,000	1,000	1,000	1,000	1,000
Lime[e]		2,200,000	2,500,000 [r]	2,200,000 [r]	2,200,000 [r]	2,200,000
Lithium, spodumene		239,528	197,482	295,000	421,391 [r]	456,921
Magnesite		126,000	344,000	275,000 [r]	300,000 [e]	300,000 [e]
Perlite, crude[e]		6,500	6,500	7,000	7,000	7,000
Phosphate rock:[e]						
Gross weight		2,950,000	2,500,000	2,600,000	2,650,000 [r]	2,600,000
P_2O_5 content		678,000	575,000	600,000	610,000 [r]	600,000
Rare earths, rare-earth oxide equivalent		--	--	--	2,188	3,222

See footnotes at end of table.

TABLE 1—Continued
AUSTRALIA: PRODUCTION OF MINERAL COMMODITIES[1]

(Metric tons unless otherwise specified)

Commodity		2008	2009	2010	2011	2012
INDUSTRIAL MINERALS—Continued						
Salt[4]	thousand metric tons	11,160	10,316	11,968	11,744	10,821
Soda ash[e]	do.	310	310	310	310	300
Stone and sand and gravel:[e]						
Construction sand	do.	37,000	34,000	21,000 [r]	24,000 [r]	25,000
Crushed and broken stone	do.	110,000	115,000	100,000	100,000	100,000
Dimension stone	do.	230	180	120 [r]	140 [r]	140
Gravel	do.	12,000	12,000	6,000 [r]	8,000 [r]	8,000
Limestone	do.	18,400	16,800	17,000	18,000	18,000
Silica in the form of quartz, quartzite, glass sand	do.	5,500	4,000 [r]	3,100 [r]	3,500 [r]	3,500 [e]
Sulfur, byproduct:[e]						
Metallurgy	do.	880	870	800	800	800
Petroleum	do.	60	60	60	60	60
Total	do.	940	930	860	860	860
Talc, chlorite, pyrophyllite, steatite[e]		120,000	90,000 [r]	100,000 [r]	120,000	120,000
MINERAL FUELS AND RELATED MATERIALS						
Coal, salable:						
Bituminous and subbituminous	thousand metric tons	332,000	348,000	356,000 [r]	348,000 [r]	365,000
Lignite[e]	do.	71,000	74,000	71,000	65,000	65,000
Total[e]	do.	403,000	422,000	427,000 [r]	413,000 [r]	430,000
Gas, natural, marketed	million cubic meters	38,256	42,345	51,868	51,253	55,970
Petroleum:						
Crude, includes condensate	thousand 42-gallon barrels	168,123	169,211	169,985	143,456	119,200
Refinery products	do.	246,717	241,233	235,971	239,618	234,734
Uranium, mine output, U_3O_8 content		9,989	7,942	7,440	6,942	6,968

[e]Estimated; estimated data are rounded to no more than three significant digits; may not add to totals shown. [r]Revised. do. Ditto. -- Zero.
[1]Table includes data available through July 9, 2013.
[2]Reported figure.
[3]Does not include tin production from heavy-mineral sands in Western Australia.
[4]Does not include production from Northern Territory and Victoria.

TABLE 2
AUSTRALIA: STRUCTURE OF THE MINERAL INDUSTRY IN 2012

(Thousand metric tons unless otherwise specified)

Commodity	Facilities, major operating companies, and major equity owners	Location of main facilities[1, 2]	Annual capacity[e]
Aluminum:			
Bauxite	Gove open pit bauxite mine [Pacific Aluminum (Rio Tinto Ltd., 100%)]	15 km southeast of Nhulunbuy, NT	8,000
Do.	Huntly open pit bauxite mine (Alcoa World Alumina Australia, 100%)	80 km south of Perth, WA	20,000
Do.	Weipa-Andoom open pit bauxite mine [Comalco Ltd., operator (Rio Tinto Alcan, 100%)]	Weipa, QLD	23,000
Do.	Willowdale open pit bauxite mine (Alcoa World Alumina Australia, 100%)	130 km south of Perth, WA	8,600
Do.	Boddington-Worsley open pit bauxite mine {Worsley Alumina Pty. Ltd., manager [BHP Billiton Ltd., 86%; Japan Alumina Associates (Australia) Pty. Ltd., 10%; Sojitz Alumina Pty. Ltd., 4%]}	14 km south of Boddington, WA	19,000
Alumina, refinery	Gladstone alumina refinery [Queensland Alumina Ltd., operator (Rio Tinto Alcan, 80%, and United Company RUSAL, 20%)]	Gladstone, QLD	3,850
Do.	Gove alumina refinery {Alcan Gove Pty Ltd. [Pacific Aluminum, 100% (Rio Tinto Ltd., 100%)]}	Nhulunbuy, Gove, NT	3,800
Do.	Kwinana alumina refinery (Alcoa World Alumina Australia, 100%)	Kwinana, WA	2,100
Do.	Pinjarra alumina refinery (Alcoa World Alumina Australia, 100%)	Pinjarra, WA	4,200
Do.	Wagerup alumina refinery (Alcoa World Alumina Australia, 60%, and Western Mining Corp., 40%)	Waroona, WA	2,600
Do.	Worsley alumina refinery {Worsley Alumina Pty. Ltd., manager [BHP Billiton Ltd., 86%, and Japan Alumina Associates (Australia) Pty Ltd., 10%]}	20 km northwest of Collie, WA	4,600
Do.	Yarwun alumina refinery (Rio Tinto Alcan, 100%)	Gladstone, QLD	3,400
Metal smelter	Bell Bay aluminum smelter [Pacific Aluminum (Rio Tinto Ltd., 100%)]	Bell Bay, TAS	160
Do.	Boyne Island aluminum smelter [Boyne Smelters Ltd., operator [Pacific Aluminum, 64% (Rio Tinto Ltd., 100%); Sumitomo Light Metal Industries Ltd., 17%; Ryowa Development Pty. Ltd., 12%; Kobe Steel Ltd., 5%; Sumitomo Chemical Co. Ltd., 2%]	Boyne Island, QLD	550
Do.	Point Henry aluminum smelter (Alcoa of Australia, 100%)	Point Henry, VIC	185
Do.	Portland aluminum smelter [Alcoa of Australia, 55%, manager; China International Trust Investment Co. (China state-owned company), 22.5%; Marubeni Australia Pty. Ltd., 22.5%]	Portland, VIC	345
Do.	Tomago aluminum smelter {Tomago Aluminium Co. Pty. Ltd., operator [Gove Aluminium Finance Ltd., 36.05%; Pacific Aluminum 51.55% (Rio Tinto Ltd., 100%); Hydro Aluminium, 12.40%]}	Tomago, NSW	525
Antimony	Costerfield underground antimony-gold mine [AGD Mining, operator (Mandalay Resources Ltd., 100%)]	50 km east and southeast of Bendigo, VIC	5
Do.	Hillgrove Mine (Straits Resources Ltd., 100%)	25 km east of Armidale, NSW	10
Bentonite	Arumpo open pit bentonite mine (Arumpo Bentonite Pty. Ltd., 100%)	95 km northeast of Mildura, NSW	10
Do.	Cedars open pit bentonite mine (PCP Douglass Pty. Ltd., 100%)	10 km southwest of Yarraman, QLD	20
Do.	Cressfield open pit bentonite mine (Unimin Australia Ltd., 100%)	20 km north of Scone, NSW	12
Do.	Mantuan Downs (Pacific Enviromin Ltd., 100%)	West of Springsure, QLD	100
Do.	Miles open pit bentonite mine (Unimin Australia Ltd., 100%)	350 km west of Brisbane, QLD	100
Cement, plant	Adelaide Brighton Cement Pty Ltd., 100%	Angaston, SA	250
Do.	do.	Birkenhead, SA	1,200
Do.	do.	Geelong, VIC	800
Do.	do.	Munster, SA	590
Do.	Blue Circle Southern Cement Ltd., 100%	Berrima, NSW	1,200
Do.	do.	Maldon, NSW	700
Do.	do.	Waurn Ponds, VIC	250
Do.	Cement Australia Pty Ltd. (Hanson Ltd. and Holcim Australia Pty Ltd.)	Brisbane, QLD	1,200
Do.	do.	Gladstone, QLD	1,700
Do.	do.	Railton, TAS	1,000
Do.	Cockburn Cement Ltd., 100%	Munster, 30 km south of Perth, WA	700
Chromite	Coobina open pit chromite mine (Palmary Enterprises Ltd., 100%)	80 km southeast of Newman, WA	250

See footnotes at end of table.

TABLE 2—Continued
AUSTRALIA: STRUCTURE OF THE MINERAL INDUSTRY IN 2012

(Thousand metric tons unless otherwise specified)

Commodity	Facilities, major operating companies, and major equity owners	Location of main facilities[1, 2]	Annual capacity[e]
Coal	Angus Place longwall coal mine (Centennial Coal Co. Ltd., 50%, and SK Corp., 50%)	16 km northwest of Lithgow, NSW	3,000
Do.	Appin longwall coal mine [Illawarra Coal Holdings Pty Ltd., operator (BHP Billiton Ltd., 100%)]	40 northwest of Wollongong, NSW	8,800
Do.	Ashton open pit/underground coal mine (Felix Resources Ltd., 60%; Chu Corp., 10%; private, 30%)	14 km northwest of Singleton, NSW	4,000
Do.	Awaba underground coal mine [Powercoal Pty. Ltd., operator (Centennial Coal Co. Ltd., 100%)]	30 km southwest of Newcastle, NSW	2,000
Do.	Baal Bone coal mine [Oakbridge Pty. Ltd., 74.1% (Xstrata plc, 100%); Sumitomo Corp., 5%; Toyota Tsusho Mining (Australia) Pty Ltd. 4.75%; private, 14.44%]	24 km northwest of Lithgow, NSW	2,500
Do.	Bengalla open pit coal mine [Coal and Allied Industries Ltd., 40%, manager; Wesfarmers Bengalla Ltd., 40%; MCDA Bengalla Investment Pty. Ltd., 10%; Taipower Bengalla Pty. Ltd., 10%]	5 km west of Muswellbrook, NSW	6,600
Do.	Blackwater open pit coal mine (includes South Blackwater) [BHP Billiton Mitsubishi Alliance, manager (BHP Billiton Ltd., 50%, and Mitsubishi Corp., 50%)]	195 km west of Rockhampton, QLD	14,000
Do.	Broadmeadow open pit/underground coal mine [BHP Billiton Mitsubishi Alliance, manager (BHP Billiton Ltd., 50%, and Mitsubishi Corp., 50%)]	30 km north of Moranbah, QLD[3]	3,000
Do.	Bulga open pit coal mine [Oakbridge Pty Ltd., manager (Xstrata plc, 68.25%; Nippon Steel Australia Pty. Ltd., 12.5%; Toyota Tsusho Mining (Australia) Pty Ltd., 4.38%; private, 13.3%]	16 km southwest of Singleton, NSW	10,000
Do.	Burton open pit coal mine (Peabody Energy Corp., 95%, and Thiess Pty. Ltd., 5%)	150 km southwest of Mackay, QLD	5,800
Do.	Callide coal mine (Anglo Coal Pty Ltd., 100%)	120 km southwest of the Port of Gladstone, QLD	10,700
Do.	Camberwell open pit coal mine [Camberwell Coal Pty. Ltd., manager [Toyota Tsusho Mining (Australia) Pty. Ltd., 90%, and Dia Coal Mining (Australia) Pty Ltd., 10%]	10 km northwest of Singleton, NSW	4,000
Do.	Clarence underground coal mine [Centennial Coal Co. Ltd., 85%, (manager) and SK Australia Pty. Ltd., 15%]	10 km east of Lithgow, NSW	2,500
Do.	Commodore open pit coal mine Roche Mining Pty. Ltd., operator [Intergen (Australia) Pty. Ltd., 100%]	80 km southwest of Toowoomba, QLD	3,600
Do.	Coppabella open pit coal mine (Macarthur Coal Ltd., 73.3%, and others, 26.7%)	140 km southwest of Mackay, QLD	4,000
Do.	Cumnock No. 1 open pit coal mine (Cumnock Coal Ltd., 100%)	28 km northwest of Singleton, NSW	3,000
Do.	Curragh open pit coal mine (Wesfarmers Ltd., 100%)	70 km east of Emerald, QLD	9,000
Do.	Dartbrook coal mine (Anglo Coal Holdings Australia Ltd., 77.3%)	70 km north of Singleton, NSW[3]	3,750
Do.	Dawson coal complex (includes Moura, Taroom, and Theodore) [Anglo American plc, 51%, and Mitsui & Co. (Australia) Ltd., 49%]	230 km west of Bundaberg, QLD	7,000
Do.	Dendrobium underground coal mine (BHP Billiton Ltd., 100%)	15 km southwest of Wollongong, NSW	5,200
Do.	Donaldson open pit coal mine (Donaldson Coal Pty Ltd., 100%)	5 km southeast of Maitland, NSW	2,500
Do.	Drayton open pit coal mine [Anglo Coal Holdings Australia Ltd., 88.2%, manager; Mitsui Coal Development Australia Pty. Ltd., 3.8%; Mitsui Mining (Australia) Pty. Ltd., 3%; others, 5%]	35 km northwest of Singleton, NSW	5,000
Do.	Duralie open pit coal mine (Gloucester Coal Ltd., 100%)	110 km of Newcastle, NSW	2,000
Do.	Elouera underground coal mine (Gujarat NRE Resources NL, 100%)	15 km southwest of Wollongong, NSW	2,000
Do.	Ensham-Yongala open pit coal mine [Idemitsu Kosan Co. Ltd., 85%; J-Power (Australia) Pty. Ltd., 10%; LG International (Australia) Pty Ltd., 5%]	40 km northeast of Emerald, QLD	9,000

See footnotes at end of table.

TABLE 2—Continued
AUSTRALIA: STRUCTURE OF THE MINERAL INDUSTRY IN 2012

(Thousand metric tons unless otherwise specified)

Commodity	Facilities, major operating companies, and major equity owners	Location of main facilities[1,2]	Annual capacity[e]
Coal—Continued	Ewington II open pit coal mine (Griffin Coal Mining Co. Pty. Ltd., 100%)	8 km east of Collie, WA	1,000
Do.	Foxleigh open pit coal mine (Foxleigh Mining Pty Ltd., 100%)	Bowen basin, QLD	3,600
Do.	German Creek and German Creek East open pit/underground coal mines [Anglo American plc, 70%, and Mitsui & Co. (Australia) Ltd., 30%]	275 km west-northwest of Rockhampton, QLD	6,000
Do.	Glennies Creek longwall coal mine (CVRD Inco Ltd., 85%; Nippon Steel Australia Pty Ltd., 5%; POSCO Australia Pty Ltd., 5%; private, 5%)	12 km north of Singleton, NSW	2,800
Do.	Goonyella-Riverside-Broadmeadow open pit coal mines (BHP Billiton Ltd., 50%, and Mitsubishi Corp., 50%)	140 km southwest of Mackay, QLD	16,000
Do.	Gregory Crinum open pit/underground coal mine [BHP Billiton Mitsubishi Alliance, manager (BHP Billiton Ltd., 50%, and Mitsubishi Corp., 50%)]	60 km north of Emerald, QLD	5,500
Do.	Hunter Valley Operations (includes Carrington Chestnut, Howick, Hunter Valley No. 1, Lemington, Riverview open pit coal mines) (Coal and Allied Industries Ltd., 100%)	10 km west and 25 km north of Singleton, NSW	15,000
Do.	Hail Creek open pit coal mine (Rio Tinto Ltd., 82%; Nippon Steel Australia Pty Ltd., 8%; Marubeni Coal Pty. Ltd., 6.66%)	100 km west of Mackay, QLD	8,000
Do.	Hazelwood open pit coal mine (International Power Hazelwood, 100%)	150 km southeast of Melbourne, VIC	20,000
Do.	Jellinbah East open pit coal mine (Queensland Coal Mine Management Pty. Ltd., 70%; Marubeni Coal Pty. Ltd., 15%; Sojitz Australia Ltd., 15%)	90 km east of Emerald, QLD	4,000
Do.	Kestrel underground coal mine [Rio Tinto Ltd., 80%, and Mitsui & Co. (Australia) Ltd., 20%]	40 km north-northeast of Emerald, QLD	5,500
Do.	Liddell open pit coal mine (Xstrata Coal Australia Pty. Ltd., 67.5%, and Mitsui Matushima Australia Pty. Ltd., 32.5%)	25 km northwest of Singleton, NSW	4,000
Do.	Loy Yang open pit coal mine (Loy Yang Power Ltd., 100%)	165 km east of Melbourne, VIC	30,000
Do.	Mondalong underground coal mine (Centennial Coal Co. Ltd., 100%)	35 km southwest of Newcastle, NSW	4,500
Do.	Moorvale open pit coal mine (Macarthur Coal Ltd., 73.3%; CITIC Resources Australia Pty Ltd., 7%; Sojtz Australia Ltd., 7%; Nippon Steel Australia Pty Ltd., 2%)	10 km south of Coppabella, QLD	3,400
Do.	Moranbah North longwall coal mine (Anglo American plc., 88%, and Nippon Steel Australia Pty. Ltd., 5%)	150 km southwest of Mackay, QLD	5,800
Do.	Mount Arthur open pit coal mine (BHP Billiton Ltd., 100%)	5 km southwest of Muswellbrook, NSW	15,000
Do.	Mount Owen open pit coal mine (Xstrata plc, 100%)	20 km northwest of Singleton, NSW	7,700
Do.	Mount Thorley open pit coal mine (Coal and Allied Industries Ltd., 80%, and POSCO Australia Pty. Ltd., 20%)	14 km southwest of Singleton, NSW	12,000
Do.	Muja open pit coal mine (The Griffin Coal Mining Co. Pty. Ltd., 100%)	18 km southeast of Collie, WA	2,000
Do.	Muswellbrook No. 2 open pit coal mine (Muswellbrook Coal Co., 100%)	4 km northeast of Muswellbrook, NSW	1,700
Do.	Myuna underground coal mine (Centennial Coal Co. Ltd., 100%)	35 km south of Newcastle, NSW	1,500
Do.	New Acland open pit coal mine (New Hope Corp. Ltd., 100%)	35 km northwest of Toowoomba, QLD	3,750
Do.	Newlands-Collinsville-Abbot Point open pit coal mine (Xstrata plc, 55%; Itochu Corp., 35%; Sumitomo Corp., 10%)	130 km west of Mackay, QLD	15,000
Do.	Newstan longwall coal mine (Centennial Coal Co. Ltd., 100%)	30 km southwest of Newcastle, NSW	4,000
Do.	North Goonyella underground coal mine (Peabody Energy Corp., 100%)	40 km north Moranbah, QLD	3,000
Do.	Norwich Park open pit coal mine (BHP Billiton Ltd., 50%, and Mitsubishi Corp., 50%)	85 km north-northeast of Emerald, QLD	5,000
Do.	Oaky Creek longwall and Alliance open pit coal mines (Xstrata plc, 55%; Sumitomo Coal Australia Pty. Ltd., 25%; Itocho Corp., 20%)	300 km west-northwest of Rockhampton, QLD	9,500
Do.	Peak Downs open pit coal mine (BHP Billiton Ltd., 50%, and Mitsubishi Development Pty. Ltd., 50%)	145 km north of Emerald, QLD	9,000
Do.	Premier open pit coal mine (Wesfarmers Premier Coal Ltd., 100%)	10 km southeast of Collie, WA	4,000

See footnotes at end of table.

TABLE 2—Continued
AUSTRALIA: STRUCTURE OF THE MINERAL INDUSTRY IN 2012

(Thousand metric tons unless otherwise specified)

Commodity	Facilities, major operating companies, and major equity owners	Location of main facilities[1,2]	Annual capacity[e]
Coal—Continued	Ravensworth-Narama open pit coal mine (includes Ravensworth East) (Xstrata Coal Australia Pty. Ltd., 100% of Ravensworth and 50% of Narama; Iluka Resources Ltd., 50% of Narama)	20 km northwest of Singleton, NSW	3,500
Do.	Rixs Creek open pit coal mine (Bloomfield Colliers Pty. Ltd., 100%)	5 km northwest of Singleton, NSW	2,000
Do.	Rolleston open pit coal mine (Xstrata plc, 75%; Itochu Corp., 12.5%; Sumitomo Corp., 12.5%)	90 km south-southeast of Emerald, QLD	8,000
Do.	Saraji open pit coal mine (BHP Billiton Ltd., 50%, and Mitsubishi Corp., 50%)	125 km north of Emerald, QLD	6,500
Do.	South Walker Creek open pit/underground coal mine (BHP Mitsui Coal Pty. Ltd., 100%)	90 km southwest of Mackay, QLD	4,300
Do.	Springvale underground coal mine (Centennial Coal Co. Ltd. 50%; SK Corp., 25%; Korea Resources Corp. Australia, 25%)	16 km northwest of Lithgow, NSW	3,000
Do.	Tahmoor longwall coal mine (includes Tahmoor North and Bargo) (Centennial Coal Co. Ltd., 85.79%, and private, 14.21%)	70 km southwest of Sydney, NSW	2,500
Do.	Tarong-Meandu open pit coal mine (Rio Tinto Ltd., 100%)	85 km north of Toowoomba, QLD	7,000
Do.	Ulan underground coal mine (Xstrata plc, 90%, and Mitsubishi Corp., 10%)	45 km northwest of Mudgee, NSW	5,000
Do.	United Collieries underground coal mine (Xstrata plc, 95%, and private, 5%)	15 km west of Singleton, NSW	3,000
Do.	Wambo open pit/underground coal mine (Peabody Energy Corp., 100%)	30 km from Singleton, NSW	6,000
Do.	West Cliff longwall coal mine (BHP Billiton Ltd., 100%)	43 km northwest of Wollongong, NSW	2,300
Do.	West Wallsend longwall coal mine (Xstrata plc, 70%; Marubeni Coal Pty Ltd., 17%; private, 13%)	25 km southwest of Newcastle, NSW	2,500
Do.	Yallourn open pit lignite mine (CLP Power Asia Ltd., 100%)	140 km southeast of Melbourne, VIC	18,000
Cobalt:			
Mine	Cawse open pit nickel-cobalt mine (OJSC MMC Norilsk Nickel, 100%)	50 km northwest of Kalgoorlie, WA	0.2
Do.	Murrin Murrin open pit nickel-cobalt mine (Minara Resources Ltd., 60%, and Glencore Australia Pty. Ltd., 40%)	60 km east of Leonora, WA	2.0
Do.	Radio Hill underground nickel-cobalt mine (Fox Resources Ltd., 100%)	35 km south of Karratha, WA	0.2
Do.	Ravensthorpe open pit mine (BHP Billiton Ltd., 100%)	155 km west of Esperance, WA	1.4
Refinery	Yabulu nickel-cobalt refinery (Nickel Consolidated Pty Ltd., Nickel House Pty, and Nickel Process Pty)	Townsville, QLD	3
Copper:			
Mine, Cu content	Boddington open pit/underground gold mine (Newmont Mining Corp., 100%)	130 km southeast of Perth, WA	35
Do.	Cadia Valley open pit/underground gold-copper mine (includes Cadia East, Cadia Hill, and Ridgeway) (Newcrest Mining Ltd., 100%)	21 km south-southwest of Orange, NSW	90
Do.	Cobar underground copper mine (Glencore International plc, 100%)	12 km northwest of Cobar, NSW	30
Do.	Doolgunna open pit/underground gold-copper mine (includes DeGrussa) (Sandfire Resources NL, 100%)	140 km north of Meekatharra, WA	300
Do.	Eloise underground copper mine (FMR Investments Pty Ltd., 100%)	60 km southeast of Cloncurry, QLD	70
Do.	Ernest Henry open pit/underground copper-gold mine (Xstrata plc, 100%)	35 km northeast of Cloncurry, QLD	115
Do.	Golden Grove underground zinc-copper mine [(MMG Ltd., operator) China Minmetals Group, 100%]	225 km east of Geraldton, WA	20
Do.	Hellyer underground zinc-lead-copper-silver mine (Bass Metals Ltd., 100%)	80 km south-southwest of Burnie, TAS	1
Do.	Lady Annie copper (solvent extraction-electrowinning) mine (CST Mining Group Ltd., 100%)	100 km north-northwest of Mount Isa, QLD	19
Do.	Leichhardt copper mine (Cape Lambert Resources Ltd., 100%)	110 km northwest of Cloncurry, QLD[3]	10
Do.	Mount Gordon open pit copper (solvent extraction-electrowinning) mine (Aditya Birla Minerals Ltd., 100%)	120 km north of Mount Isa, QLD	50
Do.	Mount Isa underground copper-lead-zinc-silver mine (also includes Enterprise, George Fisher, and Hilton Mines) (Xstrata plc, 100%)	Mount Isa, QLD	190
Do.	Mount Lyell underground copper-gold mine [Sterlite Industries (India) Ltd., 100%]	2 km northeast of Queenstown, TAS	35

See footnotes at end of table.

TABLE 2—Continued
AUSTRALIA: STRUCTURE OF THE MINERAL INDUSTRY IN 2012

(Thousand metric tons unless otherwise specified)

Commodity	Facilities, major operating companies, and major equity owners	Location of main facilities[1,2]	Annual capacity[e]
Copper—Continued:			
Mine, Cu content—Continued	Nifty open pit copper (solvent extraction-electrowinning) mine (Aditya Birla Minerals Ltd., 100%)	200 km southeast of Marble Bar, WA	25
Do.	Northparkes open pit/underground copper-gold mine (Rio Tinto Ltd., 80%; Sumitomo Metal Mining Oceania Pty. Ltd., 13.3%; SC Mineral Resources Pty. Ltd., 6.7%)	30 km northwest of Parkes, NSW	90
Do.	Olympic Dam underground copper-silver-gold-uranium mine [Olympic Dam Operations Pty. Ltd., operator (BHP Billiton Ltd., 100%)]	Roxby Downs, 80 km north of Woomera, SA	235
Do.	Osborne underground copper-gold mine (Ivanhoe Australia Ltd., 100%)	120 km northeast of Boulia, QLD	22
Do.	Peak underground gold-zinc-lead-copper-silver underground mine (includes New Cobar, New Occidental, and Perseverance) (GoldCorp Inc., 100%)	8 km south of Cobar, NSW	3
Do.	Prominent Hill open pit/underground copper-gold mine (OZ Minerals Ltd., 100%)	650 km northwest of Adelaide, SA	140
Do.	Ridgeway underground gold-copper mine (Newcrest Mining Ltd., 100%)	5 km south of Orange, NSW	30
Do.	Rosebery underground zinc-lead-silver-copper-gold mine [Minerals and Metals Group Australia Ltd., operator (China Minmetals Nonferrous Metals Co. Ltd., 100%)]	35 km north of Queenstown, TAS	2
Do.	Tritton underground mine (Straits Resources Ltd., 100%)	Nyngan, NSW	30
Smelter	Mount Isa copper smelter (Xstrata plc, 100%)	Mount Isa, QLD	250
Do.	Olympic Dam copper smelter [Olympic Dam Operations Pty. Ltd., operator (BHP Billiton Ltd., 100%)]	Roxby Downs, 80 km north of Woomera, SA	70
Do.	Port Kembla copper smelter (Furukawa Co. Ltd., 52.5%; Nittetsu Mining Co., 20%; Nissholwai Corp., 17.5%; Itochu Corp., 10%)	Port Kambla, NSW	120
Refinery	Olympic Dam copper refinery [Olympic Dam Operations Pty. Ltd., operator (BHP Billiton Ltd., 100%)]	Roxby Downs, 80 km north of Woomera, SA	235
Do.	Port Kembla copper refinery (Furukawa Co. Ltd., 52.5%; Nittetsu Mining Co., 20%; Nissholwai Corp., 17.5%; Itochu Corp., 10%)	Port Kambla, NSW	120
Do.	Townsville copper refinery (Xstrata plc, 100%)	Townsville, QLD	300
Diamond thousand carats	Argyle Mine (AK–1 lamproite pipe and alluvial diamond mines) (Rio Tinto plc, 100%)	120 km southwest of Kununurra, WA	30,000
Do. do.	Ellendale Mine (includes pipes 4 and 9) (Gem Diamond Ltd., 100%)	130 east southeast of Derby, WA	700
Do. do.	Ellendale 9 North Mine (Blina Diamond NL, 100%)	140 east of Derby, WA	500
Diatomite	Barraba open pit diatomite mine (Australia Diatomite Mining Pty. Ltd., 100%)	85 km north-northwest of Tamworth, NSW	25
Dolomite	Ardrossan metallurgical dolomite quarry (OneSteel Ltd., 100%)	Northern York Peninsula, SA	650
Do.	Cookes Hill Mine (includes Nickol River and Warrawoona) (Haoma Mining NL, 100%)	Near Port Hedland, WA	400
Feldspar	Broken Hill open pit feldspar mine (includes Bakers, Lady Beryl, and Spar Ridge) (Unimin Australia Ltd., 100%)	42 km southwest of Broken Hill, NSW	15
Garnet	Port Gregory open pit industrial garnet mine (GMA Garnet Pty. Ltd., 100%)	100 km north of Geraldton, WA	250
Gas:			
Condensate thousand 42-gallon barrels per day	North West Shelf gas operations {Woodside Petroleum Pty. Ltd., manager [BHP Petroleum Pty. Ltd., BP Australia Holdings Ltd., Chevron Asiatic Ltd., Japan Australia LNG (MIMI) Pty. Ltd., Shell Development (Australia) Pty. Ltd., and Woodside Petroleum Ltd., 16.67% each]}	130 km offshore Dampier, WA	60
Natural million cubic meters per day	do.	do.	20
Liquefied natural million metric tons	do.	Four-train liquefaction plant, Burrup Peninsula, WA	12

See footnotes at end of table.

TABLE 2—Continued
AUSTRALIA: STRUCTURE OF THE MINERAL INDUSTRY IN 2012

(Thousand metric tons unless otherwise specified)

Commodity		Facilities, major operating companies, and major equity owners	Location of main facilities[1, 2]	Annual capacity[e]
Gold:				
Mine	kilograms	Agnew open pit/underground gold mine (Gold Fields Ltd., 100%)	23 km west of Leinster, WA	5,600
Do.	do.	Boddington open pit/underground gold mine (Newmont Mining Corp., 100%)	130 km southeast of Perth, WA	31,000
Do.	do.	Bronzewing underground gold mine (includes Mount McClure, Venus, Success, Cockburn, Corboys, Mount Joel) (Audax Resources Ltd., 100%)	65 km northeast of Leinster, WA	9,000
Do.	do.	Burnside open pit mines (includes Union Reefs, Brocks Creek, North Point, Princess Louise, Rising Tide, Zapopan, Fountain Head) (Crocodile Gold Corp., 100%)	Pine Creek, NT	6,500
Do.	do.	Cadia Valley open pit/underground gold-copper mine (includes Cadia East, Cadia Hill, and Ridgeway) (Newcrest Mining Ltd., 100%)	21 km south-southwest of Orange, NSW	25,000
Do.	do.	Doolgunna open pit/underground gold-copper mine (includes DeGrussa) (Sandfire Resources NL, 100%)	140 km north of Meekatharra, WA	270
Do.	do.	Ernest Henry open pit copper-gold mine (Xstrata plc, 100%)	35 km northeast of Cloncurry, QLD	3,000
Do.	do.	Granny Smith open pit gold mine (includes Wallaby) (Barrick Gold Corp., 100%)	20 km south of Laverton, WA	16,000
Do.	do.	Gwalia underground gold mine (St Barbara Ltd., 100%)	3 km south of Leonora, WA	2,600
Do.	do.	Henty underground gold-silver mine (Barrick Gold Ltd., 100%)	30 km north of Queenstown, TAS	3,700
Do.	do.	Hillgrove Mine (Straits Resources Ltd., 100%)	25 km east of Armidale, NSW	650
Do.	do.	Jundee-Nimary open pit/underground gold mine (Newmont Mining Corp., 100%)	45 km northeast of Wiluna, WA	12,000
Do.	do.	Kalgoorlie open pit/underground gold mine [Kalgoorlie Consolidated Gold Mine Pty Ltd., operator (Barrick Gold Australia, 50%, and Newmont Mining Corp., 50%)]	600 km east Perth, WA	20,000
Do.	do.	Kanowna Belle underground gold mine (Barrick Gold Corp., 100%)	18 km northeast of Kalgoorlie, WA	7,000
Do.	do.	Lawlers underground gold mine (Barrick Gold Corp., 100%)	30 km southwest of Leinster, WA	3,000
Do.	do.	Mount Lyell underground copper-gold mine [Sterlite Industries (India) Ltd., 100%]	2 km northeast of Queenstown, TAS	1,000
Do.	do.	Mount Magnet open pit/underground gold mine (includes Hill 50 and Star) (Ramelins Resources Ltd., 100%)	2 km from Mount Magnet, WA	8,500
Do.	do.	Norseman underground gold mine (Norseman Gold Plc, 100%)	Norseman, WA	3,700
Do.	do.	Northparkes open pit/underground copper-gold mine (Rio Tinto Ltd., 80%, and Sumitomo Metal Mining Oceania Pty. Ltd., 20%)	30 km north of Parkes, NSW	155,000
Do.	do.	Osborne underground copper-gold mine (Ivanhoe Australia Ltd., 100%)	120 km northeast of Boulia, QLD	1,000
Do.	do.	Olympic Dam underground copper-silver-gold-uranium mine [Olympic Dam Operations Pty. Ltd., operator (BHP Billiton Ltd., 100%)]	Roxby Downs, 80 km north of Woomera, SA	1,500
Do.	do.	Paddington open pit/underground gold operation [Noron Gold Fields Ltd., operator (Zijin Mining Group Co. Ltd., 89%)]	35 km north of Kalgoorlie, WA	5,000
Do.	do.	Pajingo underground gold mine (includes Vera-Nancy) [North Queensland Metals Ltd. (operator), 60%, and Heemskirk Consolidated Ltd., 40%]	60 km south-southeast of Charters Towers, QLD	6,400
Do.	do.	Plutonic open pit/underground gold mine (Barrick Gold Corp., 100%)	180 km northeast of Meekatharra, WA	8,000
Do.	do.	Prominent Hill open pit copper-gold mine (OZ Minerals Ltd., 100%)	650 km northwest of Adelaide, SA	2,200
Do.	do.	Ravenswood open pit mine (includes Nolans, Sarsfield, and Mount Wright) (Resolute Mining Ltd., 100%)	100 km south of Townsville, QLD	3,000
Do.	do.	Ridgeway underground gold-copper mine (Newcrest Mining Ltd., 100%)	25 km south of Orange, NSW	10,800
Do.	do.	Rosebery underground zinc-lead-silver-copper-gold mine [Minerals and Metals Group Australia Ltd., operator (China Minmetals Nonferrous Metals Co. Ltd., 100%)]	35 km north of Queenstown, TAS	1,000
Do.	do.	Saint Ives open pit/underground gold mine (Gold Fields Ltd., 100%)	75 km south-southeast of Kalgoorlie, WA	15,000
Do.	do.	Selwyn underground copper-gold mine (Barrick Gold Corp., 100%)	160 km southeast of Mount Isa, QLD	700
Do.	do.	Stawell underground gold mine (Perseverance Corp. Ltd., 100%)	250 km west of Melbourne, VIC	3,000

See footnotes at end of table.

TABLE 2—Continued
AUSTRALIA: STRUCTURE OF THE MINERAL INDUSTRY IN 2012

(Thousand metric tons unless otherwise specified)

Commodity		Facilities, major operating companies, and major equity owners	Location of main facilities[1,2]	Annual capacity[e]
Gold—Continued:				
Mine—Continued	kilograms	Sunrise Dam open pit mine gold (includes Cleo) (AngloGold Ashanti Ltd., 100%)	55 km south of Laverton, WA	15,000
Do.	do.	Super Pit open pit gold mine (includes Fimiston) [Kalgoorlie Consolidated Gold Mines Pty. Ltd., manager (Barrick Gold Corp., 50%, and Newmont Mining Corp., 50%)]	Southeast corner of the Kalgoorlie-Boulder Township, WA	25,000
Do.	do.	Tanami open pit gold mine (includes Central Desert Joint Venture) (Newmont Gold Corp., 100%)	650 km northwest of Alice Springs, NT	15,000
Do.	do.	Telfer copper and gold mine (Newcrest Mining Ltd., 100%)	400 km east southeast of Port Hedland, WA	15,000
Do.	do.	Thunderbox gold mine (LionOre Mining International Ltd., 100%)	90 km northeast of Leonora, WA	5,000
Do.	do.	Trident gold mine (Avoca Resources Ltd., 100%)	Higginsville, WA	5,000
Do.	do.	Wattle Dam gold mine (Ramelius Resources Ltd., 100%)	70 km south of Kalgoorlie, WA	3,000
Do.	do.	Wiluna open pit/underground gold mine (Apex Minerals NL, 100%)	7 km south of Wiluna, WA	3,300
Smelter	do.	Gidji Roaster gold smelter (Kalgoorlie Consolidated Gold Mines Pty. Ltd., 100%)	Kalgoorlie, WA	24,300
Refinery	do.	Perth Refinery [AGR Management Services Ltd. (Australian Gold Alliance Pty Ltd., 40%; Western Australian Mint, 40%; Johnson Matthey (Australian) Ltd., 20%]	Newburn, WA	300,000
Gypsum		Gypsum Resources Australia Pty. Ltd., 100%	Lake MacDonnell open pit gypsum mine, near Point Thevenard, SA	1,400
Do.		Dampier Salt Ltd., 100%	Lake MacLeod salt and gypsum solar	900
Iron and steel:				
Iron ore		Channar open pit iron ore mine [Hamersley Iron Pty. Ltd., 60% (Rio Tinto Ltd., 100%), and China Iron and Steel Industry & Trade Group Corp. (SINOSTEEL) (a China state-owned company), 40%]	70 km south of Tom Price, WA	11,000
Do.		Cloudbreak iron ore mine (includes Chicester Range, Christmas Creek, WhiteKnight, Mount Lewin, Mount Nicholas, and Flinders) (Fortescue Metals Group Ltd., 100%)	Chichester Ranges, East Pilbara, WA	55,000
Do.		Cockatoo Island open pit iron ore mine (BHP Billiton Ltd., 100%)	130 km north northeast of Derby, WA	1,500
Do.		Eastern Range open pit iron ore mine [Hamersley Iron Pty. Ltd., 54% (Rio Tinto Ltd., 100%), and Shanghai Baosteel Group Corp., 46%]	10 km east of Paraburdoo, WA	10,000
Do.		Extension Hill open pit iron ore mine (Mount Gibson Iron Ltd., 100%)	85 km of Perenjori, WA	3,000
Do.		Hamersley Operations (includes Brockman No. 2, Marandoo, Mount Tom Price, Nammuldi, Paraburdoo, and Yandicoogina open pit iron ore mines) [Hamersley Iron Pty. Ltd., 100% (Rio Tinto Ltd., 100%)]	30 km to 85 km northeast, northwest, and south of Tom Price, WA	90,000
Do.		Hope Downs Mine [Hope Downs Iron Ore Pty Ltd. (Hancock Prospecting Pty Ltd. 100%), 50%, and Rio Tinto Ltd., 50%]	75 km northwest of Newman, WA	30,000
Do.		Jimblebar open pit iron ore mine {BHP Iron Ore [Jimblebar], 85% [BHP Billiton Ltd., 100%]; Mitsui Itochu Iron Pty Ltd., 10% [Mitsui & Co. (Australia) Ltd. 100%]; CI Minerals Australia Pty Ltd., 5% [Itochu Corp., 100%)]}	40 km east of Newman, WA	35,000
Do.		Karara open pit iron ore mine (Anshan Iron and Steel Group Corp., 50%, and Gindalbie Metals Ltd., 50%)	110 km south of Yalgoo, WA	8,000
Do.		Koolan Island open pit iron ore mine (Mount Gibson Iron Ltd., 100%)	140 north of Derby, WA	4,000
Do.		Koolyanobbing Central open pit iron ore mine (Portman Ltd., 100%)	50 km north-northeast of Southern Cross, WA	6,000
Do.		Mount Goldsworthy mining associates joint venture (includes Area C, Goldsworthy, and Nimingarra) [BHP Billiton Minerals Pty Ltd. (manager), 85%; ITOCHU Minerals & Energy of Australia Pty Ltd., 8%; Mitsui Iron Ore Corp. Pty. Ltd., 7%]	180 km east of Port Hedland, WA	42,000
Do.		Mount Gould open pit iron ore mine (Unimin Australia Ltd., 100%)	160 km west of Meekatharra, WA	6,000

See footnotes at end of table.

TABLE 2—Continued
AUSTRALIA: STRUCTURE OF THE MINERAL INDUSTRY IN 2012

(Thousand metric tons unless otherwise specified)

Commodity	Facilities, major operating companies, and major equity owners	Location of main facilities[1,2]	Annual capacity[e]
Iron and steel—Continued:			
Iron ore—Continued	Mount Newman open pit iron ore mine (includes Mount Whaleback, Orebody 23–25, Orebody 29, and Orebody 30–35) {BHP Billiton Minerals Pty Ltd., 85% [BHP Billiton Ltd., 100%]; Mitsui Itochu Iron Pty Ltd., 10% [Mitsui & Co. (Australia) Ltd., 100%]; CI Minerals Australia Pty Ltd., 5% [Itochu Corp., 100%]}	Within 13 km of Newman, WA	30,000
Do.	Pannawonica (includes Mesa A and J) open pit iron ore mine [Robe River Iron Associates, manager (Rio Tinto Ltd., 53%; Mitsui & Co. (Australia) Ltd., 33%; Nippon Steel Australia Pty. Ltd., 10.5%; Sumitomo Metal Australia Pty. Ltd., 3.5%]	130 km south-southwest of Dampier, WA	32,000
Do.	Savage River open pit iron ore mine (Stemcor Holdings Ltd., 100%)	100 km southwest of Burnie, TAS	2,400
Do.	Tallering Peak open pit iron ore mine (Mount Gibson Iron Ltd., 100%)	120 northeast of Geraldton, WA	3,000
Do.	Whyalla open pit iron ore mines (OneSteel Ltd., 100%)	270 km northwest of Adelaide, SA	2,600
Do.	Yandi open pit iron ore mine (BHP Billiton Minerals Pty Ltd., 85%, manager; ITOCHU Minerals & Energy of Australia Pty Ltd., 8%; Mitsui Iron Ore Corp. Pty. Ltd., 7%)	92 km north of Newman, WA	42,000
Pig iron	Hismelt pig iron plant [Hismelt Corp. Pty Ltd. (Rio Tinto Ltd., 60%; Nucor Corp., 25%; Mitsubishi Corp., 10%; and Shougang Corp., 5%]	Kwinana, WA	800
Steel	OneSteel Whyalla steelworks (OneSteel Ltd., 100%)	Whyalla, SA	1,200
Do.	Port Kembla steelworks (Blue Scope Steel Ltd., 100%)	Port Kembla, NSW	2,500
Do.	Smorgon Steel Group Ltd.	Laverton, Melbourne, VIC	700
Do.	do.	Waratch, NSW	285
Kaolin	Axedale Clays open pit kaolin mine (E Clay Pty Ltd., 100%)	18 km east of Bendigo, VIC	50
Do.	Pittong open pit kaolin mine (Imerys Minerals Australia Pty Ltd., 100%)	35 km southwest of Ballarat, VIC	110
Do.	Skardon River open pit kaolin mine (Queensland Kaolin Pty. Ltd., 96.6%, and private, 3.4%)	85 km north of Weipa, QLD	150
Lead:			
Mine, lead content	Anges zinc mine (Terramin Australia Ltd., 100%)	2 km from Strathalbyn, SA	10
Do.	Broken Hill underground silver-zinc-lead mine (Shenzhen Zhongjin Lingnan Nonfemet Co. Ltd., 50.1%, and Perilya Ltd., 49.9%)	Broken Hill, NSW	90
Do.	Cannington underground silver-lead-zinc mine (BHP Billiton Ltd., 100%)	85 km southwest of McKinlay, QLD	265
Do.	Century open pit zinc-silver-lead mine (Zinifex Ltd., 100%)	250 km north of Mount Isa, QLD	90
Do.	Endeavor underground zinc-silver-lead mine (CBH Resources Ltd., 100%)	40 km northwest of Cobar, NSW	24
Do.	Hellyer underground zinc-lead-copper-silver mine (Bae Metals Ltd., 100%)	80 km south-southwest of Burnie, TAS	44
Do.	Mount Isa underground copper-lead-zinc-silver mine (also includes Enterprise, George Fisher, and Hilton Mines) (Xstrata plc, 100%)	Mount Isa, QLD	150
Do.	Rosebery underground zinc-lead-silver-copper-gold mine [Minerals and Metals Group Australia Ltd., operator (China Minmetals Nonferrous Metals Co. Ltd., 100%)]	5 km north of Queenstown, TAS	25
Smelter	Mount Isa smelter (Xstrata plc, 100%)	Mount Isa, QLD	240
Do.	Port Pirie smelter (Nyrstar Corp., 100%)	5 km north of Queenstown, TAS	235
Magnesite	Kunwarara open pit magnesite mine (includes Marlborough) [Queensland Magnesia Pty Ltd., operator (Sibelco Group, 100%)]	70 km northwest of Rockhampton, QLD	3,000
Do.	Salt Creek open pit mine (Agricola Mining Pty Ltd., 100%)	70 km southeast of Meningie, SA	NA
Do.	Thuddungra Mine (Orind Australia Pty Ltd., 100%)	38 km northwest of Young, NSW	80

See footnotes at end of table.

TABLE 2—Continued
AUSTRALIA: STRUCTURE OF THE MINERAL INDUSTRY IN 2012

(Thousand metric tons unless otherwise specified)

Commodity	Facilities, major operating companies, and major equity owners	Location of main facilities[1,2]	Annual capacity[e]
Manganese:			
Mine, concentrate	Bootu Creek open pit manganese mine (OM Holding Ltd., 100%)	110 km north of Tennant Creek, NT	600
Do.	Groote Eylandt open pit manganese mine [Groote Eylandt Mining Co., operator (BHP Billiton Ltd., 60%, and Anglo American Corp., 40%)]	Groote Eylandt, NT	3,100
Do.	Woodie Woodie open pit manganese mine (includes Bells and East Pilbara leases) [Pilbara Manganese Pty Ltd., operator (Consolidated Minerals Ltd., 100%)]	400 southeast of Port Hedland, WA	1,000
Alloys	Bell Bay Smelter [Tasmanian Electro Metallurgical Co. Pty. Ltd., operator (BHP Billiton Ltd., 100%)]	Bell Bay, TAS	250
Mineral sands	Broken Hill region mines (Cristal Australia Pty Ltd., 100%)	120 km north of Mildura, NSW	NA
Do.	Murray Basin heavy-mineral sands mine (Iluka Resources Ltd., 100%)	80 km southeast of Mildura, VIC	NA
Do.	Perth Basin heavy-mineral sands mine (Iluka Resources Ltd., 100%)	260 km north of Perth, WA	NA
Do.	North Capel open pit heavy-mineral sands mine (Iluka Resources Ltd., 100%)	7 km north of Capel, WA	NA
Do.	North Stradbroke Island heavy-mineral sands dredge (Stradbroke Rutile Pty. Ltd., 100%)	35 km east of Brisbane, QLD	NA
Do.	Tiwest Joint Venture heavy-mineral sands dredge (Exxaro Resources Ltd., 50%, and Tronox Inc., 50%)	180 km north of Perth, WA	NA
Molybdenum metric tons	Wolfram Camp molybdenum-tungsten mine (Queensland Ore Ltd., 85%, and private, 15%)	85 km west of Cairns, QLD	120
Nickel:			
Mine, Ni content	Avebury nickel mine (includes Bison, North Avebury, Saxon, and West Viking) [Minerals and Metals Group Australia Ltd., operator (China Minmetals Nonferrous Metals Co. Ltd., 100%)]	Near Zeehan, TAS	7
Do.	Black Swan underground nickel mine (includes Silver Swan) (OJSC MMC Norilsk Nickel, 100%)	53 km northeast of Kalgoorlie, WA	10
Do.	Carnilya Hill open pit mine (Mincor Resources NL, 70%, and View Resources Ltd., 30%)	25 km northeast of Kambalda, WA	5
Do.	Cawse open pit nickel-cobalt mine (OJSC MMC Norilsk Nickel, 100%)	50 km northeast of Kalgoorlie, WA	9
Do.	Cosmos open pit nickel mine (Xstrata plc, 100%)	50 km north of Leinster, WA	13
Do.	Flying Fox underground mine (Western Areas NL, 100%)	108 km south of Marvel Loch, WA	15
Do.	Kambalda underground nickel mines (Palmary Enterprises Ltd., 100%)	5 km south of Kambalda, WA	35
Do.	Lake Johnson underground nickel mine (includes Maggie Hays, Maggie Hays Lake, and Emily Ann) (OJSC MMC Norilsk Nickel, 100%)	130 km west of Norseman, WA	12
Do.	Lanfranchi underground mine (includes Deacon, Schmitz, Tramway, and Winner) (Panoramic Resources Ltd., 100%)	42 km south of Kambalda, WA	10
Do.	Leinster open pit/underground nickel mines (BHP Billiton Ltd., 100%)	10 km north of Leinster, WA	44
Do.	Long underground mine (Independence Group NL, 100%)	Near Kambalda East, WA	10
Do.	Miitel underground nickel mine (includes Redross and Mariners) (Mincor Resources NL, 100%)	70 km south of Kambalda, WA	10
Do.	Mount Keith open pit nickel mine (includes Cliffs and Yakabindie) (BHP Billiton Ltd., 100%)	70 km south-southeast of Wiluna, WA	40
Do.	Murrin Murrin open pit nickel-cobalt mine (Minara Resources Ltd., 60%, and Glencore International plc, 40%)	60 km east of Leonora, WA	34
Do.	Radio Hill underground nickel-cobalt mine (Fox Resources Ltd., 100%)	35 km south of Karratha, WA	4
Do.	Ravensthorpe open pit mine (First Quantum Minerals Ltd., 100%)	155 km west of Esperance, WA[3]	39
Do.	Savannah underground mine (Panoramic Resources Ltd., 100%)	120 km north of Halls Creek, WA	8
Do.	Spotted Quoll nickel mine (includes Tim King and Willy Willy) (Western Areas NL, 100%)	114 km south of Marvel Loch, WA	10
Do.	Waterloo underground nickel mine (includes Amorac) (OJSC MMC Norilsk Nickel, 100%)	90 km north of Leonora, WA	5

See footnotes at end of table.

TABLE 2—Continued
AUSTRALIA: STRUCTURE OF THE MINERAL INDUSTRY IN 2012

(Thousand metric tons unless otherwise specified)

Commodity	Facilities, major operating companies, and major equity owners	Location of main facilities[1,2]	Annual capacity[e]
Nickel—Continued:			
Smelter	Kalgoorlie nickel smelter (BHP Billiton Ltd., 100%)	Kalgoorlie, WA	100
Refinery	Kwinana nickel refinery (BHP Billiton Ltd., 100%)	Kwinana, WA	67
Do.	Murrin Murrin nickel refinery (Minara Resources Ltd., 60%, and Glencore International plc, 40%)	Murrin Murrin, WA	45
Do.	Yabulu nickel-cobalt refinery (Nickel Consolidated Pty Ltd., Nickel House Pty Ltd., and Nickel Process Pty Ltd.)	Townsville, QLD	40
Opal	Many small producers	Andamooka and Coober Pedy areas, SA; Lightning Ridge area, NSW	NA
Petroleum thousand 42-gallon barrels per day	Exxon Mobil Corp., 100%	Altona Refinery, VIC	120
Do. do.	Bulwer Island Refinery [BP Amoco Refinery (Bulwer Island) Pty. Ltd., 100%]	Bulwer Island, QLD	69.3
Do. do.	Clyde Refinery [Shell Refining (Australia) Pty. Ltd., 100%]	Clyde, NSW	85
Do. do.	Geelong Refinery [Shell Refining (Australia) Pty. Ltd., 100%]	Geelong, VIC	110
Do. do.	Kurnell Refinery (Caltex Australia Ltd., 100%)	Kurnell, NSW	114
Do. do.	Kwinana Refinery [BP Amoco Refinery (Kwinana) Pty. Ltd., 100%]	Kwinana, WA	138
Do. do.	Lytton Refinery (Caltex Australia Ltd., 100%)	Lytton, QLD	106
Do. do.	Port Stanvac Refinery (Exxon Mobil Corp., 100%)	Port Stanvac, SA	69
Phosphate rock	Phosphate Hill-Duchess open pit phosphate mine (Incitec Pivot Ltd., 100%)	140 km northwest of Mount Isa, QLD	2,200
Rare earths, rare-earth oxide	Mount Weld Mine (Lynas Corp. Ltd.)	Mount Weld, WA	1,100
Salt	Dampier solar evaporation salt pans (Dampier Salt Ltd., 100%)	Near Dampier, WA	4,000
Do.	Lake MacLeod solar salt and gypsum evaporation pans (Dampier Salt Ltd., 100%)	65 km north of Carnarvon, WA	900
Do.	Port Hedland solar salt fields (Dampier Salt Ltd., 100%)	Port Hedland, WA	3,000
Silica	Itochu Corp., 50%, and Tochu Corp., 50%	Kemerton silica sands dredge, 25 km northeast of Bunbury, WA	450
Silver:			
Mine, kilograms Ag content	Broken Hill underground silver-zinc-lead mine (Shenzhen Zhongjin Lingnan Nonfemet Co. Ltd., 50.1%, and Perilya Ltd., 49.9%)	Broken Hill, NSW	81,200
Do. do.	Cannington underground silver-lead-zinc mine (BHP Billiton Ltd., 100%)	85 km southwest of McKinlay, QLD	700,000
Do. do.	Century open pit zinc-silver-lead mine [Minerals and Metals Group Australia Ltd., operator (China Minmetals Nonferrous Metals Co. Ltd., 100%)]	250 km north of Mount Isa, QLD	3,000
Do. do.	Pasminco Ltd., 100%	Cockle Creek silver smelter, NSW	85,000
Do. do.	Endeavor underground zinc-silver-lead mine (CBH Resources Ltd., 100%)	40 km northwest of Cobar, NSW	35,000
Do. do.	Hellyer underground zinc-lead-copper-silver mine (Intec Ltd., 50%, and Polymetals Mining Services Pty Ltd., 50%)	80 km south-southwest of Burnie, TAS	60,000
Do. do.	Henty underground gold-silver mine (Barrick Gold Ltd., 100%)	30 km north of Queenstown, TAS	1,100
Do. do.	Mount Isa underground copper-lead-zinc-silver mine (also includes Enterprise, George Fisher, and Hilton Mines) (Xstrata plc, 100%)	Mount Isa, QLD	375,000
Do. do.	Olympic Dam underground copper-silver-gold-uranium mine [Olympic Dam Operations Pty. Ltd., operator (BHP Billiton Ltd., 100%)]	Roxby Downs, 80 km north of Woomera, SA	27,000
Do. do.	Peak underground gold-zinc-lead-copper-silver underground mine (includes New Cobar, New Occidental, and Perseverance), (GoldCorp Inc., 100%)	8 km south of Cobar, NSW	6,000

See footnotes at end of table.

TABLE 2—Continued
AUSTRALIA: STRUCTURE OF THE MINERAL INDUSTRY IN 2012

(Thousand metric tons unless otherwise specified)

Commodity		Facilities, major operating companies, and major equity owners	Location of main facilities[1, 2]	Annual capacity[e]
Silver:				
Mine, Ag content—Continued	kilograms	Rosebery underground zinc-lead-silver-copper-gold mine [Minerals and Metals Group Australia Ltd., operator (China Minmetals Nonferrous Metals Co. Ltd., 100%)]	5 km north of Queenstown, TAS	35,000
Smelter	do.	Port Pirie smelter (Nyrstar Corp., 100%)	do.	450,000
Refinery	do.	Perth Refinery [AGR Management Services Ltd. (Australian Gold Alliance Pty Ltd., 40%; Western Australian Mint, 40%; and Johnson Matthey (Australian) Ltd., 20%]	Newburn, WA	81,000
Spodumene		Greenbushes open pit/underground tantalite-spodumene mine {Windfield Holding Pty Ltd., operator [Chengdu Tianqi Industry (Group) Co. Ltd., 100%]}	70 km southeast of Bunbury, WA	260
Do.		Mount Cattlin spodumene mine (Galaxy Resources Ltd., 100%)	2 km north of Ravensthorpe, WA	140
Talc		Three Springs open pit talc mine (Imerys SA, 100%)	330 km north of Perth, WA	150
Tantalum, tantalite, Ta_2O_5 content	metric tons	Greenbushes open pit/underground tantalite-spodumene mine (Global Advanced Metals Ltd., 100%)	70 km southeast of Bunbury, WA	550
Do.	do.	Bald Hill tantalite mine (Haddington Resources Ltd., 100%)	60 km southeast of Kambalda, WA[3]	100
Do.	do.	Wodgina open pit tantalite mine (Global Advanced Metals Ltd., 100%)	70 km southeast of Bunbury, WA[3]	250
Tin:				
Mine, Sn content	do.	Collingwood underground tin mine (Metals X Ltd., 100%)	35 km south of Cooktown, QLD[3]	3,000
Do.	do.	Greenbushes open pit/underground tantalite-spodumene mine (Global Advanced Metals Ltd., 100%)	70 km southeast of Bunbury, WA[3]	1,000
Do.	do.	Mount Bischoff open pit mine (Metals X Ltd., 50%; L'sea Resources International Holdings Ltd. and YT Parksong Australia Holdings Pty Ltd., 50%)	55 km southwest of Burnie, TSA[3]	6,000
Do.	do.	Renison Bell underground tin mine (Metals X Ltd., 50%; L'sea Resources International Holdings Ltd. and YT Parksong Australia Holdings Pty Ltd., 50%)	136 km south of Burnie, TAS	4,000
Smelter	do.	Greenbushes smelter (Global Advanced Metals Ltd., 100%)	70 km southeast of Bunbury, WA[3]	1,000
Tungsten, W content	do.	Kara magnetite and scheelite mine (Tasmania Mines Ltd., 100%)	30 km south of Burnie, TAS	50
Do.	do.	Mount Carbine tungsten mine (Carbine Tungsten Ltd., 100%)	75 km west of Cairns, QLD	4,000
Do.	do.	Wolfram Camp molybdenum-tungsten mine (Deutsche Rohstoff AG, 100%)	85 km west of Cairns, QLD	500
Uranium, U_3O_8 content	do.	Beverley in situ leach uranium operation (Heathgate Resources Pty. Ltd., 100%)	300 km northeast of Port Augusta, SA	1,000
Do.	do.	Honeymoon uranium mine (UraniumOne Inc., 100%)	75 km northwest of Broken Hill, SA	400
Do.	do.	Olympic Dam underground copper-silver-gold-uranium mine [Olympic Dam Operations Pty. Ltd., operator (BHP Billiton Ltd., 100%)]	Roxby Downs, 80 km north of Woomera, SA	4,400
Do.	do.	Ranger open pit uranium mine (Energy Resources of Australia Ltd., 100%)	230 km east of Darwin, NT	5,000
Vanadium, V_2O_5 content	do.	Windimurra open pit mine vanadium (Precious Metals Australia Ltd., 90%, and Noble Group Ltd., 10%)	100 km east-southeast of Mount Magnet, WA3	8

See footnotes at end of table.

TABLE 2—Continued
AUSTRALIA: STRUCTURE OF THE MINERAL INDUSTRY IN 2012

(Thousand metric tons unless otherwise specified)

Commodity	Facilities, major operating companies, and major equity owners	Location of main facilities[1, 2]	Annual capacity[e]
Zinc:			
Mine, Zn content	Anges zinc mine (Terramin Australia Ltd., 100%)	2 km from Strathalbyn, SA	24
Do.	Broken Hill underground silver-zinc-lead mine (Shenzhen Zhongjin Lingnan Nonfemet Co. Ltd., 50.1%, and Perilya Ltd., 49.9%)	Broken Hill, NSW	360
Do.	Cannington underground silver-lead-zinc mine (BHP Billiton Ltd., 100%)	85 km southwest of McKinlay, QLD	100
Do.	Century open pit zinc-silver-lead mine [(MMG Ltd., operator) China Minmetals Group, 100%]	250 km north of Mount Isa, QLD	500
Do.	Endeavor underground zinc-silver-lead mine (CBH Resources Ltd., a subsidiary of Toho Zinc Co. Ltd. of Japan, 100%)	40 km northwest of Cobar, NSW	44
Do.	Golden Grove underground zinc-copper mine [(MMG Ltd., operator) China Minmetals Group, 100%]	225 km east of Geraldton, WA	150
Do.	Hellyer underground zinc-lead-copper-silver mine (Intec Ltd., 50%, and Polymetals Mining Services Pty Ltd., 50%)	80 km south-southwest of Burnie, TAS[3]	130
Do.	Jaguar underground mine (Jabiru Metals Ltd., 100%)	250 km north of Kalgoorlie, WA	420
Do.	McArthur River open pit mine [McArthur River Mining Pty Ltd., operator (Xstrata plc, 100%)]	60 km southwest of Borroloola, NT	143
Do.	Mount Isa underground copper-lead-zinc-silver mine (also includes Enterprise, George Fisher, and Hilton Mines) (Xstrata plc, 100%)	Mount Isa, QLD	175
Do.	Peak underground gold-zinc-lead-copper-silver underground mine (includes New Cobar, New Occidental, and Perseverance) (GoldCorp Inc., 100%)	8 km south of Cobar, NSW	8
Do.	Rosebery underground zinc-lead-silver-copper-gold mine [Minerals and Metals Group Australia Ltd., operator (China Minmetals Nonferrous Metals Co. Ltd., 100%)]	35 km north of Queenstown, TAS	100
Smelter	Port Pirie smelter (Nyrstar Corp., 100%)	5 km north of Queenstown, TAS	45
Do.	Hobart smelter (Nyrstar Corp., 100%)	Hobart, TAS	320
Refinery	Sun Metals zinc refinery [Sun Metals Corp. Pty. Ltd., operator (Korea Zinc Co., 100%)]	Townsville, QLD	170

[e]Estimated; estimated data are rounded to no more than three significant digits. Do., do. Ditto. NA Not available.

[1]Abbreviations used for States and Territories in this table include the following: NSW—New South Wales; NT—Northern Territory; QLD—Queensland; SA—South Australia; TAS—Tasmania; VIC—Victoria; WA—Western Australia.

[2]Abbreviation(s) used for unit(s) of measure in this table include the following: km—kilometer.

[3]On care-and-maintenance status; expansion project development decision pending.

TABLE 3
AUSTRALIA: RESERVES OF MAJOR MINERAL COMMODITIES IN 2012

Commodity		Reserves[1]
Antimony, Sb content	thousand metric tons	105
Bauxite	million metric tons	5,670
Coal:		
Black:		
In situ	billion metric tons	70
Recoverable	do.	57
Brown:		
In situ	do.	49
Recoverable	do.	44
Cobalt, Co content	thousand metric tons	1,200
Copper, Cu content	million metric tons	86
Diamond	million carats	2,700
Gold, Au content	metric tons	9,100
Iron ore	billion metric tons	38
Lead, Pb content	million metric tons	36
Lithium, Li content	thousand metric tons	1,000
Magnesite ($MgCO_3$ content)	million metric tons	330
Manganese ore	do.	200
Mineral sands:		
Ilmenite	do.	190
Rutile	do.	27
Zircon	do.	46
Molybdenum, Mo content	thousand metric tons	167
Nickel, Ni content	million metric tons	20
Niobium (columbium) and tantalum:		
Niobium (columbium), Nb content	thousand metric tons	200
Tantalum, Ta content	do.	62
Platinum-group metals (Pd, Pt)	metric tons	4
Rare earths (REO plus Y_2O_3)	thousand metric tons	2,000
Silver, Ag content	do.	88
Tin, Sn content	do.	240
Tungsten, W content	do.	370
Uranium, U content	do.	1,200
Vanadium, V content	do.	1,520
Zinc, Zn content	million metric tons	68

do. Ditto.

[1]Economic demonstrated resources. Data are rounded to no more than three significant digits.

Source: Geoscience Australia, 2010, Australia's identified mineral resources 2012: Canberra, Australian Capital Territory, Australia, Geoscience Australia, p. 5.

THE MINERAL INDUSTRY OF BANGLADESH

By Yolanda Fong-Sam

In 2012, the mineral industry of Bangladesh produced mainly coal, natural gas, salt, and stone. The country lacks reserves of metallic minerals but is prospective for additional reserves of natural gas. According to preliminary data from the Central Bank of Bangladesh, in fiscal year 2012 (July 2012 through June 2013), mining and quarrying accounted for about 1.1% of the country's gross domestic product (GDP), which stayed the same compared with that of 2011; the construction sector accounted for about 8.3% of the country's GDP in 2012 compared with 8% in 2011. Total foreign trade increased by 5.6% to $55.9 billion from $52.9 billion in 2011. The country's total exports were valued at about $24 billion compared with $22.6 billion in 2011, which was an increase of 6.2%; similarly, the total value of imports in 2012 increased by 5.2% to $31.9 billion from $30.3 billion in 2011. Imports of iron, steel, and other base metals (unspecified) were valued at $2.2 billion, which was an increase of 11% compared with that of 2011 (Central Bank of Bangladesh, 2013, tables I.1, I.2, IV.3).

The Bangladesh Oil, Gas and Mineral Corp. (Petrobangla) is the Government entity that is responsible for the exploration for, production of, and transmission and distribution of natural gas and oil in Bangladesh. In recent years, Petrobangla continued its campaign for the discovery of new gasfields to increase the country's natural gas reserves by performing exploratory drillings and conducting seismic surveys. Petrobangla was also in charge of the development of some of the country's nonfuel mineral deposits, which included deposits of industrial minerals and coal that had been determined to be economically feasible. Oversight of the exploration for minerals, however, was the responsibility of the Geological Survey of Bangladesh (Bangladesh Oil, Gas and Mineral Corp., 2012, p. 8, 15–16).

In June, the Ministry of Finance submitted the National Budget Proposal to the Parliament for fiscal year 2013 that outlined a budget of about $24.3 billion, which was equivalent to about 18% of the country's GDP. The proposal targeted economic growth of more than 7% per fiscal year and identified the power and energy sector as a priority by allocating 17.3% of the national budget to the development of that sector. In its budget proposal, the Government indicated that it was committed to solving the country's energy crisis as a measure to achieve GDP growth by gradually decreasing the deficits in energy production and the power supply and by improving the energy infrastructure. The Government projected a GDP growth rate of approximately 8% by fiscal year 2015. In addition, the proposal mentioned the 20-year railway master plan, which was under Government consideration and covered the implementation of 231 projects to connect the Bangladesh railway to subregional, regional, and international railway networks. The proposed railway master plan had an estimated cost of $16 billion (Xinhuanet.com, 2012).

Bangladesh and neighboring countries Burma (to the east) and India (to the west) were involved in a maritime boundary dispute concerning their respective sovereignty in the Bay of Bengal. For many years, these countries had attempted to negotiate and delimit their claims in the disputed area. In December 2009, Bangladesh and Burma accepted the jurisdiction of the International Tribunal for the Law of the Sea (ITLOS) for the settlement of their boundary delimitation. ITLOS is an independent judicial body established by the United Nations Convention on the Law of the Sea (UNCLOS) that has jurisdiction to arbitrate disputes arising out of the interpretation and application of the Law of the Sea. UNCLOS establishes a legal framework to regulate ocean space and its resources and uses (International Tribunal for the Law of the Sea, 2010).

In September 2011, representatives from Bangladesh and Burma met with the ITLOS in Germany for a final round of arguments regarding the maritime boundary. During the final judgment on March 14, 2012, ITLOS dealt with the delimitation in three parts—the territorial sea, the exclusive economic zones and continental shelf within 200 nautical miles, and the continental shelf beyond 200 nautical miles. ITLOS rendered its judgment regarding the territorial sea by drawing an equidistant line from the countries' baselines. For the exclusive economic zones and continental shelf within 200 nautical miles, the tribunal drew a provisional equidistant line that adjusts to the concavity of the coast of Bangladesh. For the delimitation of the continental shelf beyond 200 nautical miles, the tribunal concluded that it should not differ from that for within the 200 nautical miles and should continue in the same direction beyond the limit of Bangladesh. With the resolution of the boundary dispute, the countries involved would be able to redefine their exploration areas offshore. The maritime dispute with India was set to be resolved in the near future (International Law Observer, 2012; International Tribunal for the Law of the Sea, 2012).

Production

Bangladesh produced small amounts of industrial minerals and processed products. The level of production in 2012 remained similar to that of 2011 (table 1).

Structure of the Mineral Industry

The major public gas-producing companies, both of which were under Petrobangla, were Bangladesh Gas Fields Co. Ltd. (BGFCL) and Sylhet Gas Fields Ltd. (SGFL). Together, the companies produced about 44% of the country's total gas production for fiscal year 2012. In addition to exploring for, producing, and distributing oil and gas, Petrobangla also explored for and produced coal and granite through its subsidiaries Barapukuria Coal Mining Co. Ltd. (BCMCL) and Maddhapara Granite Mining Company Ltd., respectively. Eastern Refinery Ltd. (ERL), which was a subsidiary of

Government-owned Bangladesh Petroleum Corp. (BPC), was Bangladesh's sole petroleum refining company. The refinery, which was located in Chittagong, produced petroleum products from imported crude oil. Output from the refinery met about 10% of the country's demand for petroleum products (Bangladesh Oil, Gas and Mineral Corp., 2012, p. 24–25; Daily Sun, 2012). Table 2 is a list of major mineral industry facilities.

Commodity Review

Industrial Minerals

Cement.—In February, HeidelbergCement Bangladesh Ltd., which was a subsidiary of HeidelbergCement Group of Germany, inaugurated a new cement mill at its Chittagong grinding plant, which had a designed capacity to produce 0.8 million metric tons per year (Mt/yr) of cement. The mill was constructed at a cost of $16 million (International Cement Review, 2012).

Premier Cement Mills Ltd. announced the completion of an expansion project at the company's facility in West Mukterpur in Munshigonjto, central Bangladesh, to increase the capacity to 2.4 Mt/yr of cement from 1.2 Mt/yr. The estimated cost for the expansion was $13.56 million (International Cement Review, 2013).

Stone, Crushed.—Maddhapara Granite was responsible for the production of granite at Petrobangla's underground mine in the District of Dinajpur. The facility had the capacity to produce about 1.65 Mt/yr of construction aggregate. For fiscal year 2011 (the most recent year for which reported data were available), the Maddhapara Mine produced 360,071 metric tons (revised) of stone, of which the majority was sold domestically for use as construction material, such as aggregates (Bangladesh Oil, Gas and Mineral Corp., 2012, p. 31).

Mineral Fuels

Coal.—As of 2012, five coalfields had been discovered in the country, which included the Barapukuria, the Dighipara, the Jamalganj, the Khalashpir, and the Phulbari fields, most of which are located in northern Bangladesh. The combined total of probable reserves in the coalfields was estimated to be 3.3 billion metric tons of coal. The Barapukuria coal mine, which was managed by BCMCL, was the first and only operating coal mine in Bangladesh. Coal produced from the mine was being used for power generation by two 150-megawatt (MW)-capacity coal-fired thermal powerplants located near the mine in the District of Dinajpur (Bangladesh Oil, Gas and Mineral Corp., 2012, p. 41).

In 2012, the United Kingdom-based company Global Coal Management Resources Plc. (GCM) through its subsidiary Asia Energy Corp. (Bangladesh) Pty Ltd. was still awaiting approval from the Government to develop the Phulbari coal project located in the northwestern region of Phulbari, Dinajpur District, which is 350 kilometers (km) from Dhaka. Asia Energy held a contract with the Government for the exploration and mining of coal in northern Bangladesh. As part of the application process for the project's mining permit, the company completed a feasibility study and submitted a proposal for mine development to the Government. The project had already obtained the environmental clearance, and according to the company, if the project is approved, the development of the open pit mine could start immediately and production could start within 3 to 4 years. The Phulbari coal project, which had a coal resource of 572 million metric tons (Mt) and a proposed annual capacity to produce 15 Mt/yr, could potentially alleviate the country's electricity shortages by providing coal to support up to 4,000 MW of power-generating capacity. The project could potentially contribute an amount equal to about 1% of the GDP to the country's revenue; in addition, it could generate approximately 17,000 jobs (CGM Resources, 2012, p. 2, 3; PhulbariCoal.com, 2013).

Natural Gas.—In 2012, Bangladesh's electricity-generating capacity was insufficient to meet the country's increased demand for energy, and power shortages continued to affect the country. Natural gas accounted for about 73% of commercial energy in the country; gas supplies were scarce, which forced authorities to ration the supply between businesses, households, and industries. Since 2009, Petrobangla had been carrying out extensive exploration and drilling activities to increase gas reserves. By 2012, a total of 25 gasfields had been discovered in the country for an overall recoverable proven and probable reserve of 765.7 billion cubic meters, of which about 40% had already been produced as of December. In 2012, the sole national exploration company, Bangladesh Petroleum Exploration and Production Co. Ltd. (BAPEX), which was a subsidiary of Petrobangla, for the first time started running a three-dimensional (3–D) seismic program to survey and delineate the extension of discovered fields, increase the success rate of exploration and development of wells, and estimate the content of gas-in-place in the discovered fields. BAPEX planned to survey about 1,950 square kilometers between 2012 and 2017 using the 3–D technology. According to Petrobangla's annual report, BAPEX announced the discovery of two new gasfields, the Sundolpur and the Srikail fields, during a well-drilling program in 2012 (Bangladesh Oil, Gas and Mineral Corp., 2012, p. 8–9, 17).

As of June 2012, the total gas pipeline network in Bangladesh encompassed nearly 21,380 km, of which 15,036 km was feeder and service lines; 2,431 km was distribution lines; 2,110 km was transmission lines; and 1,800 km was undefined. By the end of the year, a total of 393 km of new transmission pipelines had been installed across the country to contribute to a larger flow of gas in the system. As the gas flow progressively increased in the country (surpassing 62.3 million cubic meters per day by the end of 2012), the Government lifted the ban that it had imposed in July 2010 on the installation of new gas connections from the transmission pipelines to households. According to Petrobangla, a project to expand the gas distribution network in the western and southwestern parts of the country was expected to be commissioned in December 2013. Gas Transmission Co. Ltd. (GTCL) was the state company in charge of the transport of high-pressure gas in the country (Bangladesh Oil, Gas and Mineral Corp., 2012, p. 8–9, 18–19).

By the end of 2012, the following areas had active production-sharing contracts in place between the Government and international oil companies—Block 9 (Bangora gasfield) operated by Tullow Oil plc of United Kingdom;

Block 12 (Bibiyana gasfield), Block 13 (Jalalabad gasfield), and Block 14 (Maulavibazar gasfield) operated by Chevron Corp. of the United States; Block 16 (Sangu and Magnama gasfields) operated by Santos International Holdings Pty. Ltd. of Australia; and Block 17 and Block 18 operated by PTT Exploration & Production Public Co. Ltd. of Thailand. Chevron's Bibiyana gasfield, which had the capacity to produce 22.7 million cubic meters per day of gas, was the leading natural gas producer in the country; Chevron was planning to increase the field's production capacity to 31.1 million cubic meters per day of gas by 2014 (Bangladesh Oil, Gas and Mineral Corp., 2012, p. 20–21).

ConocoPhillips Co. of the United States explored two offshore blocks (Block 10 and Block 11) that covered an area of approximately 566,600 hectares (1.4 million acres) in the Bay of Bengal. In 2012, the company performed seismic surveys on both blocks, and it planned to continue with the analysis of the collected data in 2013 to define a possible drilling location. The company held a 100% interest in both blocks. The Government had awarded ConocoPhillips the two blocks in 2009, and in 2011, the company received approval from the Government for a production-sharing contract, which determines each party's share in the project (Bangladesh Oil, Gas and Mineral Corp., 2012, p. 20; ConocoPhillips Co., 2012, p. 44, 48).

Chevron held 98% interest in its three gasfields in Bangladesh, which were operated under two production-sharing contracts, one for Block 12, and one for Block 13 and Block 14. In April, Chevron announced that it had completed the installation of two turbine compressors at the Muchai compression station to support additional production capacity of 2.3 million cubic meters per day of natural gas from the company's three gasfields. In July, the company announced plans to implement the Bibiyana expansion project, which would include an expansion of the gas plant, development of additional wells, and other upgrades to increase the daily production capacity by 8.5 million cubic meters of natural gas and 4,000 barrels of condensate. The expansion project was expected to be commissioned by 2014 (Chevron Corp., 2012, p. 27).

Petroleum.—During the last quarter of 2012, BPC announced plans to expand the refining capacity at ERL. The expansion, which was projected to cost $1 billion, would increase the refinery's total capacity by 3 Mt/yr to a total of about 4.5 Mt/yr of crude oil. BPC asked ERL to prepare a project development proposal. ERL is located in Chittagong in southeastern Bangladesh. On average, the facility currently refined a total of 1.4 Mt/yr of crude oil, although the capacity was 1.5 Mt/yr. No expected date was set for completion of the expansion project. In April, BPC submitted a proposal to Kuwait Petroleum International (KPI) that the two companies form a joint venture for the construction and installation of a second oil refinery in the country. KPI requested a feasibility study. The proposed refinery would have a refining capacity of about 5 Mt/yr of crude oil (Daily Sun, 2012).

Outlook

The Government has made investment in the country's power generation platform a priority as a way to help resolve the country's energy crisis. The Government expects to decrease the deficits in the power supply and energy production through the gradual expansion of the gas distribution network and gas transmission pipelines and the implementation of other infrastructure improvements, such as bridges and a railway system to support these plans. The gap between the demand for and supply of natural gas is expected to be further narrowed as the Government starts the implementation of projects undertaken to increase the production of natural gas. In addition, because the country's dependence on natural gas has skyrocketed in recent years, the Government is considering the use of domestically produced coal as an alternative fuel. The development of the coal sector has been continuing at a slow pace, however, as the Government is still working on the development of the country's coal policy.

In the near future, and taking into consideration the many active development projects in the country, an increase in the demand for building materials, such as cement and crushed stone, is expected. The country is expected to become less dependent on imported industrial materials, such as cement, as expansion projects in the sector have reached the production stage. The fast-growing infrastructure development and the increased demand for materials will most likely influence trading between neighboring countries.

References Cited

Bangladesh Oil, Gas and Mineral Corp., 2012, Annual report 2012: Dhaka, Bangladesh, Bangladesh Oil, Gas and Mineral Corp., 56 p. (Accessed November 22, 2013, at http://www.petrobangla.org.bd/annualreport2012pb.pdf.)

Central Bank of Bangladesh, 2013, Bangladesh Bank quarterly—January–March 2013: Dhaka, Bangladesh, Central Bank of Bangladesh, v. X, no. 3, 19 p. [tables only]. (Accessed June 4, 2013, http://www.bangladesh-bank.org/pub/quaterly/bbquarterly/jan-mar2013/bbquarterly.php.)

CGM Resources, 2012, Annual report & accounts 2012: London, United Kingdom, CGM Resources, November, 52 p. (Accessed November 7, 2013, at http://www.gcmplc.com/ir/gcm/pdf/Annual_Report_2012.pdf.)

Chevron Corp., 2012, Supplement to the annual report—2012: San Ramon, California, Chevron Corp., 64 p. (Accessed December 9, 2013, at http://www.chevron.com/documents/pdf/chevron2012annualreportsupplement.pdf?.)

ConocoPhillips Co., 2012, Annual report—2012: Houston, Texas, ConocoPhillips Co., 214 p. (Accessed December 9, 2013, at http://www.conocophillips.com/investor-relations/Company Reports/2012_Annual_Report_CR.pdf.)

Daily Sun, 2012, Govt. decides to establish 2nd unit of Eastern Refinery: Daily Sun [Dhaka, Bangladesh], September 15. (Accessed November 7, 2013, at http://www.daily-sun.com/details_yes_15-09-2012_Govt.-decides-to-establish-2nd-unit-of-Eastern-Refinery_262_1_0_3_4.html.)

International Cement Review, 2012, Heidelberg inaugurates mill: International Cement Review, February, p. 8.

International Cement Review, 2013, Premier completion: International Cement Review, January, p. 14.

International Law Observer, 2012, Judgment in Bangladesh-Myanmar maritime boundary dispute: International Law Observer, March 15. (Accessed January 15, 2013, at http://www.internationallawobserver.eu/2012/03/15/judgment-in-bangladesh-myanmar-maritime-boundary-dispute/.)

International Tribunal for the Law of the Sea, 2010, Dispute concerning delimitation of the maritime boundary between Bangladesh and Myanmar in the Bay of Bengal: WorldCourts.com, January 28, Case no. 16, 3 p. (Accessed June 21, 2011, at http://www.worldcourts.com/itlos/eng/decisions/2010.01.28_Bangladesh_v_Myanmar.pdf.)

International Tribunal for the Law of the Sea, 2012, Dispute concerning delimitation of the maritime boundary between Bangladesh and Myanmar in the Bay of Bengal: International Tribunal for the Law of the Sea, press release 175, March 14, 5 p. (Accessed January 15, 2013, at http://www.itlos.org/fileadmin/itlos/documents/press_releases_english/pr175_engf.pdf.)

PhulbariCoal.com, 2013, Phulbari coal project—Project at a glance: PhulbariCoal.com. (Accessed April 2, 2014, at http://www.phulbaricoal.com/?page_id=54.)

Xinhuanet.com, 2012, Bangladesh unveils 24.30 bln-USD proposed budget targeting 7.2 pct economic growth: Xinhuanet.com, June 7. (Accessed March 31, 2014, at http://news.xinhuanet.com/english/business/2012-06/07/c_131638485.htm.)

TABLE 1

BANGLADESH: ESTIMATED PRODUCTION OF MINERAL COMMODITIES[1, 2]

(Metric tons unless otherwise specified)

Commodity[3]		2008	2009	2010	2011	2012
Cement, hydraulic[4]		5,000,000	5,000,000	5,000,000	NA [r]	NA
Clays, kaolin[4]		8,500	8,500	8,500	-- [r]	--
Coal, bituminous[4]		857,648 [5]	730,866 [5]	790,579 [r, 5]	820,437 [r, 5]	820,000
Gas, natural, marketed[4, 6]	million cubic meters	18,511 [5]	19,919 [5]	20,312 [r, 5]	20,951 [r, 5]	21,000
Iron and steel, metal, steel products[4]		60,000	60,000	60,000	-- [r]	--
Nitrogen, N content of urea, ammonia, ammonium sulfate		1,300,000	1,300,000	1,300,000	-- [r]	--
Petroleum:						
Crude	thousand 42-gallon barrels	1,800	1,800	1,800	-- [r]	--
Refinery products	do.	9,500	9,500	9,500	NA [r]	NA
Salt, marine[4]		1,368,323 [5]	1,388,557 [5]	1,409,239 [5]	1,410,000	1,400,000
Stone, crushed, granite		267,434 [5]	290,187 [r, 5]	300,000	360,071 [r, 5]	360,000

[r]Revised. do. Ditto. NA Not available. -- Zero.

[1]Estimated data are rounded to no more than three significant digits.

[2]Table includes data available through February 21, 2014.

[3]In addition to the commodities listed, construction materials, such as limestone, sand and gravel and other varieties of stone, are known to have been produced, but available information is inadequate to make reliable estimates of output.

[4]Data are for fiscal year ending June 30 of following year.

[5]Reported figure.

[6]Gross production is not reported; the quantity vented, flared, or reinjected is thought to be negligible.

TABLE 2
BANGLADESH: STRUCTURE OF THE MINERAL INDUSTRY IN 2012

(Thousand metric tons unless otherwise specified)

Commodity		Major operating companies and major equity owners	Location of main facilities	Annual capacity^e
Cement		Bangladesh Oil, Gas and Mineral Corp. (Petrobangla)	Chittagong	1,000.
Do.		do.	Sylhet	1,100.
Do.		Cemex Cement Bangladesh Ltd.	Mahmudnagar	600.
Do.		HeidelbergCement Bangladesh Ltd.	Chittagong and Narayangonj (near Dhaka)	2,000.
Do.		Holcim (Bangladesh) Ltd.	Bagerhat and Narayanganj	1,300.
Do.		Lafarge Surma Cement Ltd. (Lafarge Group and Cementos Molins S.A.)	Chhatak, Sunamganj	1,500 (1,150 clinker).
Do.		Meghna Cement Mills Ltd. (an enterprise of the Bashundhara Group of Bangladesh)	Mongla Port Industrial Zone and Pashur River Bank facility	1,000.
Do.		Premier Cement Mills Ltd.	West Mukterpur in Munshigonjto	2,400.
Do.		Shah Cement Industries Ltd.	Dhaka	1,860.
Do.		Unique Cement Industries Ltd.	Chittagong, Dhaka, and Sylhet	1,440.
Do.		Various	18 additional facilities	5,240.
Coal		Barapukuria Coal Mining Co. Ltd. (BCMCL) [Bangladesh Oil, Gas and Mineral Corp. (Petrobangla), 100%]	Barapukuria	1,000.
Fertilizer		Bangladesh Chemical Industries Corp.	Auganish	560.
Do.		do.	Fenchugani	100.
Do.		do.	Ghorasai	600.
Gas, natural	million cubic meters per day	Bangladesh Gas Fields Co. Ltd. (BGFCL) [Bangladesh Oil, Gas and Mineral Corp. (Petrobangla), 100%]	Bakhrabad, Habigangj, Kamta, Meghna, Narsingdi, and Titas	22.
Do.	do.	Bangladesh Petroleum Exploration and Production Co. Ltd. (BAPEX) [Bangladesh Oil, Gas and Mineral Corp. (Petrobangla), 100%]	Fenchuganj and Saldanadi	2.
Do.	do.	Chevron Corp.	Bibiyana gasfield (Block 12)	23.
Do.	do.	do.	Jalalabad gasfield (Block 13)	7.
Do.	do.	do.	Maulavibazar gasfield (Block 14)	2.
Do.	do.	Niko Resources Ltd.	Bibiyana and Feni	6.
Do.	do.	Santos International Holdings Pty. Ltd.	Sangu gasfield (Block 16)	3.
Do.	do.	Sylhet Gas Fields Ltd. (SGFL) [Bangladesh Oil, Gas and Mineral Corp. (Petrobangla), 100%]	Beanibazar, Haripur, Kailashtila, and Rashidpur	5.
Do.	do.	Tullow Oil plc, 30%	Bangora gasfield (Block 9)	3.
Do.	do.	PTT Exploration & Production Public Co. Ltd.	Block 17 and Block 18	NA.
Petroleum:				
Crude	42-gallon barrels per day	Santos International Holdings Pty. Ltd.	Sangu	30,000.
Refined	do.	Eastern Refinery Ltd. (Bangladesh Petroleum Corp.)	Chittagong	34,000.
Steel, crude		Bangladesh Steel and Engineering Corp.	do.	20.
Stone, crushed, granite		Maddhapara Granite Mining Co. Ltd. [Bangladesh Oil, Gas and Mineral Corp. (Petrobangla)]	Maddhapara, District of Dinajpur	1,650 (hard rock).

^{e}Estimated; estimated data are rounded to no more than three significant digits. Do., do. Ditto. NA Not available.

THE MINERAL INDUSTRIES OF BHUTAN AND NEPAL

By Lin Shi

BHUTAN

Bhutan's economy was significantly influenced by its monetary and trade links mainly with India. Bhutan's economy continued to grow steadily in 2012, although the real growth rate of the gross domestic product (GDP) slowed to about 8% from about 10% (revised) in 2011. The GDP based on purchasing power parity was about $5 billion, and the average inflation rate was about 11%. In addition to receiving financial assistance from India, Bhutan generated income by selling hydroelectric power to India (Asian Development Bank, 2013a; U.S. Central Intelligence Agency, 2013).

The mineral industry of Bhutan was small in scale relative to other sectors and not a significant contributor to the country's economy. The country produced and exported cement, copper wire, ferrosilicon, and manganese. Other mineral production included coal, dolomite, and limestone. Detailed or updated mining and mineral production data have not been available in recent years in the country's official reports, and the production estimates in table 1 are based on historic production data and information contained in public media reports. Bhutan and Nepal were mineral trade partners (table 1; Republica, 2013; U.S. Central Intelligence Agency, 2013).

Bhutan has the potential for 26,760 megawatts (MW) of hydroelectric generating capacity. In 2012, only about 6% (1,500 MW) of that potential capacity was actually installed. About 80% of the electricity generated by hydropower was exported to India. The 126-MW-capacity Dagachhu hydropower development project was started in 2008, and the Government aimed to export power from Bhutan to India through the existing river grid beginning at yearend 2012. As a low-carbon energy source, the exported hydropower was prompting regional economic cooperation among the neighboring countries and would provide clean energy to the coal-dominated India power market. The project was operated by a joint venture of Druk Green Power Corp. in Bhutan and Tata Power Co. in India through a public-private partnership. Financially supported by the Asian Development Bank, the Dagachhu project was nominated for a U.S. Department of Treasury Impact Honor for international clean and renewable energy development projects in 2012. Bhutan's economic growth for 2013 was expected to be about 9% because of the country's strategic economic investments in, mainly, hydropower and tourism development projects (Asian Development Bank, 2013b).

References Cited

Asian Development Bank, 2013a, Economy, Bhutan: Asian Development Bank. (Accessed August 22, 2013, at http://www.adb.org/countries/bhutan/economy.)

Asian Development Bank, 2013b, Projects, Bhutan: Asian Development Bank. (Accessed August 22, 2013, at http://www.adb.org/projects/37399-013/details.)

Republica, 2013, Araniko cement starts production from new plant: Republica, April 1st. (Accessed April 1, 2013, at http://www.myrepublica.com/portal/index.php?action=news_details&news_id=52461.)

U.S. Central Intelligence Agency, 2013, Bhutan, *in* The world factbook: U.S. Central Intelligence Agency, August 13. (Accessed August 23, 2013, at https://www.cia.gov/library/publications/the-world-factbook/geos/bt.html.)

NEPAL

Nepal mined mostly for industrial minerals, which were used for domestic construction. Cement and brick production were Nepal's main mineral-related industries. The country also produced red clay, coal, limestone, marble, and rolled steel. Nepal's mineral resources were mostly unexploited, and its mineral industry was not significant to the country's economy. Nepal has significant economically feasible hydropower potential.

Nepal's real GDP growth rate in 2012 was 4.6%. This increase was attributed to a favorable monsoon season, growth in the service sector, moderated inflation, a surplus in remittances, and reduced imports. In 2012, Nepal's exports accounted for 10% of the GDP, and 55.7% of the exports went to India; 10.1%, to the United States; and 4.4%, to Germany. Nepal's imports came mainly from India (51%) and China (34.5%). The major imports included gold, electrical goods, machinery and equipment, and petroleum products (table 1; Asian Development Bank, 2013; U.S. Central Intelligence Agency, 2013).

Nepal's cement industry has about 5.6 million metric tons (Mt) of annual production capacity. The country imported cement from Bhutan and India. There were about 45 cement factories in Nepal, of which only a few produced clinker. The country previously depended on India for 90% of its clinker needs. During fiscal year 2012, Nepal's cement imports decreased by 25% compared with those of fiscal year 2011 owing to the increase in domestic cement production. More grinding factories set up clinker production units, which began to reduce the country's dependency on clinker imports. Also, Nepal's limestone quarries were being acquired by cement producers, as limestone is a key raw material in the production of cement (Bell, 2012, p. 109; Republica, 2013).

The Nepal Bureau of Standards & Metrology (NBSM) closed two cement plants—the Butwal Cement Mills and the Shubha Shree Jagadamba—because the cement products produced at the plants failed to meet the NBSM's standard. An additional 16 cement plants were facing bans from the market because they had not acquired the NBSM's standard mark (Global Cement, 2013).

The Asian Development Bank projected that Nepal's 2013 GDP growth rate would slow to 3.5% because of the negative effects of the upcoming monsoon season, low investor and business confidence, the continuing domestic political uncertainties, and the slowdown in economic growth in India (Asian Development Bank, 2013).

References Cited

Asian Development Bank, 2013, Economy, Nepal: Asian Development Bank. (Accessed August 22, 2013, at http://www.adb.org/countries/nepal/economy.)

Bell, Peter, 2012, Emerging market challenges: International Cement Review, July, p. 109.

Global Cement, 2013, Nepal Bureau of Standards & Metrology closes two cement plants: Global Cement, August 21. (Accessed August 22, 2013, at http://www.globalcement.com/news/itemlist/tag/Nepal.)

Republica, 2013, Araniko cement starts production from new plant: Republica, April 1. (Accessed April 1, 2013, at http://ww.myrepublica.com/portal/index.php?action=news_details&news_id=52461.)

U.S. Central Intelligence Agency, 2013, Nepal, in The world factbook: U.S. Central Intelligence Agency, August 13. (Accessed August 22, 2013, at https://www.cia.gov/library/publications/the-world-factbook/geos/np.html.)

TABLE 1

BHUTAN AND NEPAL: PRODUCTION OF MINERAL COMMODITIES[1]

(Metric tons unless otherwise specified)

Country and commodity[2]		2008	2009	2010	2011	2012
BHUTAN						
Cement[e]	thousand metric tons	180	180	200	544 [r]	521
Coal, bituminous		123,704	48,545	87,814	108,904 [r]	98,731
Dolomite		1,247,568	1,028,993	1,192,374	1,082,300 [r]	1,499,534
Ferrosilicon, exports		30,824 [r]	90,798 [r]	97,528 [r]	79,181 [r]	67,900
Granite	square meters	199	217	18,731	462 [r]	1,806
Gypsum		248,445	299,735	344,034	352,233 [r]	313,172
Limestone		583,707	591,027	715,956	649,291 [r]	677,128
Marble	square meters	1,143	31	--	71,582 [r]	59,541
Quartzite		94,688	82,578	104,580	95,015 [r]	88,630
Slate	square meters	764	1,765	--	--	--
Stone		408,945	475,614	716,760	1,842,678 [r]	1,494,467
Talc		56,077	64,381	26,302	8,562 [r]	16,062
NEPAL						
Cement[e]	thousand metric tons	295	295	295	3,900 [r]	3,900
Clay, red	cubic meters	14,135	8,950	9,000	6,705 [r,3]	9,066
Coal:						
Bituminous		13,845	14,819	16,000	3,391 [r,3]	10,904
Lignite		60	NA	--	--	--
Total		13,905	14,819	16,000	3,391 [r,3]	10,904
Gemstones, quartz	kilograms	930	826	1,000	560 [r,3]	839
Quartzite[e]		--	--	3,000	3,000	3,000
Steel, rolled[e]	thousand metric tons	85	85	85	80	80
Stone:						
Limestone		701,950	582,999	580,000	580,000	1,276,452
Marble:						
Chips		441	1,047	900	1,330 [r,3]	1,969
Slab, cut	cubic meters	1,781	426	500	NA	NA
Craggy	do.	--	8,062	--	--	--
Talc		7,996	6,601	9,000	1,655 [r,3]	6,935

[e]Estimated; estimated data are rounded to no more than three significant digits; may not add to totals shown. [r]Revised. do. Ditto. NA Not available. -- Zero.

[1]Table includes data available through September 24, 2013.

[2]In addition to the commodities listed, metallic commodities, such as copper wire, ferrosilicon, and manganese; and crude construction materials, such as sand and gravel and a variety of stone, presumably are produced in Bhutan and Nepal, but information is inadequate to make reliable estimates of output.

[3]Reported figure.

TABLE 2
BHUTAN AND NEPAL: STRUCTURE OF THE MINERAL INDUSTRIES IN 2012

(Thousand metric tons unless otherwise specified)

Country and commodity	Major operating companies and major equity owners	Location of main facilities	Annual capacity[e]
BHUTAN			
Cement	Dungsam Cement Construction Ltd. (DCCL)	Nganglam, Pemagatshel District	1,300
Do.	Penden Cement Authority Ltd.	Gomtu, Samtse District	348
Dolomite	Jigme Mining Corp. Ltd.	do.	2,000
Ferrosilicon	Bhutan Ferro Alloys Ltd. (Government of Bhutan, Marubeni Co., and Tashi Commercial Co.)	Phuentsholing	34
NEPAL			
Cement	Dang Cement Industries (a subsidiary of Ambuja Cements of India)	NA	NA
Do.	Lhaki Cement Pvt. Ltd.	Bhawani Khola	660
Do.	Hetauda Cement Industries Ltd.	Hetauda	260
Do.	Himal Cement Co. Ltd.	Chobhar	130
Do.	Manasa Cement Industry	Chandragadhi, Jhapa	37
Do.	More than 40 cement manufacturers under the Cement Manufacturers Association of Nepal (CMAN)	NA	4,500
Lead and zinc	Nepal Metal Co. Ltd.	Lari	NA
Magnesite metric tons	Nepal Orind Magnesite Ltd.	Dolkha	50
Marble	Godavari Marble Industries Ltd.	Latitpur	1

[e]Estimated. Do., do. Ditto. NA Not available.

THE MINERAL INDUSTRY OF BURMA

By Yolanda Fong-Sam

In April 2012, Burma (also known as Myanmar) started seeing the easing and, in some cases, the suspension of decades of sanctions imposed by Western countries after more than 50 years of military rule in Burma. The countries that agreed to suspend sanctions were the United States, Australia, Canada, the European Union (EU), and Japan. The decision was made as Burma's Government started to implement democratic reforms. With the suspension of sanctions, companies from the countries mentioned above were allowed to start investing in Burma. The United States suspended the sanctions in July 2012 (Pawlak and Moffett, 2012; Spetalnick, 2012).

In 2012, Burma produced a variety of mineral commodities, including cement, coal, copper, lead, natural gas, petroleum, petroleum products, precious and semiprecious stones, tin, tungsten, and zinc. On November 11, 2012, a 6.8-magnitude earthquake struck the country. The epicenter was located in central Burma near the town of Shwebo, 60 kilometers (km) northwest of Burma's second largest city, Mandalay. Damage to many buildings was reported, including hospitals, monasteries, and schools, mainly in the villages of Male, Mandalay, Mogok, and Shwebo. Reports also indicated that miners were trapped in a gold mine in the Singgu area in Mandalay. On March 24, 2011, a previous earthquake of 6.9 magnitude had struck the eastern part of the country just north of Tachileik town in Shan State close to the border with Laos and Thailand. Production of such commodities as brine salt and some semiprecious stones dipped during the period following the 2011 earthquake, but the mineral industry in general was not affected (table 1; U.S. Geological Survey, 2011, 2012; Tun, 2012b).

Bangladesh, Burma, and India were involved in maritime boundary disputes concerning their respective sovereignty in the Bay of Bengal. For many years, these countries had attempted to negotiate and delimit their claims in the disputed area. In December 2009, Bangladesh and Burma accepted the jurisdiction of the International Tribunal for the Law of the Sea (ITLOS) for the settlement of their boundary delimitation. ITLOS is an independent judicial body established by the United Nations Convention on the Law of the Sea (UNCLOS) that has jurisdiction to arbitrate disputes arising out of the interpretation and application of the Law of the Sea. UNCLOS establishes a legal framework to regulate ocean space and its resources and uses (International Tribunal for the Law of the Sea, 2010).

In September 2011, representatives from Bangladesh and Burma met with the ITLOS in Germany for a final round of pleadings regarding the maritime boundary. The final ITLOS judgment of March 14, 2012, dealt with the delimitation in three parts—the territorial sea, the exclusive economic zones and continental shelf within 200 nautical miles, and the continental shelf beyond 200 nautical miles. ITLOS rendered its judgment in relation to the territorial sea by drawing an equidistant line from the countries' baselines. For the exclusive economic zones and continental shelf within 200 nautical miles, the tribunal decided to draw a provisional equidistant line that adjusts to the concavity of the coast of Bangladesh. For the delimitation of the continental shelf beyond 200 nautical miles, the tribunal concluded that it should not differ from that of within the 200 nautical miles and should continue in the same direction beyond the limit of Bangladesh (International Law Observer, 2012; International Tribunal for the Law of the Sea, 2012).

Government Policies and Programs

During fiscal year 2012–13, which started on April 1, Burma began implementing a reform of its exchange rate system. The reform consisted of a managed float of the currency with the objective of ending the fixed-rate currency system, unifying or consolidating the existing dual exchange rate system, and stabilizing domestic prices by creating an interbank money market. In recent years, on average, the official exchange rate of one U.S. dollar (US$) bought a little more than six Burmese kyat (MMK); on the other hand, the unofficial exchange rate was about US$1=MMK800. The official rate was typically used for Government revenue and for imports by some state-owned enterprises. The foreign currency exchange is regulated by the Foreign Exchange Management Act of 2012 (FEMA), which was enacted by Burma's Parliament in November 2012. FEMA allows for a transparent trading of the currency by lifting some trading restrictions. Under the exchange rate reform, foreign banks are to be allowed to form joint ventures in Burma beginning in 2014. In addition, banks from member countries of the Association of Southeast Asian Nations (ASEAN) will be allowed to open branch offices in the country. By 2015, ASEAN countries are planning to integrate and stimulate southeastern countries' economies by forming an economic block to facilitate immigration, increase trading of products, and improve the economy of the region overall (Raybould, 2012; Szep and Tun, 2012).

The Myanmar Foreign Investment Law of 2012 was signed into law on November 2, 2012, and promulgated on January 31, 2013. The new investment law offers tax breaks for the first 5 years of operation, allows foreign firms to fully own ventures, allows registered investors to execute leases of up to 5 years in duration, and provides for the possibility of two 10-year extensions on existing leases. Tax relief of up to 50% may be granted to foreign manufacturing companies on profits made from exports under the condition that the profits are reinvested in the business within 1 year. Joint ventures between foreign investors and Burmese nationals and (or) the Government are allowed with any stake ratio agreed to between the parties. (Previously, foreign investors were required to supply a minimum of 35% of the capital for joint ventures.) Under certain circumstances and restrictions, foreign investors are not required to have a local partner, which allows foreign investors to own a 100% interest in their businesses (Finch, undated; Tun, 2012a).

The Government of Burma was seeking to encourage the participation of foreign and local investors in part to draw in industry experts who have the knowledge of how to develop the country's mineral industry. The Union of Myanmar's Mineral Law went into effect in September 1994, and the rules related to the law were implemented in December 1996. The Ministry of Mines is the Government entity responsible for implementing the Government's mineral policy, for planning, and for enforcing the laws and regulations related to the mineral sector. The Ministry evaluates and processes all license applications for the prospecting for and production and beneficiation of minerals in accordance with the Mineral Law and regulations; it also monitors production operations and promotes investment in the mineral sector. According to the Mineral Law, any naturally occurring minerals found on or under Burmese soil and on Burma's Continental Shelf belong to the state (Ministry of Mines, undated a, b).

Production

During 2012, the availability of data on Burma's mineral industry statistics was limited compared with that of previous years. The data shown in table 1 for year 2012 are estimates based on data for 2011 unless otherwise stated. In the metals mining sector, production increases listed for copper, lead, and zinc were reported; the decreases listed for manganese and tin were estimated (table 1).

Mineral Trade

The latest period for which comprehensive data reported by the Government of Burma through its Central Statistical Organization were available was from January to September 2011. For reference purposes only, Burma's total foreign trade turnover for the first 9 months of 2011 was $14.11 billion,[1] of which exports totaled $7.05 billion and imports totaled $7.07 billion. The currency exchange system in Burma changed in April 2012 (Central Statistical Organization, 2011, p. 1, 50).

In 2011, Thailand remained Burma's primary export partner, followed by China, India, Hong Kong, and Singapore. China remained Burma's major import source, followed by Singapore and Thailand (Central Statistical Organization, 2011, p. 9–10, 18–19).

Structure of the Mineral Industry

The mineral sector in Burma includes mining and mineral processing industries, which are mainly Government owned. Table 2 is a list of Burma's major mineral industry facilities.

Commodity Review

Metals

Copper.—On March 19, 2012, the Chinese company China North Industries Corp. (NORINCO) announced the commencement of the construction phase of the Monywa copper project. The Monywa copper project, which was Burma's main copper asset, was located in the town of Monywa in Sagaing Region in the northwestern part of the country. The copper project, which was estimated to have a life of 30 years, consisted of the Letpadaung Mine (L Mine), the Sabetaung Mine (S Mine), the Sabetaung South Mine (SS Mine), and the Kyisintaung Mine (K Mine). The L Mine was scheduled to start operations in March 2012, but owing to protests from local villagers, operations were suspended. On July 24, Wanbao Mining Copper Co. Ltd. (Wanbao) and its local partner Union of Myanmar Economic Holdings (UMEHL) revised their profit-sharing agreement with the Ministry of Mines. Under the new contract, the Government would earn 51% of the project's profits and the remaining 49% would belong to Wanbao and UMEHL. Wanbao's obligations included spending $1 million each year on corporate social responsibility, $2 million on environmental protection and conservation, and 2% of earnings on local development once the project is in operation. The company planned to hire about 3,500 workers for the L Mine, and 1,300 workers for the S and K Mines combined. Additional details on the terms of the contract were not available (MCC8 Group Co. Ltd., 2012; Burmanet.org, 2013).

Nickel.—On August 13, China ENFI Engineering Corp. (ENFI) announced that it had put into operation the first ferronickel furnace in the Tagaung Taung nickel project, which is located in Thabeikying, Mandalay Region, in central Burma. The furnace was under a trial production phase during 2012. The plant, which had a design capacity of 25,000 metric tons per year of nickel, was operated by China Nonferrous Metal Mining Co. Ltd. (CNMC) under a production-sharing contract with state-owned Mining Enterprise No. 3; ENFI had been in charge of the design phase of the project (Csteellnews.com, 2012).

Zinc.—On December 31, South East Asia Metals Co., Ltd. (SEAMET) (which was a subsidiary of Padaeng Properties Co., Ltd. of Thailand), and Mayflower Mining Enterprises Ltd. (MME) agreed to terminate exploration operations in the area near the town of Mawkhi in Burma. Exploration activities did not identify any commercially recoverable zinc resource. Since July 2012, however, Padaeng had intensified its search for other zinc deposits in the country and identified several zones that would require further exploration to assess the resource potential (Padaeng Industry Public Co., Ltd., 2012, p. 25, 66).

Industrial Minerals

Cement.—Based on industry estimates, Burma has the capacity to produce approximately 3.46 million metric tons per year (Mt/yr) of cement, although most of the facilities lack the ability to operate at full capacity mainly because of unreliable energy sources and a lack of proper infrastructure. In 2011, four cement plant proposals were approved for development by the Myanmar Investment Commission (MIC). Three out of the four projects were still under construction in 2012; no information was available on the construction status of the fourth facility. The MIC is under the Ministry of National Planning and Economic Development and is the Government agency responsible for evaluating domestic and foreign

---

[1]Where necessary, values have been converted from Burmese kyat (MMK) to U.S. dollars (US$) at the rate of MMK5.38=US$1.00 for 2011.

investment proposals. Each of the four projects was planned to have the capacity to produce 1,000 metric tons per day of cement. Two of the plants were to be located in central Burma at Pyinyaung, Mandalay Region, and were owned by Htoo Cement Co. Ltd. and Shwe-Taung Cement Co. Ltd., respectively. Another plant was proposed to be built between the towns of Hopone and Taungyi in Shan State by Kanbawza Cement Ltd.; this plant was expected to be commissioned in early 2013. The fourth plant was proposed to be built in Naungcho in northern Shan State (about 100 km east of Mandalay) by Ngwey Yi Pale Mining Co. Several other cement project proposals were under consideration by the MIC (International Cement Review, 2012b).

On February 7, Siam City Cement Plc. (SCCC) of Thailand announced that it was still considering the construction of a cement plant in Burma after it carried out a feasibility study to determine the viability of the project. No new details were released as to the location or the capacity of the project. PT Semen Gresik of Indonesia was also considering building a cement plant in Burma with a capacity of 2.5 Mt/yr. No details were released as to when the company was planning to start construction (International Cement Review, 2012a; Myanmar Business Network, 2012).

Outlook

Burma's economic future seems promising, as many economic reforms were approved and implemented during 2012. In the near future, an increase in foreign direct investment is expected in the wake of the exchange rate reform that started in 2012, and the easing and (or) temporary suspension of sanctions by the United States, Australia, Canada, the EU, and Japan, which will allow companies to invest in Burma. The country is in a political and economic transition that is opening opportunities for business competition.

As the economy of the country gets stronger and investments increase, an appreciation in the Burmese currency (kyat) is also expected. The floating of currency could encourage investors to consider Burma as a serious investment partner as the Government continues its program of reforming and modernizing the economy and creating an atmosphere favorable for foreign investment. As Burma's economy grows, the other economies in the Southeast Asia region are also expected to grow, especially as the effort led by ASEAN to create an economic block to integrate and stimulate the region's economies materializes by 2015.

The exploration for and production of metals and industrial minerals is expected to increase as new developments and expansion projects progress and mines and plants start being commissioned in the next few years. This increase will be particularly noticeable in the cement production sector, as much foreign direct investment has been approved by the Government for the construction of new plants. The increase in cement demand will likely be directly influenced by the Government's infrastructure plans, which include the construction of oil and gas pipelines to China, highways and transportation projects, and a number of deep-sea ports. The development of other mining projects will also be subject to the continuing demand for mineral commodities from neighboring countries and world market prices.

References Cited

Burmanet.org, 2013, Build Monywa copper mine into a model project in Myanmar: Burmanet.org, September 20. (Accessed December 13, 2013, at http://www.burmanet.org/news/2013/09/20/guangming-daily-build-monywa-copper-mine-into-a-model-project-in-myanmar/.)

Central Statistical Organization, 2011, Selected monthly economic indicators: Nay Pyi Taw, Myanmar, Central Statistical Organization, September, 73 p.

Csteellnews.com, 2012, ENFI designed No. 1 ferro-nickel furnace commissioned in Tagaung Taung: Csteellnews.com, August 16. (Accessed May 17, 2013, at http://www.csteelnews.com/csteelnews/ztbd/cmn/EquipmentTechnology/201208/t20120816_72152.html.)

Finch, James, [undated], Legal provisions for foreign investors: IFLR1000.com, Phnom Penh, Cambodia. (Accessed December 12, 2013, at http://www.iflr1000.com/LegislationGuide/964/Legal-provisions-for-foreign-investors.html.)

International Cement Review, 2012a, Myanmar—Building bridges: International Cement Review International Report, April, p. 118–126.

International Cement Review, 2012b, Semen Gresik to build 2.5Mta plant in Myanmar: Cemnet.com, March 12. (Accessed March 13, 2012, at http://www.cemnet.com/News/story/149207/semen-gresik-to-build-2-5mta-plant-in-myanmar.html.)

International Law Observer, 2012, Judgment in Bangladesh-Myanmar maritime boundary dispute: International Law Observer, March 15. (Accessed January 15, 2013, at http://www.internationallawobserver.eu/2012/03/15/judgment-in-bangladesh-myanmar-maritime-boundary-dispute/.)

International Tribunal for the Law of the Sea, 2010, Dispute concerning delimitation of the maritime boundary between Bangladesh and Myanmar in the Bay of Bengal: WorldCourts.com, January 28, Case no. 16, 3 p. (Accessed June 21, 2011, at http://www.worldcourts.com/itlos/eng/decisions/2010.01.28_Bangladesh_v_Myanmar.pdf.)

International Tribunal for the Law of the Sea, 2012, Dispute concerning delimitation of the maritime boundary between Bangladesh and Myanmar in the Bay of Bengal: International Tribunal for the Law of the Sea, press release 175, March 14, 5 p. (Accessed January 15, 2013, at http://www.itlos.org/fileadmin/itlos/documents/press_releases_english/pr175_engf.pdf.)

MCC8 Group Co. Ltd., 2012, NORINCO Monywa Copper Project commencement ceremony held: MCC8 Group Co. Ltd. (Accessed January 29, 2013, at http://chinamcc8.com/en/disInteraction.asp?id=169&fid=&lid=22.)

Ministry of Mines [Myanmar], [undated]a, Department of Mines: Ministry of Mines. (Accessed September 9, 2010, at http://www.mining.com.mm/imis/Mines/d_mines.asp.)

Ministry of Mines [Myanmar], [undated]b, Government policy and legislation on investment in minerals: Ministry of Mines. (Accessed September 9, 2010, at http://www.mining.com.mm/imis/Mines/pltim.asp.)

Myanmar Business Network, 2012, Thai cement company eyeing Myanmar: Myanmar Business Network, February 7. (Accessed January 31, 2013, at http://www.myanmar-business.org/2012/02/thai-cement-company-eyeing-myanmar.html.)

Padaeng Industry Public Co., Ltd., 2012, Annual report 2012: Bangkok, Thailand, Padaeng Industry Public Co., Ltd., 106 p. (Accessed January 31, 2013, at http://www.padaeng.com/files/en/report/2013_09/pdf/PDI_Annual_Report2012_Eng.pdf.)

Pawlak, Justyna, and Moffett, Sebastian, 2012, EU suspends most Myanmar sanctions, not arms ban: Thomson Reuters, April 23. (Accessed March 18, 2014, at http://www.reuters.com/article/2012/04/23/us-eu-myanmar-sanctions-idUSBRE83M0CK20120423.)

Raybould, Alan, 2012, Exclusive—Myanmar to float currency in 2012/13, unify FX rates: Thomson Reuters, March 6. (Accessed November 13, 2013, at http://www.reuters.com/assets/print?aid=USTRE8250CT20120306.)

Spetalnick, Matt, 2012: Obama eases U.S. sanctions on Myanmar: Thomson Reuters, March 18. (Accessed November 13, 2013, at http://www.reuters.com/article/2012/11/11/us-myanmar-quake-idUSBRE8AA00L20121111.)

Szep, Jason, and Tun, A.H., 2012, Myanmar's central bank aims for weaker currency: Thomson Reuters, May 18. (Accessed November 13, 2013, at http://www.reuters.com/assets/print?aid=USBRE84H05V20120518.)

Tun, A.H., 2012a, Myanmar state media details new foreign investment law: Thomson Reuters, November 3. (Accessed November 13, 2013, at http://www.reuters.com/assets/print?aid=USBRE8A204F20121103.)

Tun, A.H., 2012b, Strong quake in central Myanmar kills at least six: Thomson Reuters, November 11. (Accessed November 13, 2013, at http://www.reuters.com/article/2012/11/11/us-myanmar-quake-idUSBRE8AA00L20121111.)

U.S. Geological Survey, 2011, Magnitude 6.9—Myanmar: U.S. Geological Survey Preliminary Earthquake Report. (Accessed November 21, 2014, at http://earthquake.usgs.gov/earthquakes/eqinthenews/2011/usc0002aes/#details.)

U.S. Geological Survey, 2012, M6.8 Burma earthquake of 11 November 2012: U.S. Geological Survey Earthquake Summary Poster, November 14. (Accessed November 21, 2014, at http://earthquake.usgs.gov/earthquakes/eqinthenews/2011/usc0002aes/#details.)

TABLE 1
BURMA: PRODUCTION OF MINERAL COMMODITIES[1]

(Metric tons unless otherwise specified)

Commodity[2]		2008	2009	2010	2011[e]	2012[e]
METALS						
Copper:						
Mine output, Cu content		--	3,500	9,000 [r]	9,000 [r]	19,000
Metal, refined		--	3,500	9,000 [r]	9,000 [r]	19,000
Lead:						
Mine output, Pb content[e, 3]		1,000	5,000	7,000	8,700 [4]	9,800 [4]
Metal, refined		202	200	--	--	200 [4]
Manganese, mine output, Mn content		142,600	242,900	299,900	234,400 [4]	114,500 [4]
Nickel, mine output, Ni content[e]		10	10	--	800	5,000 [p]
Silver, mine output, Ag content[3]	kilograms	--	249	--	--	--
Tin, mine output, Sn content:[5, 6]						
Of tin ores and concentrates		800	1,000	4,000	11,000	10,600
Metal, refined[e]		30	30	30	30	30
Total		830	1,030	4,030	11,000 [r]	10,600
Tungsten, mine output, W content:[3]						
Of tungsten concentrate		5	4	2	--	--
Of tin-tungsten concentrate		131	83	161	140 [e]	140
Total		136	87	163	140 [e]	140
Zinc, mine output, Zn content[e, 3]		7,000	6,000	8,600 [r]	9,300 [r, 4]	10,000
INDUSTRIAL MINERALS						
Barite		5,679	7,623	8,975	30,000	30,000
Cement, hydraulic		675,788	669,941	534,034	538,000	540,000
Gypsum		82,224	97,518	81,051	50,000	50,000
Precious and semiprecious stones:						
Jade	kilograms	30,896,440	25,427,237	38,990,035	45,000,000	45,000,000
Ruby	do.	1,868,696	1,674,579	1,612,070	870,000	900,000
Sapphire	do.	1,129,039	795,228	1,311,327	1,500,000	1,500,000
Spinel	do.	572,308	296,956	618,730	620,000	620,000
Salt, brine		54,355	133,358	97,136	100,000	100,000
Stone:						
Dolomite		4,264	4,390	3,119	2,000	2,000
Limestone, crushed and broken[e]	thousand metric tons	4,000	4,000	3,200	3,200	3,000
MINERAL FUELS AND RELATED MATERIALS						
Coal, lignite		249,442	245,418	217,650	300,000	300,000
Gas, natural, marketed	million cubic meters	12,445	11,555	12,425	12,500	12,500
Petroleum:						
Crude	thousand 42-gallon barrels	7,242	6,881	6,806	6,400	6,500
Refinery products[7]	do.	4,661	4,139	4,851	5,000	5,000

[e]Estimated; estimated data are rounded to no more than three significant digits; may not add to totals shown. [p]Preliminary. [r]Revised. do. Ditto. -- Zero.

[1]Table includes data available through February 21, 2014.

[2]In addition to the commodities listed, bentonite clay, copper matte, construction aggregates, diamond, feldspar, gold, iron and steel, lead (antimonial), nitrogen (ammonia), sand and gravel, and silica sand are produced, but available information is inadequate to make reliable estimates of output.

[3]Data are for the production by the state-owned mining enterprises under the Ministry of Mines.

[4]Reported figure.

[5]Production of tin, mine output, Sn content production as reported by the Government was, in metric tons, 2008—499; 2009—518; 2010—374; 2011—350, and 2012—350 (estimated).

[6]Data compiled from the United Nations Comtrade database for tin ores and concentrates imported from Burma by China, Malaysia, and Thailand.

[7]Includes diesel, distillate fuel oil, gasoline, jet fuel, kerosene, and residual fuel oil.

Sources: Ministry of Mines and Central Statistical Organization (Yangon), Statistical Yearbook 2009; Selected Monthly Economic Indicators, May 2008, January 2009, January 2010, December 2010, and September 2011.

TABLE 2
BURMA: STRUCTURE OF THE MINERAL INDUSTRY IN 2012

(Metric tons unless otherwise specified)

Commodity		Major operating companies and major equity owners	Location of main facilities	Annual capacity[e]
Cement		AAA Cement International Co. Ltd.	Cement plant in Kyaukse, Mandalay Region	180,000.
Do.		Dragon Cement	Cement plant in Pinlaung, Shan State	180,000.
Do.		Mandalay Cement Industries Co. Ltd.	Cement plant in Kyaukse, Mandalay Region	135,000.
Do.		Max Cement	Cement plant in Aung Nan Cho Village, Lewe Naypyidaw Township, Mandalay Region	150,000.
Do.		Myanma Ceramic Industries	Cement plant in Kyangin, Ayeyarwady	363,000.
Do.		do.	Cement plant in Kyaukse, Mandalay Region	120,000.
Do.		do.	Cement plant in Thayet, Magway Region	170,000.
Do.		Myanmar Economic Co. Myaing Galay 1	Cement plant in Hpa An, Kayin State	240,000.
Do.		Myanmar Economic Co. Myaing Galay 2	do.	1,200,000.
Do.		Naypyidaw Development Committee	Cement plant in Naypyidaw Township, Mandalay Region	150,000.
Do.		Tiger Head Cement (Myanmar)	Cement plant in Kyaukse, Mandalay Region	90,000.
Do.		Union of Myanmar Economic Holdings Ltd. Sin Min 1	do.	330,000.
Do.		Union of Myanmar Economic Holdings Ltd. Sin Min 2	do.	NA.
Do.		Yangon City Development Committee	Myodaw cement plant in Thazi, Mandalay Region	150,000.
Coal		Mining Enterprise No. 3 (ME–3)	Kalewa coal mine near Kalewa, Sagaing Region	13,000.
Copper		Mining Enterprise No. 1, Myanmar Yang Tse Copper Ltd.	Monywa copper project, S&K Mine, and Monywa solvent extraction electrowinning plant in Monywa region, central Burma	40,000.
Fertilizer, N content		Myanma Petrochemical Enterprise (Government, 100%)	No. 1 fertilizer plant at Sales, 190 kilometers southwest of Mandalay	94,900.
Do.		do.	No. 2 fertilizer plant at Kyun Chaung, central Burma	75,600.
Do.		do.	No. 3 fertilizer plant at Kyaw Zwar, central Burma	219,000.
Natural gas	million cubic meters	Total E&P Myanmar, 31.2%; Chevron Corp., 28.26%; PTT Exploration and Production Public Co. Ltd. (PTTEP), 25.5%; Myanma Oil and Gas Enterprise (MOGE), 15%	Yadana gasfield in Moattama, Gulf of Martaban	7,300.
Do.	do.	Petronas Carigali Myanmar Inc., 40.91%; Myanma Oil and Gas Enterprise (MOGE), 20.45%; PTT Exploration and Production Public Co. Ltd. (PTTEP), 19.32%; Nippon Oil Exploration (Myanmar) Ltd., 19.32%	Yetagun gasfield in Tanintharyi, Gulf of Martaban	4,600.
Do.	do.	Myanmar Petroleum Resources Ltd. and Myanma Oil and Gas Enterprise (MOGE)	Mann oilfield, south of Yangon	40.

See footnotes at end of table.

TABLE 2—Continued
BURMA: STRUCTURE OF THE MINERAL INDUSTRY IN 2012

(Metric tons unless otherwise specified)

Commodity		Major operating companies and major equity owners	Location of main facilities	Annual capacity^e
Nickel		China Nonferrous Metal Mining Group Co. Ltd. of China and Taiyuan Iron and Steel Co. (TISCO) of China	Tagaung Taung nickel ore project (mine and smelter) at Thabeikying, Mandalay Region	20,000 (nickel); 85,000 (ferronickel).
Petroleum:				
Crude	thousand 42-gallon barrels	Myanmar Petroleum Resources Ltd. and Myanma Oil and Gas Enterprise (MOGE)	Mann oilfield, south of Yangon	876.
Refined	do.	Myanma Petrochemical Enterprise (Government, 100%)	No. 1 refinery at Thanlyin (near Yangon)	9,490.
Do.	do.	do.	No. 2 refinery at Chauk, central Burma	2,190.
Do.	do.	do.	No. 3 refinery at Thanbayakan, central Burma	9,130.
Steel		POSCO, 70%	POSCO steel plant at Yangon	30,000.

^{e}Estimated. Do., do. Ditto. NA Not available.

The Mineral Industry of Cambodia

By Yolanda Fong-Sam

In 2012, industrial minerals dominated the production of minerals in Cambodia. These included gravel, sand, and stone, which were mostly consumed domestically by the construction industry, and limestone, which was used for the production of cement. The gross domestic product (GDP) at constant 2000 prices was $8.7 billion in 2012, which was an increase of 6.8% compared with that of 2011. In 2012, the National Bank of Cambodia (NBC) reported that the country's total trade increased by 12.3% to $12.8 billion from $11.4 billion in 2011. The country's domestic exports were valued at $5.50 billion compared with $4.93 billion in 2011, which was an increase of about 12%; also, the total value of imports in 2012 increased by 13% to $7.32 billion from $6.49 billion in 2011. Imports of petroleum and electricity were valued at $1.13 billion and $145.7 million, respectively, and represented 15% and approximately 2% of the country's total import value, respectively. Cambodia's exports to the United States were valued at $2.7 billion. Imports from the United States were valued at about $226 million and included aluminum and alumina, fertilizers, nonmonetary gold, nonferrous metals, nonmetallic minerals, precious metals, and petroleum products valued at about $802,000. In 2012, the NBC reported that it had issued four licenses for the trade of metals and precious stones, which brought the total number of valid licenses to 19 (table 1; National Institute of Statistics of Cambodia, 2012; National Bank of Cambodia, 2013, p. 11, 14, 41; U.S. Census Bureau, 2013a, b).

Government Policies and Programs

The Ministry of Industry, Mines and Energy (MIME), which implements Cambodia's mineral law and policy, is the Government entity that leads the effort to promote and develop the industrial sector in the country. The Department of Geology and Mines and the Department of Energy, which are both under the MIME, coordinate the development of the energy and mineral sectors (Investincambodia.com, 2009).

The Law on Management and Exploitation of Mineral Resources, promulgated on July 13, 2001, regulates the exploitation of mineral resources in Cambodia. The law stipulates that all mineral resources found in, on, or under the islands, land, seabed, territorial sea, and water bodies within the sovereignty of the Kingdom of Cambodia belong to the state. All raw mineral resources that are mined in the country are banned from export and are reserved for local use to manufacture finished products. Only finished products are allowed to be exported (General Department of Mineral Resources of Cambodia, 2013, p. 3, 6).

The law defines the following six categories of mineral licenses: artisanal mining license, gem mining license, industrial mining license, mineral exploration license, mineral (gemstone) cutting license, and pits and quarries mining license. To obtain a mineral license, the applicant must submit an application to the MIME for review, and approval is granted based on the applicant's technical and financial capability to manage the license. The ministry has 45 days to reach a decision. Then, the mining license is issued after the Council for the Development of Cambodia (CDC) approves the applicant's mining proposal. Exploration permits are granted for a period of 6 years, and mining permits are granted for up to 30 years followed by two 5-year extensions (General Department of Mineral Resources of Cambodia, 2013, p. 4, 8).

The policy governing foreign direct investment (FDI) in Cambodia is based on the Law on Investment of the Kingdom of Cambodia, which was enacted in 1994. The objective of the law is to restructure, simplify, and more efficiently manage foreign investment in the country. These efforts include modernizing the legal structure that supports commercial activities, strengthening and organizing Government departments, and enacting laws and regulations to support business activities in Cambodia. The law provides competitive concessions and incentives for FDI and also provides comparable treatment to domestic and foreign investors, with the exception of land ownership, which is based on the country's Constitution and allows only foreign investors to lease the land for a period of up to 70 years with the option to renew from then on. The investment law also allows foreign investors to own 100% of their mining investments in the country, and guarantees that foreign-owned assets will not be nationalized and that the prices for goods produced and services provided by the investors will not be regulated. Under the investment law, the Government created the Council for the Development of Cambodia (CDC), which oversees the investment policy and assists investors in all the aspects related to doing business in the country. The CDC, through the Cambodian Investment Board, processes and evaluates all proposals for exploration and is required to respond to each application within 45 days of submittal, after which time the CDC either grants the exploration license or rejects the proposal. If exploration is successful, the CDC is responsible for granting the mining license to the investors; a mining project plan or proposal is required as a condition for acquiring a mining license (Investincambodia.com, 2010).

Production

The General Department of Mineral Resources of Cambodia reported modest increases in the output of industrial minerals in 2012 compared with the output in 2011. The production of laterite increased by 16.2% to about 412,500 metric tons (t) in 2012 from 355,000 t (revised) in 2011; crushed stone (used as construction material) increased to an estimated 5.85 million metric tons (Mt) in 2012 from about 5.22 Mt (revised) in 2011, which represented an increase of approximately 12%. The production of gravel was estimated to have increased by about 12% to 43,000 t from a reported 38,438 t (revised) in 2011 (table 1).

Structure of the Mineral Industry

In 2012, the MIME reported that a total of 91 domestic and foreign companies held mining and exploration licenses in Cambodia. A total of 139 exploration projects were authorized under the licenses granted, out of which 13 had been licensed to conduct mining projects. Table 2 is a list of major mineral industry facilities (General Department of Mineral Resources of Cambodia, 2013, p. 2).

Commodity Review

Metals

Gold.—In May 2012, Renaissance Minerals Ltd. of Australia announced that it had finalized the purchase of OZ Minerals of Australia's assets in Cambodia for a cash cost of $7.8 million and 26.4 million shares. The license for the Okvau gold deposit, which is located in Mondulkiri Province approximately 265 kilometers (km) northeast of the capital city of Phnom Penh, covered an area of approximately 800 square kilometers (km^2). The independent Joint Ore Reserves Committee (JORC) resource estimate for the deposit was 15.6 Mt at a grade of 2.4 grams per metric ton (g/t) gold for an estimated resource of 37,300 kilograms of gold (reported as 1.2 million troy ounces) (Australia's Paydirt, 2012; Gold Mining Journal, 2012; Renaissance Minerals Ltd., 2013).

Southern Gold Ltd. of Australia held a total of seven exploration licenses for gold and base metals in an area covering 1,500 km^2 located in several Provinces in eastern Cambodia. The tenements were distributed within five main projects—the Kratie North, the Kratie South, the Memot, the Phnum Romdul, and the Srae Pok. On July 5, Southern Gold announced that it had finalized an earn-in agreement with Mekong Minerals Ltd. of Australia. The agreement entitled Mekong to earn an interest in Southern Gold's subsidiary Southern Gold Asia Pty. Ltd. (SG Asia) and to manage all the projects under SG Asia and its wholly owned subsidiary Southern Gold Cambodia Ltd. (SG Cambodia). With an investment of $5.7 million towards funding the projects at SG Asia and SG Cambodia, Mekong would have the exclusive right to earn up to 70% interest in SG Asia. Once Mekong had earned a 70% interest, the company would have the option to purchase Southern Gold's remaining interest (30%) in SG Asia (Southern Gold Ltd., 2012).

Iron Ore.—The Government of Cambodia announced that it had granted Cambodia Iron & Steel Mining Industry Group of China (CISMIG) a planning permit to build a steel plant in Preah Vihear Province in northern Cambodia; no additional details were available about the steel plant. The Government also announced that CISMIG had contracted China Railway Group to build a 405-km railway to connect the proposed steel plant to the seaport of Kaoh Kong in southern Cambodia by way of the Provinces of Kampong Chnang, Kampong Speu, and Kampong Thom. The contract was valued at an estimated $11 billion. The construction of the railway was expected to start in the summer of 2013, and the railway would be commissioned by 2017 (Railwaygazette.com, 2013).

Industrial Minerals

Cement.—In December, Siam City Cement Co. (SCCC) of Thailand announced that it was still considering plans to invest in a cement plant in Cambodia through a joint venture with a local partner. The company was planning to reach a final decision about the investment in 2013. If approved, the plant would have an estimated cost of $150 million (GlobalCement.com, 2012; WorldCement.com, 2012).

Mineral Fuels

Natural Gas and Petroleum.—Chevron Corp. of the United States through its subsidiary Chevron Overseas Petroleum Cambodia Ltd. operated the offshore oilfield Block A, which is located about 200 km off the coast of Cambodia in the Gulf of Thailand. Chevron Overseas held a 30% interest in the block, which covers 4,709 km^2. According to the company, meetings held with the Government during the course of 2012 brought progress in their discussion regarding the production permit to develop Block A. Chevron Overseas was waiting for a finalized agreement on the commercial terms and the Government's approval of the production permit before making a final investment decision to start development (Chevron Corp., 2013).

On January 31, Japan Oil, Gas, and Metals National Corp. (JOGMEC) of Japan announced the beginning of a 4-month seismic survey focused on oil and gas exploration in onshore Block 17. The block was located in Preah Vihear Province about 500 km northwest of Phnom Penh. In May 2010, the Cambodian National Petroleum Authority and JOGMEC signed a memorandum of understanding to cooperate in the oil and gas sector. Under the agreement, JOGMEC would carry out a feasibility study in an area that covers 6,500 km^2, which includes the Provinces of Kampong Thom, Preah Vihear, and Sem Preap (Xinhuanet.com, 2012b).

On December 28, the Government of Cambodia granted an oil refinery license to Cambodian Petrochemical Co. (CPC) to build the country's first oil refinery, which was expected to generate about 4,000 jobs. The refinery was a joint project between CPC and Sinomach China Perfect Machinery Industry Corp. The project, which had an estimated cost of $2.3 billion, was located in an 80-hectare area along the borders of Kampot Province and Preah Sihanouk Province. The oil refinery would have a designed production capacity of 5 million metric tons per year. CPC contracted Cambodia's National Petroleum Authority to develop the project, which was to be commissioned by the end of 2015 (Phnom Penh Post, The, 2012; Xinhuanet.com, 2012a).

Outlook

In recent years, Cambodia has signed significant bilateral cooperation agreements with neighboring countries, as well as with other Asian countries. The cooperation efforts could potentially open the country to more opportunities for investment in the near future. When completed, the country's first oil refinery is expected to contribute greatly to the development and strengthening of Cambodia's economy. It is also expected to reduce the country's dependence on imported oil as the country transitions from being petroleum import dependent to being an oil producer and exporter. As the economy of Cambodia and those of neighboring countries are developed, the demand for cement and other industrial mineral commodities will most likely increase, both domestically and regionally.

The lack of suitable infrastructure in Cambodia, such as reliable electricity sources and paved roads, continue to present challenges for the Government to attract foreign investors. The situation is expected to improve in the near future as the Government continues the implementation of reforms to tackle these challenges.

References Cited

Australia's Paydirt, 2012, Cambodian gold attracts cash: Australia's Paydirt, November, v. 1, no. 200, p. 76.

Chevron Corp., 2013, Cambodia factsheet: Chevron Corp., April, 4 p. (Accessed October 17, 2013, at http://www.chevron.com/documents/pdf/CambodiaFactSheet.pdf.)

General Department of Mineral Resources of Cambodia, 2013, Current status of mining in Cambodia: Ministry of Industry, Mines and Energy of the Kingdom of Cambodia, March 6, 12 p.

GlobalCement.com, 2012, Siam City to fire up closed kiln: GlobalCement.com, December 12. (Accessed March 19, 2014, at http://www.globalcement.com/news/item/1307-siam-city-to-fire-up-closed-kiln.)

Gold Mining Journal, 2012, Renaissance gives Cambodia fresh impetus: Gold Mining Journal, v. 1, no. 108, July–September, p. 52–53.

Investincambodia.com, 2009, Mineral resources: Invest in Cambodia. (Accessed June 22, 2010, at http://www.investincambodia.com/minerals.htm.)

Investincambodia.com, 2010, Investment information: Invest in Cambodia. (Accessed September 9, 2011, at http://www.investincambodia.com/investing/investing.htm.)

National Bank of Cambodia, 2013, Annual report 2012: Phnom Penh, Cambodia, National Bank of Cambodia, December 25, 41 p. (Accessed July 17, 2013, at http://www.nbc.org.kh/download_files/publication/annual_rep_eng/AnnualReport2012.pdf.)

National Institute of Statistics of Cambodia, 2012, National accounts 2012: Phnom Penh, Cambodia, National Institute of Statistics of Cambodia. (Accessed March 19, 2014, at http://www.nis.gov.kh/nis/NA/NA2012.html.)

Phnom Penh Post, The, 2012, New oil refinery joint venture: The Phnom Penh Post, December 31. (Accessed March 20, 2014, at http://www.phnompenhpost.com/business/new-oil-refinery-joint-venture.)

Railwaygazette.com, 2013, Railway planned to link steel plant and port: Railwaygazette.com, January 4. (Accessed January 11, 2013, at http://www.railwaygazette.com/news/single-view/view/railway-planned-to-link-steel-plant-and-port.html.)

Renaissance Minerals Ltd., 2013, Cambodian gold project: Perth, Western Australia, Australia, Renaissance Minerals Ltd. (Accessed March 19, 2014, at http://www.renaissanceminerals.com.au/index.php/projects/cambodian-gold-project-update.)

Southern Gold Ltd., 2012, Cambodia: Southern Gold Ltd. (Accessed December 10, 2013, at http://www.southerngold.com.au/projects/cambodia/.)

U.S. Census Bureau, 2013a, U.S. exports to Cambodia from 2004 to 2013 by 5-digit end-use code: U.S. Census Bureau. (Accessed March 19, 2014, at http://www.census.gov/foreign-trade/statistics/product/enduse/exports/c5550.html.)

U.S. Census Bureau, 2013b, U.S. imports from Cambodia from 2004 to 2013 by 5-digit-end-use code: U.S. Census Bureau. (Accessed March 19, 2014, at http://www.census.gov/foreign-trade/statistics/product/enduse/imports/c5550.html.)

WorldCement.com, 2012, Cement industry trends—Asia: WorldCement.com, September 7. (Accessed October 29, 2013, at http://www.worldcement.com/news/cement/articles/Cement_market_export_expansion_Asia_1160.aspx#.Uyse2z8VAQp.)

Xinhuanet.com, 2012a, Cambodian, Chinese firms unveil 1st oil refinery project in Cambodia: Xinhuanet.com, December 28. (Accessed March 20, 2014, at http://news.xinhuanet.com/english/china/2012-12/28/c_132069193.htm.)

Xinhuanet.com, 2012b, Japan's firm launches onshore oil, gas exploration in Cambodia: Xinhuanet.com, January 31. (Accessed March 20, 2014, at http://news.xinhuanet.com/english/business/2012-01/31/c_131384271.htm.)

TABLE 1
CAMBODIA: PRODUCTION OF MINERAL COMMODITIES[1]

(Metric tons)

Commodity[2]	2008	2009	2010	2011	2012
Cement	772,029	933,900	789,025	906,711 [r]	980,000 [e]
Gravel	37,500 [e]	41,875	82,500	38,438 [r]	43,000 [e]
Laterite, blocks	454,750	631,000	1,612,500	355,000 [r]	412,500
Salt	78,000 [e]	NA	NA	NA	NA
Sand, construction material	6,581,500	14,035,775	38,367,500	8,296,590 [r]	8,800,000 [e]
Stone, crushed	2,039,336	2,819,804	6,331,000	5,223,345 [r]	5,850,000 [e]

[e]Estimated; estimated data are rounded to no more than three significant digits. [r]Revised. NA Not available.

[1]Table includes data available through February 21, 2014.

[2]In addition to the commodities listed, clay, gemstones, gold, iron ore, and limestone are presumably produced, but available information is inadequate to make reliable estimates of output.

Source: Cambodia's Ministry of Industry, Mines and Energy, General Department of Mineral Resources; U.S. Geological Survey Minerals Questionnaires for Cambodia 2007–12.

TABLE 2
CAMBODIA: STRUCTURE OF MINERAL INDUSTRY IN 2012

(Thousand metric tons)

Commodity	Main operating companies and main equity owners	Location of main facilities	Annual capacity[e]
Cement	Kampot Cement Co. Ltd. (a joint venture between Siam City Cement Co. of Thailand, 90%, and Khaou Chuly Group, 10%)	Touk Meas District, Kampot Province	1,000
Gemstones	Jirech International Cambodia Construction and Import-Export Co. Ltd.	Anlong Krapeu-Preak Bey, Samlot District, Batdambang Province	NA
Do.	Seoul Digem Cambodia Co. Ltd.	Loamphat District, Rotanokiri Province	NA
Do.	Sonuba Cahm Industries Co. Ltd.	Phnom Trop, Pailin City	NA
Do.	Ultra Marine Kiri (Cambodia) Ltd.	Pating Thom Village, Ting Chak Commune, Borkeo District, Rotanokiri Province	NA
Do.	do.	Sen Chauv, Samlot District, Batdambang Province	NA
Granite	TTY-Rithy Mexico	Svay Chreas District, Kracheh Province	NA
Limestone	Kampot Cement Co. Ltd.	Tatung, Kampot Province	1,500
Do.	Thai Boon Roon Cement Co. Ltd.	Phnom Laang Laan Commune, Donghtung District, Kampot Province	NA
Do.	United International (Cambodia) Foreign Investment Group Co. Ltd.	East Touk Meas, Banteay Meas District, Kampot Province	NA

[e]Estimated. Do., do. Ditto. NA Not available.

THE MINERAL INDUSTRY OF CHINA

By Pui-Kwan Tse

China ranked second behind the United States as the world's leading economic power; it was also one of the world's leading mineral producing and consuming countries. In 2012, China's economy remained strong compared with most of the developed countries in the West. During the past two decades, China's economic growth was the result of a combination of trade and investment, and this growth greatly affected the global commodity market. The country's demand for energy, metals, and minerals was particularly strong. China imported significant amounts of raw materials and transformed the materials into products for export. During the 1980s and 1990s, China's commodity exports went primarily to Europe and the United States; however, during the past several years, intraregional trade within the Asia and the Pacific region increased significantly, which benefited other emerging economies in Asia, especially those that exported raw materials to China (Asian Development Bank, 2013, p. 14–16).

During the past decade, the Chinese Government's economic policy was to prevent economic slowdown and fight inflation. In November 2012, the 18th National Congress of the Communist Party of China (CPC) was convened and new leaders were elected. The CPC aimed to increase the gross domestic product (GDP) by 2020 to double that of 2010 and to continue to transform China into a modern industrialized country. The Chinese Government set the country's targeted economic growth rate at 7% in its 12th 5-year plan (which covers the years 2011 through 2015). In 2012, the GDP growth rate was 7.8% compared with 9.3% in 2011. The growth rate was the lowest of the past 13 years but was higher than the Government's target of 7.5%, which was set at the beginning of the year. Economic activity, especially manufacturing, was slow during the first half of the year but started to rebound in September. The slowdown in fixed-asset investment was a major reason for the slower economic growth rate. During the first half of the year, the Government restricted purchases of investment property as a result of decreases in real estate investment and raw material consumption. Also, the value of the industrial sector, which was one of the leading contributors to the GDP growth rate, decreased by 2.2% compared with that of the previous year. In 2012, the consumer price index increased by 2.6% compared with an increase of 5.4% in 2011. The country's fixed-asset investment increased by 20.3% to $6.0 trillion, which was the smallest growth rate of the past 3 years. The mining sector (including coal, gas, and oil) received $211 billion of the total investment, which was an increase of 11.8% from that of 2011. China's long-term challenge was to continue to foster economic growth at a sustainable rate (Citigroup Global Market Inc., 2012, p. 2; National Bureau of Statistics of China, 2013, p. 1–10).

The size of China's labor force was expected to shrink during the next decade because of the Government's adoption of a one-child policy in the 1980s. In the past, the decrease in the labor force has tended to raise wage levels and increase social welfare costs. The working-age population decreased by 3.5 million in 2012. The labor force in the mining sector was 6.11 million, or 4.2% of the country's total workforce in 2011.

In the manufacturing sector, rising production and environmental costs, overcapacity, and uncertain external demand were likely to affect the sector's profit margin. Month-on-month production growth in the ferrous metals sector decreased significantly to 19.1% in November from 23.2% in August, and the trend toward decreased production was mirrored in the cement sector as well. During the past decade, the Chinese Government has tried to restrict the expansion capacity in such energy-intensive sectors as aluminum, cement, and iron and steel. Capacity reduction in these sectors could have a significant effect on investment and employment in the future (Citigroup Global Market Inc., 2013a, p. 3; National Bureau of Statistics of China, 2013, p. 1–10).

Minerals in the National Economy

China is rich in mineral resources and was the world's leading producer of aluminum, antimony, barite, bismuth, cement, coal, fluorspar, gold, graphite, iron and steel, lead, magnesium, mercury, molybdenum, phosphate rock, rare earths, salt, talc, tin, tungsten, and zinc in 2012. China ranked among the top three countries in the world in the production of many other mineral commodities. China was the leading exporter of antimony, barite, fluorspar, graphite, indium, rare earths, and tungsten in the world. The country's consumption of chromium, cobalt, copper, iron ore, manganese, nickel, petroleum, platinum-group metals, and potash exceeded domestic production, and imports were estimated to account for more than 40% of total domestic consumption. Mineral trade accounted for about 25% of the country's total trade. China was one of the few countries whose domestic supply of and demand for a variety of mineral commodities affected the world mineral market (National Bureau of Statistics of China, 2012, p. 130).

Government Policies and Programs

The 12th National People's Congress (NPC) adopted the recommendations from the CPC to consolidate the Government's bureaucracy by combining regulators into several larger Ministries. The existing Government structure made responding quickly to shifting economic and social demands difficult. The State Council hoped that the reform would build a more transparent, service-oriented Government. The State Council would have 25 Ministries and Commissions compared with the current 27. The Ministry of Railways was eliminated, and its railway planning and development functions were placed under the Ministry of Transportation. The State Administration of Press and Publication and the State Administration of Radio, Film, and Television were merged. The Ministry of Health and National Population and Family Planning Commission were merged to become the National Health and Family Planning Commission. The Government also reduced its 1,700 requirements by one-third, subject to Central Government approval (State Council, The, 2013a, b).

China's rapid capital-intensive, export-oriented growth had been successful during the past three decades; however, the global markets it relied on were expected to be weaker in the future. The country's economic pattern of growth was energy- and natural-resource intensive and environmentally unsustainable. The constrained supply of major mineral commodities and environmental degradation were limiting the country's ability to maintain its past level of economic growth. The Government indicated that it intended to support and build a more energy-efficient and ecologically friendly society by upgrading the value chain in manufacturing while enhancing innovation and promoting the development of new strategic industries. Its plan to reduce carbon emissions was to be focused on the energy-intensive sectors, such as cement, chemicals, iron and steel, and nonferrous metals. The Government stated that the country's economic growth should be less dependent on export markets, such as Europe and the United States, and thus was planning to transform the economy from one that is export focused to one that is consumer driven (Zhonghua Renmin Gongheguo Guowuyuan Gongbao, 2012a, b).

The Government also planned to deepen administrative reform and strengthen efforts to combat corruption. The 12th 5-year plan aimed to strengthen accountability and public financial management. The Central and local government roles and responsibilities were expected to be defined more clearly. The Government planned to develop an effective anti-corruption system, including improving transparency, reporting, and enforcement (Citigroup Global Market Inc., 2013b, p. 6–10).

During the past decade, the output capacity of China's manufacturing industry had increased rapidly, including in the ferrous and nonferrous metals, industrial, and new energy (solar photovoltaic and wind power generation) sectors. Owing to a decrease in global demand for these products since the financial crisis starting in 2008, the domestic capacity utilization ratio decreased sharply. As a result, cumulative financial losses in these sectors became a significant issue for the Government. Excess capacity had been a long chronic problem for China's manufacturing industries. During the past decade, the Government repeatedly issued guidelines to restrict the construction of new plants and expansion of facilities in these sectors. The Central Government's effort to reduce excess output capacity was undermined, however, by local governments and state-owned enterprises that were keen to expand and the performances of which were based on how much revenue they could generate (China Metal Bulletin, 2013g).

In 2011, the ferrous and nonferrous metals sectors each received $61 billion in fixed–asset investment and the nonmetallic sector received $166 billion. Domestic analysts estimated that fixed-asset investment in these sectors was about 20% more in 2012 than in 2011. Since the global financial crisis in 2008, most recent investments have been supported by local governments.

The Ministry of Finance (MOF) announced that the Government would disband the investment tax for domestic investors and companies beginning on January 1, 2013. In 1983, the Government had used the construction tax as a way to manage and control the country's investment conditions and subsequently changed it to an investment adjustment tax in 1991. The investment tax applied only to domestic investors. The investment tax was adopted under a planned economy and did not fit within the country's current "social market economic state" (Ministry of Finance, 2012).

In 2012, the Ministry of Industry and Information Technology (MIIT) amended regulations that were issued in 2010 on the development of the iron and steel sector. The new regulations require iron and steel producers to produce products that are within the guidelines of the country's iron and steel products qualification standards. Iron and steel producers are forbidden to use obsolete technology to produce iron and steel products. The emission of powder dust should be less than 1.19 kilograms (kg); sulfur dioxide emission, 1.63 kg; and water consumption, 4.1 cubic meters per metric ton of steel products. For each ton produced, the energy consumption of the blast furnace for ironmaking should be less than 446 kg of standard coal [5,500-kilocalorie (kcal) coal]; sintering, 56 kg of standard coal; electric arc furnace (EAF), 92 kg of standard coal; EAF for specialty steel, 171 kg of standard coal; and coking, 155 kg of standard coal. The new regulations also specify the minimum capacities of equipment for iron and steel production including blast furnace volume, more than 400 cubic meters; EAF, 30 metric tons (t) (15,000-kiloampere capacity or higher); and EAF for ferroalloys, 10 t. The minimum required production capacity for an individual iron and steel production plant is 1 million metric tons per year (Mt/yr), and that for specialty steel plants is 300,000 metric tons per year (t/yr). The new regulations took effect on October 1, 2012. The tougher regulations are intended to curb excess capacity and to force iron and steel producers to produce more high-quality and environmentally friendly products (Ministry of Industry and Information Technology, 2012a).

As part of the Government's efforts to reduce the excess supply of nonferrous metals in the domestic market and to provide financial assistance to domestic nonferrous metal producers, the State Reserve Bureau (SRB) announced in late 2012 that the Government had started stockpiling aluminum, copper, indium, and zinc. In 2008 and 2009, the Government stockpiled 400,000 t of aluminum, 165,000 t of copper, 150,000 t of zinc, and 30 t of indium from the domestic market to stabilize metal prices in the domestic market, and local governments also stockpiled more than 1 million metric tons (Mt) of nonferrous metals during the same period. Also, the SRB took advantage of low metal prices in the international markets to purchase strategic metals, of which the country had limited resources. Domestic analysts estimated that the SRB purchased (through metal traders) more than 60,000 t of nickel from the international markets in 2012. The Government also stockpiled rare-earth products domestically (China Metals, 2012d).

The MIIT issued industrial policy guidelines for the development of the graphite sector. According to the guidelines, the Government prohibits mining within 1 kilometer of the boundary of environmentally protected areas, water resources, urban areas, and nonindustrial zones. The guidelines urge graphite producers to adopt advanced processing technology in their operations. The guidelines set the minimum processing capacity for flake graphite plants at 20,000 t/yr and the minimum recovery rate at 80% for ore grading 5% carbon, 85% for ore grading 8% carbon, and 90% for ore grading 10% carbon. The minimum processing capacity for an amorphous graphite plant is set at 150,000 t/yr, and the minimum recovery rate is set at 85%. The Government's guidelines also require all graphite producers to meet energy and water consumption limits. The Government planned to publish a list of graphite producers that had met the policy and environmental guidelines at yearend 2014 (Ministry of Industry and Information Technology, 2012c).

The MIIT and 11 ministries and commissions jointly issued policy guidelines for industry mergers. Two basic policies for saving natural resources and protecting the environment were introduced. Important objectives noted in the guidelines are reducing production and transportation costs; upgrading technological processes; improving energy development; and adjusting the raw material supply infrastructure in China. By 2015, the top iron and steel producers would account for 60% of total output. About three to five domestic iron and steel companies are projected to have the capability to compete with international iron and steel companies in the global market. The Government industry merger guidelines also urge companies to merge across regions. According to the guidelines, the top 10 aluminum producers are expected to produce about 90% of the country's total output in 2015, and producers of aluminum, coal, and power are encouraged to form joint ventures. The guidelines stress the importance of upgrading technology and efficiency through mergers, improving the quality of products, controlling production capacity, and reducing competition in nine industrial sectors, including automobile and electronics manufacturing, cement and rare-earth production, and shipbuilding (Ministry of Industry and Information Technology, 2013a).

Production

China was one of the world's leading producers of aluminum, antimony, barite, bismuth, cement, coal, copper, fluorspar, gold, graphite, indium, iron and steel, lead, lime, magnesium, manganese, molybdenum, phosphate rock, rare earths, salt, silver, talc, tin, tungsten, and zinc. The country's output quantities of these mineral commodities were sufficient to have a significant effect on world markets. In 2012, China's production of alumina, aluminum, bauxite, cement, coal, copper, gold, graphite, iron and steel, lead, phosphate rock, silver, titanium sponge, tungsten, and zinc increased compared with that of 2011 (table 1).

China's reform priorities were to improve the efficiency of resource allocation and to boost economic growth. The Government understood that the unbalanced growth of consumption, investment, and net exports could not continue unabated forever. During the past several years, the Government reduced the export tax rebates on ferrous and nonferrous metal products, increased the export duties on energy-intensive metals, and encouraged producers to produce high-value-added products. Owing to increasing domestic and overseas demand, China's minerals and metals output was expected to continue to increase.

Structure of the Mineral Industry

China's mining industry is highly fragmented and has had a poor safety record. Several companies often mined in a single mining area. The State Council approved a mining consolidation plan that had been proposed jointly by the Ministry of Land and Resources (MLR), the National Development and Reform Commission (NDRC), and other agencies. Fifteen mineral commodities—antimony, bauxite, coal, copper, gold, iron ore, lead, manganese, molybdenum, phosphorus, potassium, rare earths, tin, tungsten, and zinc—were on the consolidation plan. The Central Government worked with local governments to implement the plan. Small mine operators were targeted to be integrated into large operators through such means as acquisition or joint-management agreements. The State-Owned Assets Supervision and Administration Commission would transfer state-owned assets of these small operators to the large operators. The Government would not allow any expansion of mining boundaries during the consolidation period. The Government would not issue mining operation permits to uncooperative mine operators. Local governments were required to submit their consolidation plans to the MLR for recording. During the past several years, the Government enabled state-owned enterprises to diversify their core business into other sectors, such as by allowing Aluminum Corporation of China (Chinalco) to be a major shareholder of copper companies in the Provinces of Hebei and Yunnan and rare-earth companies in Jiangsu Province and Guangxi Zhuangzu Autonomous Region. Baoshan Iron and Steel (Group) Corp. invested in coal mining in Shanxi Province, and Jiangxi Copper Co. Ltd. took charge of the consolidation of rare-earth mining activities in Sichuan Province. Minmetal Group Co. took charge of consolidation of the Hunan Nonferrous Metal Co. and invested in rare-earth separation plants in Jiangxi Province.

Mineral Trade

China was one of the most important producing and consuming countries in the world. According to customs statistics, China's total trade was valued at $3.87 trillion in 2012, which was an increase of 6.2% compared with that of 2011. The value of exports increased by 7.9% to $2.05 trillion. The United State replaced the European Union (EU) as the leading destination for China's exports followed by Hong Kong and Japan. The value of China's imports increased by 4.1% to $1.82 trillion. The EU was China's leading source of imports followed by Japan, the Republic of Korea, and the United States. Imports of raw materials, such as chromium ore, coal, copper ore and concentrate, iron ore, nickel ore, and oil, increased sharply. In 2012, the total value of mineral and metal products trade was $991.9 billion. China's main exports were low-end and semimanufactured goods. Large amounts of capital, designs, technologies, and even raw materials were coming from abroad. Consequently, China posted a trade surplus with countries that consumed manufactured goods, such as the United States and the countries of the EU, and trade deficits with such countries as Australia, Brazil, Chile, and Indonesia, which produced and exported fuels and minerals (General Administration of Customs of the People's Republic of China, 2013; Ministry of Land and Resources, 2013, p. 12).

The Ministry of Commerce (MOC) issued circular No. 97, which details the mineral commodities that are under the Government's monitoring list for export. The commodities are ammonium paratungstate; bauxite and refractory clay; coal; coke; concentrates of antimony, cobalt, gold, molybdenum, silver, tin, tungsten, and zinc; dolomite; fluorspar; magnesite; oxides of antimony, magnesium, and tungsten; rare earths; silicon carbide; silver; talc; and unwrought metal and alloys of antimony, beryllium, bismuth, copper, gallium, germanium, nickel, niobium, platinum-group metals, tantalum, tin, and zirconium. Of these exported mineral commodities, antimony and its products, coal, petroleum and its products, silver, and tungsten and its products were under state management. In 2012, the Government encouraged the import of raw materials, such as concentrates of chromite, nickel, niobium, tantalum, titanium, and uranium; antimony concentrates with metal content higher than 30%; copper concentrates with metal content higher than 20%; cobalt concentrates with cobalt content higher than 6%; lead concentrates with lead content higher than 55%; molybdenum concentrates with metal content higher than 51%; zinc concentrates with zinc content higher than 40%; and ferronickel. Beginning on January 1, 2012, the tariff rate on imports of rare-earth compounds was reduced to zero from 5.5%. The MOF announced that the export tariff rates would be changed, including those for many minerals and metals (Ministry of Commerce, 2012a, p. 1–20).

The Government adjusted the 2013 export quota for magnesia to 1.67 Mt; phosphate rock, to 1.0 Mt; talc, to 750,000 t; antimony and antimony products (metal content), to 59,400 t; molybdenum, to 25,000 t; tin and tin products (metal content), to 17,000 t; tungsten and tungsten products (metal content), to 15,400 t; silver, to 5,387 t; and indium, to 231 t. Bauxite (alumina clay) and silicon carbide were not included in the export quota control system after the World Trade Organization's (WTO's) appellate body upheld the panel's decision that China's export restrictions on these raw materials were inconsistent with its obligations in 2012. The 2013 export quotas for magnesia and tin decreased compared with those of 2012. The first batch of export quotas for mineral products usually accounted for 60% of the total annual export quota. Analysts predicted that exports of rare metals would decrease gradually at a rate of 2% to 3% per year in the future. A planned reduction of the value-added tax rebate and reduced export quotas on energy-intensive products were expected to force producers to reduce their output, which was, in turn, expected to help protect and conserve mineral resources and minimize environmental damage. Although the annual export quotas for coal, coking coal, and rare earths were not publicly available, the announcement of the changes in the export allocations and an increase in tariffs for those commodities indicate that the export volumes of the commodities would likely be at the same level in 2013 as in 2012 (Ministry of Commerce, 2012b).

The MOC allocated a total rare-earth export quota of 30,996 t for 2012, which was a slight increase from the quota of 30,184 t in 2011. The Government specified how much light or middle and heavy rare earths each company was allocated in 2012. This policy was different than in previous years, when the Government had assigned export quotas without specification. The Government withheld export quotas for companies that did not meet the environmental protection guidelines. Of the 27 companies and traders, only 9 were cleared to export rare-earth products in February, and the total rare-earth export volume was 10,546 t. The rremaining 18 companies met the environmental protection guidelines in mid-year. The MOC announced a first-batch rare-earth export quota of 15,501 t for 2013, of which light rare earths was 13,563 t and middle and heavy rare earths was 1,938 t (Ministry of Commerce, 2012c, d).

Commodity Review

Metals

Aluminum.—China's aluminum production continued to increase in 2012. During the first half of 2012, aluminum smelters, which were located in central and southern areas of the country, reduced their output. Aluminum smelters gradually restarted their operations because local governments provided subsidies. Also, the commissioning of newly built aluminum smelters in the northwestern part of the country resulted in several facilities starting commercial operations in the second half of the year. Provinces in the northwest and Shandong Province were the major contributors to the country's growth in aluminum production. In 2012, China remained a net importer of aluminum. The net trade volume of unwrought aluminum increased to 392,913 t in 2012 from 143,172 t in 2011 but was still less than the 1.44 Mt traded in 2009. China's unwrought aluminum imports came mainly from (in descending order of volume) Russia, Australia, Oman, India, and South Africa, and the country's exports went to (in descending order of volume) the Republic of Korea and Japan (Alumina and Aluminum Monthly, 2013d).

The aluminum price in China resembled the London Metal Exchange price in 2012. The domestic market price of aluminum decreased to 15,169 yuan ($2,446) per metric ton in December 2012 from 16,002 yuan ($2,540) per metric ton in December 2011. Without the Government decision to stockpile about 160,000 t of aluminum during the last quarter of 2012, the price of aluminum might have been lower at yearend. The average market price of aluminum for the year was 15,706 yuan ($2,533) per metric ton on the Shanghai Metal Exchange in 2012. In 2012, about 78,500 t of production capacity, mainly in the central and southern parts of the country, was closed down because of the high price of electricity. About 2 Mt of new capacity was installed, mainly in the northwestern part of the country, especially in Xinjiang Uygur Autonomous Region, where the price of electricity was the lowest in the country in 2012. China's average production cost per metric ton of aluminum was about 16,000 yuan ($2,580), and electricity accounted for about 43% of the total production cost. Each ton of aluminum output consumed about 1,400 kilowatthours (kWh) of electricity. In Xinjiang, each kWh cost about 0.2 yuan ($0.03), whereas in the Provinces of Henan and Shandong, the cost was 0.5 yuan ($0.08) from company-owned powerplants. As a result, many aluminum investors built their aluminum smelters in Qinghai Province and Xinjang Uygur Autonomous Region. The country's aluminum output capacity reached 27.6 Mt/yr in 2012 and 30.2 Mt/yr in 2013. The construction sector was the leading consumer of aluminum and accounted for about 39% of total consumption followed by transportation, 18%; electronics, 9%; machinery and household appliances, 8% each; packaging, 7%; and others, 11%. China consumed about 21 Mt of aluminum in 2012 (China Metal Bulletin, 2012d, 2013e; Alumina and Aluminum Monthly, 2013b, p. 2–10).

Diaspore (orthorhombic hydrous aluminum oxide) accounted for more than 90% of China's bauxite deposits; the remaining 10% was of the gibbsite (monoclinic aluminum hydroxide) type. China followed Australia as the second-ranked bauxite-producing country in the world. Owing to the expansion of alumina production during the past 10 years, the country required extensive imports of bauxite to meet the demand from its alumina refineries, and this dependence was expected to continue into the future. China imported 39.6 Mt of bauxite in 2012 compared with 44.8 Mt in 2011, and bauxite imports from Indonesia and Australia accounted for 70.2% and 24.0% of the total, respectively, compared with 79.7% and 18.8%, respectively, in 2011. The decrease in bauxite imports was the result of the Government of Indonesia's restriction on raw materials exports, which the Government of Indonesia announced would continue. The restrictions on Indonesia's raw materials exports would affect the coastal Province of Shandong's alumina refineries in the future because these refineries produced alumina solely from imported bauxite. China's imports of bauxite from India were expected to increase during the next few years. The Government encouraged enterprises to explore for bauxite resources in African countries and in Australia. A few Chinese companies, including Chalco, planned to build alumina refineries in Indonesia that would use Indonesian bauxite resources (Alumina and Aluminum Monthly, 2013c).

China's output of alumina increased by more than 20% in 2012 compared with that of 2011, but China continued to experience a shortage of alumina. To support the aluminum sector, the country imported large quantities of alumina to meet the demand. In 2012, China imported 5.0 Mt of alumina compared with 1.9 Mt in 2011, of which about 92% was from Australia. China consumed about 42.4 Mt of alumina in 2012, of which 40.6 Mt was for metallurgical use and 1.8 Mt was for nonmetallurgical use. The balance was about 300,000 t of surplus alumina. By yearend 2012, more than 5 Mt/yr of alumina output capacity was installed, which increased the country's alumina output capacity to 57 Mt/yr. The additional alumina capacity was from greenfield and brownfield projects, including those of Shanxi Jiaokou Feime Aluminum Co. (1.2 Mt/yr), Guangxi Jinjiang Tiandong Co. (1 Mt/yr), Shandong Weiqiao Aluminum Co. (1 Mt/yr), Guizhou Qiya Aluminum Co. (600,000 t/yr), Shanxi Xinfa Xiaoyi Alumina Plant (600,000 t/yr), Chalco Shanxi Co. (500,000 t/yr), Bosai Group Nanchuan Pioneer Aluminum Co. (300,000 t/yr), and Shanxi Zhaofeng Aluminum Co. As a result, imports of alumina were expected to decrease in the future (Alumina and Aluminum Monthly, 2013a).

Shanxi Senze Coal Aluminum (Group) Co. Ltd. extracted alumina from coal gangue at its pilot plant in Liulin, Shanxi Province. The extracting technology was jointly developed by Northeast University, Shanxi Senze, and Shanxi University. The pilot plant had the capacity to process 50,000 t/yr of coal gangue to produce 12,600 t/yr of alumina. Shanxi Senze owned several bauxite mines that had a combined output capacity of 2 Mt/yr. The company's 700,000-t/yr alumina refinery was put into operation in 2012 (Alumina and Aluminum Monthly, 2013e).

Despite the Government's macroeconomic policy on investment in some commodities, the output of aluminum metal continued to increase rapidly. In 2005, the State Council issued a development policy "in principle" for the aluminum sector and assigned the NDRC to work with relevant agencies to prepare a plan for sustainable aluminum development. According to the MIIT-issued 12th 5-year development plan for the aluminum sector, the country was expected to produce 24 Mt of primary aluminum in 2015, and the top 10 smelters would account for 90% of the total output. To take advantage of investment incentives offered by local governments, many aluminum companies, including Chalco, China Power Investment Corp., Shandong Xinfa Group, Tiashan Aluminum-Power Co. Ltd., and Zhonghe Aluminum Co. Ltd., moved some of their operations to the northwestern part of the country. The government of Xinjiang Uygur Autonomous Region urged enterprises to develop an integrated coal-power-metallurgy industry in the region. Primary aluminum output capacity in Xinjiang increased to 2.3 Mt/yr in 2012 from 50,000 t/yr in 2007; Qinghai Province, to 2.2 Mt/yr from 1 Mt/yr; Gansu Province, to 2.1 Mt/yr from 96,000 t/yr; and Shandong Province, to 4.9 Mt/yr from 2.1 Mt/yr. Domestic analysts estimated that about 5.3 Mt of output capacity would be installed in 2013, of which Xinjiang would add 2.1 Mt and Shandong would add 1 Mt. By yearend 2015, the aluminum output capacity in Xinjiang was expected to increase to 13 Mt/yr. In April, the MIIT and eight other Government agencies jointly issued a circular to urge local governments to stop providing preferential policies to aluminum producers. The rapid expansion of aluminum output capacity in the western part of country would contribute to serious financial losses for existing producers in the country. It also would create infrastructure bottlenecks, as alumina would need to be transported from the eastern and southern parts of the country to the northwest, and finished products would need to be shipped to consumers in the coastal areas (China Nonferrous Metals Monthly, 2012e, 2013; China Metal Bulletin, 2013c).

Antimony.—China was the leading antimony producing country in the world. Changes in the volume of China's production and exports could affect prices of antimony in the world market. China's antimony resources are located in the Provinces of Guangdong, Guangxi, Hunan, Jiangxi, and Yunnan. In 2012, Guangxi, Hunan, Jiangxi, and Yunnan were the top mined antimony producing Provinces in China and accounted for more than 90% of the country's total. Guangxi, Hunan, Jiangxi, and Yunnan were also the top antimony metal producing Provinces in the country.

During the past several years, owing to environmental and safety problems, the Government shut down many illegal mining and smelting activities in the Provinces of Guangxi, Hunan, and Yunnan. The Government also monitored illegal exporting activities through Vietnam to other countries. Owing to the expansion of smelting capacity during the past decade, the supply of domestic antimony concentrates was insufficient to meet the smelters' demand; therefore, the country imported large quantities of antimony concentrates. In 2012, China imported 68,577 t of antimony concentrates, mainly from Australia, Burma, Kazakhstan, Russia, and Thailand. The country exported 40,598 t of antimony oxide, mainly to Japan, Taiwan, and the United States. In 2012, 9,583 t of unwrought antimony was exported mainly to Japan, the Netherlands, and the United States (Minor Metals Monthly, 2013b).

The Chinese Government considered antimony to be one of the protected and strategic minerals, and exploitation and production of antimony was strictly controlled. In 2012, the MLR allocated a total production quota of 105,000 t (metal content) of mined antimony, of which 75,360 t was assigned to individual Provinces and 29,640 t was recovered as byproduct or coproduct, which the MLR withheld. The MLR continued to refuse any exploration and exploitation applications in 2012. Many antimony producers in Lenshuijiang, Hunan Province, and in the Guangxi Zhuang Autonomous Region shut down their operations temporarily in 2012. Domestic analysts estimated that reported antimony metal output data might be double counted because the reported production of antimony metal was much higher than the supply of antimony in concentrates in the domestic market. China consumed about 60,000 t of antimony, and the Government stockpiled about 4,500 t in 2012. The flame retardant sector was the leading consumer of antimony and accounted for about 50% of the total followed by battery alloys, 17%; plastic stabilizers, 15%; glass, 10%; and others, 8% (Minor Metals Monthly, 2013a).

Copper.—Because it has limited copper resources, China imported a considerable amount of copper concentrates, scrap, anode, and refined metal from overseas markets. Domestic copper mines supplied less than 30% of the country's requirements for copper concentrates. In 2012, China imported 7.83 Mt of copper concentrates, which was 22.8% more than in 2011. Copper concentrates were imported from Chile (24.0%), Peru (19.0%), Australia (7.7%), Mexico (7.5%), Mongolia (6.8%), Canada (4.7%), the United States (4.5%), and others (25.8%). China imported 4.89 Mt of copper scrap from the United States (20.8%), Hong Kong (16.6%), Germany (8.4%), Australia (6.4%), Spain (6.2%), Japan (5.8%), the Netherlands (5.5%), the United Kingdom (4.7%), France (3.9%), Malaysia (3.3%), and others (18.4 %); and 3.40 Mt of refined copper from Chile (36.7%), India (7.9%), Japan (6.6%), the Republic of Korea (4.5%), Belgium (4.4%), Australia (4.2%), Zambia (3.7%), and others (32.0%). In 2012, imports of refined copper and copper scrap increased by 20.0% and 3.7%, respectively, compared with those of 2011. During the same period, China exported 274,014 t of refined copper, which was 75.3% more than in 2011. Even though the production of refined copper was at record-high levels and demand for refined copper in the domestic market was weak, the record-high import volume was attributed to the appreciation of the Chinese currency against the U.S. dollar, and traders took advantage of it for their financial purpose. The price of refined copper in the international market decreased to an average of $7,949 per metric ton in 2012 compared with $8,823 per metric ton in 2011. Also, some traders were hedging against projected higher demand in 2013. As a result, domestic analysts estimated that about 1 Mt of imported copper was stored in the bond warehouses. The availability of copper scrap in the international market was low; therefore, many downstream copper products producers used refined copper instead of copper scrap as raw material. In 2012, the apparent consumption of refined copper was about 8.8 Mt; however, producers and traders stockpiled more than 1 Mt of copper in their warehouses, and real copper consumption was about 7.7 Mt, which was about 5% higher than that of 2011. The power sector was the leading consumer of copper and accounted for about 47% of the total followed by household appliances, 15%; transportation, 10%; construction, 9%, electronics, 7%; and others, 12% (China Metals 2013c; Copper Monthly, 2013a).

Owing to domestic smelter and refinery expansions, China's copper output increased sharply during the past several years. China's copper production continued to expand despite the constrained supply of copper concentrates on the world market. In 2012, China's copper smelting and refining output capacities increased by 350,000 t and 930,000 t, respectively, and reached 4.4 Mt/yr and 7.9 Mt/yr, respectively. The output of domestic mined copper was expected to increase. Copper resources discovered in the western part of the country in Xinjiang Uygur Autonomous Region and Xizang Autonomous Region were expected to be put into development during the next several years. Copper resources in the Gangdise metallogenic belt in Xizang and the Tishan area in Xinjiang would be developed during the next several years. Significant copper resources were discovered recently in the southwestern part of the country and were expected to help replace depleted copper resources in the eastern part of the country (Recycling Resources, 2012).

The MOF and State Administration of Taxation issued a circular about value-added tax (VAT) and consumption tax policies on export goods and labor services in 2012. The new policy indicates that the tax on processing trade of copper would be changed. Under the current trade policy, copper producers paid a 17% VAT on imported copper concentrate, and the tariff rate on imported copper cathode was between 5% and 10%, depending on the purity of the copper cathode. The proposed new policy would reduce the VAT to 3% on imported copper concentrate for processing trade. Domestic copper mines supplied only about 30% of the country's copper demand and this new policy would reduce the tax burden for many domestic copper producers. Detailed guidelines on the implementation of the new policy were planned to be released later (China Nonferrous Metals Monthly, 2012b).

According to the Government's 12th 5-year plan, the output of secondary refined copper would account for 40% of the total copper output in 2015, although China's recycling sector remained fragile. More than 5,000 enterprises were involved in nonferrous metal recycling, and most of them used obsolete technology or manual labor. China imported more than 7 Mt/yr of nonferrous metal scrap, and domestic households generated about 2 Mt/yr of nonferrous metal scrap. Domestic analysts estimated that the secondary nonferrous metals output capacity was about 20 Mt/yr. Jiangxi Copper Co. Ltd., which was the leading copper producer in China, had the capacity to produce 1 Mt/yr of refined copper, of which more than 350,000 t was from copper scrap. Other Chinese copper producers also expanded their secondary copper capacities in the coastal areas. In 2012, China imported 4.89 Mt of copper scrap, and the average copper content was about 30%; however, copper content in domestic scrap was only one-third of imported scrap. According the MIIT guidelines on the secondary copper sector, greenfield secondary copper smelters were required to have 100,000 t/yr of capacity by 2015, and secondary copper producers would be required to shut down existing smelters, that had less than 50,000 t/yr of capacity. The Government also provided guidelines on energy consumption and the recycling rate requirements for greenfield and brownfield copper smelters. The MIIT also published a draft on conditions under which companies would be allowed entry into the copper sector. China's copper sector was expected to continue to be dependent on imported copper concentrate and scrap to meet the country's demand (China Nonferrous Metals Monthly, 2012a; Ministry of Industry and Information Technology, 2013b).

The MIIT published a list of obsolete production technologies and ordered producers to close down those production plants that used these production technologies in 2012. Copper plants that used small reverberatory or electric furnaces were included on the list.

Several greenfield and brownfield copper projects were recently completed or under construction. Tiandilong Copper Co. Ltd. completed the construction of its 150,000-t/yr copper plant in Guangfeng, Jiangxi Province. Baiyin Nonferrous Metals Co. started the construction of a 200,000-t/yr copper plant in Baiyin, Gansu Province. Duobaoshan Copper Co. Ltd. [a joint venture between Zijing Mining Co. Ltd. and Heilongjiang Mining (Group) Co. Ltd.] completed the construction of the Duobaoshan copper mine, which had the capacity to produce 30,000 t/yr of copper in concentrates in 2012, and was scheduled to complete the construction of a 100,000-t/yr copper smelter in 2015. Zijing planned to expand its refined copper capacity to 300,000 t/yr from 200,000 t/yr in Fujian Province (China Metal Bulletin, 2012a; Copper Monthly, 2012; Ministry of Industry and Information Technology, 2012b).

Guangxi Nonferrous Metal Group Co. Ltd. completed its 300,000-t/yr-capacity secondary copper cathode project at Wuzhou Recycling Industrial Park in Guangxi Zhuang Autonomous Region. The first phase of the project included a 200,000-t/yr-capacity tilting furnace and a 100,000-t/yr-capacity Kaldo smelting furnace. Shangrao Hefeng Copper Co. Ltd. planned to put its 50,000-t/yr-capacity high-purity copper cathode plant, which was located in Shangrao, Jiangxi Province, into operation in early 2013. Jiangxi Copper and Hangzhou Fuchunjiang Smelting Co. Ltd. agreed jointly to invest $1.3 billion in a copper project in Fuyang, Zhejiang Province. The project was designed to produce 200,000 t/yr of blister copper and 370,000 t/yr of refined copper. The project was scheduled to be completed in late 2013 (China Metal Bulletin, 2012b; Copper Monthly, 2013b).

Germanium.—About 90% of China's identified germanium resources are located in the Provinces of Guangdong, Guangxi, Guizhou, Jilin, Shanxi, Sichuan, and Yunnan, and the Nei Mongol Autonomous Region. In China, germanium was produced as a byproduct of lead and zinc operations or was extracted from coal. Germanium was used mainly in the chemical, optical, and solar-cell-battery sectors. In early 2012, demand for germanium was sluggish, and the domestic market price for germanium decreased to 7,200 yuan ($1,161) per kg in May from 8,600 yuan ($1,387) per kg in January. In December, after Yunnan Lincang Xinyuan Germanium Industrial Co. Ltd.'s subsidiary, Dongcheng Metal Co. Ltd., announced that the company had signed a contract to supply a total of 375 t of germanium oxide to Hanenergy Group between 2012 and 2018, the price of germanium subsequently increased to 10,750 yuan ($1,734) per kg. In 2012, the country consumed about 75 t of germanium; the optical sector accounted for 30% of the total and the solar-cell-battery and chemical sectors consumed about 25% each. The consumption of germanium by the optical sector was expected to increase by between 10% and 15% during the next several years, especially as the country had ungraded its germanium tetrafluoride production technology. The Government stockpiled 20 t of germanium in 2012 (Zhang and Xu, 2012; China Metal Bulletin, 2013d).

Sparton Resources Inc. of Canada announced that the company had signed an agreement for the sale of its subsidiary, Lincang Linxiang 306 Huajun Coal Industry Co. Ltd. (located in Lincang, Yunnan Province), to Yunnan Lincang Xinyuan Germanium Industrial Co. Ltd., which included all Huajun's germanium assets. The sale of Huajun was in response to the Chinese Government's decision to consolidate all small coal mining operations into larger coal mining operations within the next 2 years. The government of Lincang appointed Xinyuan to purchase all the smaller coal (germanium) operations in the Lincang area, including the Huajun assets (Sparton Resources Inc., 2013).

Iron Ore and Iron and Steel.—China was the world's leading iron and steel producer and accounted for more than 57% of the world's pig iron production and 45% of the world's crude steel production in 2012. The output of the country's iron and steel sector continued to increase in 2012. Production of pig iron and crude steel increased by about 3% or more each in 2012 compared with production in 2011. The rate of growth was the slowest of the past several years owing to weak demand for steel products in both the domestic and international markets. In 2012, the total fixed-asset investment in the iron and steel sector was $106.2 billion, which was an increase of 3.0% from that of 2011; of that amount of investment, iron ore mining accounted for $24.7 billion of the total. The ironmaking and steelmaking output capacity increased by 50 Mt and 45 Mt, respectively, in 2012. The ironmaking capacity reached 878 Mt in March 2012 and was expected to be about 1 billion metric tons (Gt) by 2015. The Provinces of Hebei, Jiangsu, and Liaoning ranked as the top three Provinces for fixed-asset investment in the iron and steel sector in the country. Other Provinces and cities, such as Chongqing, Henan, Hunan, and Shandong, also increased investments in the iron and steel sector. Private funds accounted for about 80% of the investment. Apparent consumption of crude steel was 659 Mt, which was 3.1% higher than that of 2011. Domestic analysts estimated that producers and traders stockpiled more than 20 Mt of steel products in their warehouses. In 2013, the China Iron and Steel Association projected that the country's crude steel consumption would increase by about 3.1% in 2012. In 2012, China exported

55.7 Mt of steel products, which was an increase of 0.4% from that of 2011, and imported 13.6 Mt of steel products, which was a decrease of 12.3% from that of 2011. China was a net exporter of 44.1 Mt of steel products. This indicated that more than 50% of the increase in crude steel output was targeted for exported steel products. Exports of China's steel products to Asian countries accounted for 59% of the total exports. Steel products imports from, in descending order of export value, Japan, the Republic of Korea, Taiwan, and the EU accounted for 90% of the total imports (Chen, 2013; Hu, 2013; Wang, 2013).

During the past several years, the country's iron ore production increased sharply; however, the percentage of iron ore (iron content) supplied by domestic producers remained at less than 50% of demand in 2012, and China continued to depend on iron ore imports to fill the gap. China had iron ore reserves of about 19.3 Gt; however, owing to the low iron content and high impurities of domestic ore, pig iron producers preferred imported ore to domestic ore. Also, the production cost for salable iron ore was about $90 to $100 per metric ton in China compared with about $40 to $50 per metric ton in Australia and Brazil. Imports of iron ore increased to more than 743 Mt in 2012 from 686 Mt in 2011 and 618 Mt in 2010. Australia remained the leading iron-ore-supplying country and accounted for 47% of the total followed by Brazil, 22%; South Africa, 5%; India, 4%; and others, 22%. The total amount of iron ore stockpiled at China's 25 major ports was more than 80 Mt at yearend 2012. About 50% of seaborne iron ore in the world was destined for China. The country was expected to import about 770 Mt of iron ore in 2013 (China Metals, 2012c; Li, 2012; General Administration of Customs of the People's Republic of China, 2013).

The Chinese Government was considering adjusting the total tax levied on iron ore companies. Currently, iron ore companies paid about 10 different types of taxes, including the VAT and resource taxes, and the total tax rate was about 25%. The MIIT and the MOF submitted a joint proposal to reduce the tax rate to between 10% and 15% to the State Council for consideration (China Metals, 2012b).

The Government hoped that the consolidation of the iron and steel industry would help the sector's efficiency, increase its bargaining power with suppliers of raw materials, and reduce competition within the sector. The Government also urged iron and steel producers to create transregional enterprises. If Government targets are met, the top 10 iron and steel producers would account for 60% of the country's total iron and steel output in 2015 compared with 46% in 2012. Because of environmental concerns, the Central and local governments urged iron and steel producers to relocate their facilities away from the cities and urban areas. Shoudu Iron and Steel (Group) Co. (Shougang) moved its production facilities to Hebei Province from Shijiangshan, Beijing. In 2012, the government of Shanghai issued the 12th 5-year plan for iron and steel development in Shanghai. The city government planned to shut down 5.8 Mt of ironmaking capacity and 6.6 Mt of steelmaking capacity within the city limits. Baoshan Iron and Steel (Group) Corp. (Baogang), which was located in Shanghai, planned to increase the company's crude steel output capacity to 66 Mt/yr in 2015 from about 43 Mt/yr in 2011, of which about 16 Mt/yr of crude steel capacity would be produced from steel plants in Shanghai. The company would shut down at least 6 Mt/yr of output capacity to reach the target of 16 Mt/yr in Shanghai by 2015. Steel facilities in Shanghai would concentrate on producing high-value-added steel products for automobile manufacturing, shipbuilding, and special steel production. The construction of Baogang's 10-Mt/yr iron and steel complex in Zhanjiang, Guangdong Province, was underway. The steel output capacity in Baogang's subsidiaries was expected to increase within the next 5 years (China Metals, 2012a; 2013d).

The Government continued its effort to curb the fast growing output capacity in the country, including by ordering iron and steel producers to phase out obsolete facilities. Major iron and steel enterprises continued to expand their output capacity, however, either by replacing smaller furnaces with larger furnaces or by relocating to coastal areas and building steel complexes with larger output capacity. Anshan Iron and Steel (Group) Co. built greenfield iron and steel production facilities at Bayuquan in Yingkou and at Chaoying to add output capacity of more than 8.5 Mt/yr during the past 5 years. Shougang and Tangshan Iron and Steel Co. jointly built a 10-Mt/yr iron and steel complex in Hebei Province. The Government approved Baogang and Wuhan Iron and Steel Group's construction of two iron and steel complexes in the Provinces of Guangdong and Guangxi, respectively. The Government also approved the merger of 12 privately owned iron and steel producers in Hebei Province to form a large iron and steel enterprise, Tangshan Bohai Iron and Steel (Group) Co. Ltd., which proposed to build a 15-Mt/yr iron and steel complex in the coastal area of Hebei. The government of Shandong planned to build a 17.5-Mt/yr iron and steel complex at Dongjiakou, Shandong Province (Shandong Provincial Government, 2012; China Metals, 2013a).

Rhenium.—Rhenium was produced as a byproduct of molybdenite concentrates from porphyry copper-molybdenum ore. Rhenium sulfide was collected as dust from molybdenum roaster-flue gas from copper-molybdenum ore; rhenium sulfide was then oxidized into rhenium oxide. Dirhenium heptoxide is readily dissolved in water, but rhenium dioxide has a low solubility in water. Rhenium oxide was usually leached out with acid and then purified by an anion exchange and solvent extraction process.

In China, the Huanglong polymetallic mine at Luonan County in Shaanxi Province contained more than 300 grams of rhenium per metric ton of ore. Western Xinxing Metal Materials Co. Ltd. constructed a plant that had the capacity to produce 10,000 t/yr of molybdenum concentrates and to recover 200 kilograms per year of ammonium perrhenate. Jiangxi Copper Co. Ltd.'s Guixi Smelter had the capacity to produce about 3 t/yr of ammonium perrhenate mainly from imported copper concentrates. Zhuzhou Kete Industries Co. Ltd. at Zhuzhou in Hunan Province was also reported to have the capacity to produce high-purity rhenium products (Ma and others, 2012).

Industrial Minerals

Lithium.—China has abundant salt lakes that contain lithium, and these lakes accounted for 80% of the country's total lithium resources. Most of the lakes are located in the western part of the country, however, and the infrastructure in these areas was relatively undeveloped. Spodumene resources were located mainly in the Altai and the Keketuohai areas in Xinjiang Uygur Autonomous Region and in the Kangding area in Sichuan Province. Lepidolite resources were found at the Yichun area in Jiangxi Province. Brine resources were found in salt lakes at the Qaidam basin in Qinghai Province and Zabuye Lake in Xizang Autonomous Region. The country's primary lithium products, which were lithium carbonate, lithium hydroxide, and lithium salt, were produced from domestic and imported raw materials. These compounds could be further processed into lithium metal or lithium cobalt oxide. In 2012, China's lithium carbonate output was estimated to be about 35,000 t; lithium hydroxide, 18,000 t; and lithium metal, 2,000 t. The total output capacity of lithium salt was about 100,000 t/yr in 2011. In 2012, lithium production from spodumene and lepidolite increased to about 160,000 t of spodumene concentrates, but lithium from brine decreased because producers were working on technology to increase the efficiency to extract lithium from brine. Output of lithium from brine was expected to increase during the next several years. Xizang planned to expand the capacity of its Jieze Chaka salt lake operation. China Minmetals expanded its output capacity in the Qaidam basin (China Metal Bulletin, 2013h).

Sichuan Tianqi Lithium Industry Co. Ltd., which used imported spodumene to produce lithium compounds (including carbonate, chloride, and hydroxide), received Government approval to mine spodumene resources at Yajiang in Sichuan Province. The Cuola lithium deposit had resources of 20 Mt grading 1.3% lithium oxide. The company planned to design a mine to produce 1.2 Mt/yr of ore for about 20 years. The deposit also contained beryllium, niobium, and tantalum. The company planned to increase its total output capacity of lithium carbonate, anhydrous lithium chloride, and lithium hydroxide to about 20,000 t/yr. The country depended on imported spodumene resources to meet its demand. In 2012, Tianqi invested $850 million to acquire the outstanding shares and stock options for Talison Lithium. Talison operated the Greenbushes Mine in Australia, which supplied about 80% of China's lithium needs (China Metal Bulletin, 2013i).

Rare Earths.—China was rich in rare-earth resources, and the country produced different kinds of rare-earth products. China's rare-earth production accounted for about 90% of the world total, and the volume of exports had a significant effect on the world markets. During the past several years, the Government adjusted the rare-earth production and export quota to protect the domestic resources and the environment. In 2012, the State Council issued "Situation and Policies of China's Rare Earth Industry," which provided a plan for the development of the country's rare-earth industry during the next several years. The goal was to establish orderly rare-earth operations, including the development of rare-earth resources, followed by separation, smelting, and marketing. The Government would control unregistered exploitation, environmental damage, unapproved production capacity expansion, and illegal trading. According to a 2009 MLR rare-earth resources survey, China's rare-earth reserves were 18.59 Mt of rare-earth oxide (REO) equivalent content, which accounted for 23% of the world's total reserves of 81.1 Mt. This estimate of China's reserves was much lower than the previous MLR estimate. In 2011, the country had 126 rare-earth companies with a combined total output capacity of 320,000 t of REO. In 2012, the MLR issued rare-earth mining production was 93,800 t. Rare-earth consumption in China had increased steadily, and the country consumed more than 84,000 t/yr of REO in 2012 compared with 19,300 t in 2000. To maintain the same volume of rare-earth production, the Government decided to reduce the rare-earth export quota to less than 31,000 t in 2010, 2011, and 2012 from 65,000 t in 2005. China, however, continued to have difficulty controlling global rare-earth prices because a significant volume of rare-earth products was exported through unofficial channels. Domestic analysts estimated that about 20,000 t/yr of rare-earth products was unaccounted for (Ma, 2012; State Council, The, 2012, p. 1–5; China Metal Bulletin, 2013b).

In 2012, the MLR issued a total 67 rare-earth mining licenses, which was 46 fewer than before. Jiangxi Province had 45 mining licenses, of which Ganzhou Rare Earth (Group) Co. Ltd. received 43 of the total. The MLR assigned seven mining licenses to seven companies in Sichuan Province. Five mining companies in Fujian Province received a total five mining licenses. In Guangdong Province, Guangdong Rising Group Co. Ltd.'s subsidiaries had three mining licenses. Baotou Iron and Steel was the only company to receive two mining licenses in Nei Mongol Autonomous Region. Yunnan Province had two mining licenses, and the Provinces of Guangxi, Hunan, and Shandong each had one. The MLR also issued 10 rare-earth exploration licenses (Ministry of Land and Resources, 2012).

The consolidation of the rare-earth industry continued in 2012. The Government of Nei Mongol Autonomous Region assigned Baotou Iron and Steel [through its subsidiary Baotou Rare Earth Hi-Tech Holding Co. Ltd. (Baotou Hi-Tech)] to be the sole producer and manager of all rare-earth mining, separation, and trading activities in the Autonomous Region. Those domestic producers in the Autonomous Region that met the asset transfer requirement were required to sell their assets to Baotou Hi-Tech. The remainder would be shut down without compensation. The consolidation of 35 rare-earth producers was set to be completed in June 2011; however, many rare-earth producers refused to shut down. In December 2012, Baotou Hi-Tech signed tentative agreements with 12 rare-earth companies—Baotou Damao Rare Earth Co. Ltd., Baotou Feida Rare Earth Co. Ltd., Baotou Hongtianyu Rare Earth Magnet Co. Ltd., Baotou Jinmeng Rare Earth Co. Ltd., Baotou Sanlong Rare Metal Material Co. Ltd., Baotou Shengyou Rare Earth Co. Ltd., Baotou Xijun Rare Earth Co. Ltd., Baotou Xinye New Material Co. Ltd., Baotou Xinyuan Rare Earth Hi-Tech Material Co. Ltd., Inner Mongolia Jinxia Chemical Engineering Co. Ltd., Inner Mongolia Shengyilum Rare Earth Co. Ltd., and Wuyuan Runzhe Rare Earth Co. Ltd. Each company transferred 51% of its shares to Baotou Hi-Tech without compensation. An official agreement would be signed within 1 year. Otherwise, agreements would be voided after the expiration date. The integration of 12 rare-earth companies into Baotou Hi-Tech appeared to depend on the establishment of state-owned China Northern Rare Earth (Group) Hi-Tech Co. Ltd., which would include rare-earth producers from the Provinces of Shandong and Sichuan. The company would invest in the development of innovative technologies to improve mining, smelting, and product development. This would further consolidate the rare-earth sector in the northern part of China, where most of the light rare-earth producers were located (China Metals, 2013b).

The Provincial government of Guangdong established a Guangdong Rare Earth Industry Group to manage the rare-earth sector in Guangdong. Guangdong-based Guangdong Rising Group Co. Ltd., which was the parent company of Guangdong Rising Nonferrous Metals Co. Ltd. and which had three rare-earth mining licenses in Guangdong, was a member of the Guangdong Rare Earth Industry Group. The Guangdong Rare Earth Industry Group would be in charge of the integration of exploration, mining, and smelting, and the development of high-value-added products in the Province. The Provincial government of Fujian established an integrated rare-earth enterprise, Fujian Rare Earth Group Co. Ltd., in the Province in 2012. Province-owned Fujain Metallurgy Holdings Co. Ltd. transferred 33% of its shares in Xiamen Tungsten Co. Ltd. to Fujian Rare Earth. Xiamen Tungsten would be transformed into a leading company to perform prospecting, mining, smelting, and downstream product development. The Provincial government of Fujian would assist the acquisition by local rare-earth companies in Longyan and Sanming within the next 2 years (China Metal Bulletin, 2012c; China Nonferrous Metals Monthly, 2012c).

Dingnan Nanfang Rare Earth Co. Ltd., Ganzhou Rare Earth, and Xunwu Nanfang Rare Earth Co. Ltd. reached an agreement to merge. Ganzhou Rare Earth also acquired Longnan Wanbao Rare Earth Separating Co. Ltd. and Longnan Kaisheng Nonferrous Metal Co. Ltd. Ganzhou Rare Earth became an integrated rare-earth enterprise in Jiangxi Province. The government of Hunan Province announced its intent to establish an integrated rare-earth company, Hunan Rare Earth Group. State-owned Hunan Jinxin Gold Group Co. Ltd. would be the leading shareholder of Hunan Rare Earth. It appeared that local governments wanted to protect their own rare-earth resources and had no intention of letting state-owned enterprises participate in the reform of their rare-earth sectors. Two state-owned companies, Chalco and Minmetals, established their integrated rare-earth companies in Guangxi Zhuang Autonomous Region and Hunan Province, respectively (China Nonferrous Metals Monthly, 2012d; China Metal Bulletin, 2013f).

Strontium.—China's strontium sector developed quite rapidly during the past decade, and the country became one of the leading producers of strontium carbonate in the world. Strontium carbonate was produced from celestite and strontianite by either the carbon reduction method or the decomposition method. China's celestite resources are located in the Provinces of Hubei, Jiangsu, Qinghai, Sichuan, and Yunnan; Xinjiang Uygur Autonomous Region; and Chongqing City. Strontianite resources are located in Chongqing City. Compared with that of such countries as Iran, Mexico, and Spain, the strontium sulfate content in Chinese celestite and strontianite was lower. Dafengshan Celestite Mine in Qinghai Province, which is the largest celestite mine in China, contains between 38% and 40% strontium sulfate, and concentrates contain about 70% strontium sulfate. Chongqing's Dazu and Tongliang Mines were the country's leading celestite and strontianite mining sites, and the combined output capacity of the two mines was 300,000 t/yr of ore to produce about 100,000 t/yr of concentrates. After more than 20 years of exploitation, most of the high-grade resources had been depleted, and production costs were increasing. Other significant celestite mining areas included the Hubei Strontium Mine in Hubei, the Jiangsu Strontium Mine in Jiangsu, the Xinjiang Strontium Mine in Xinjiang, and the Lanping Strontium Mine in Yunnan (Yang, Wu, and Xu, 2012).

In the early 2000s, numerous strontium carbonate producers in China had a total combined output capacity of 350,000 t/yr. The television cathode ray tube (CRT) glass sector was the leading consumer of strontium carbonate. Owing to the development of new generation televisions, the demand for CRT glass had decreased significantly during the past several years. As a result, strontium carbonate producers closed down their operations partially during the past few years, and the total combined output capacity had decreased to about 170,000 t/yr in 2011. The Government urged celestite producers to mine low- and high-grade ore together and to increase the recovery efficiency and strontium sulfate content in concentrates (Yang, Wu, and Xu, 2012).

Mineral Fuels

Coal.—China had undergone significant economic reform and had one of the world's fastest growing economies. Coal consumption had increased to meet the high demand for industrial production and power generation. Coal was the country's primary source of energy—two-thirds of the country's electricity was produced by coal-fired powerplants. About 50% of the country's total coal output was consumed by the power sector. Even though China's coal production continued to increase in 2012 because of an increase in demand for coal by every industrial sector, China became a net coal importing country. In 2012, the country imported a total of 289 Mt of coal, which was an increase of 59% from that of 2011. China exported a total of 14.7 Mt of coal, which was the same as in 2011. The increase in coal imports was the result of the price of coal on the international market being lower than that on the domestic market. Also, the Government had gradually eliminated import tariff rates on coal during the past several years and imposed tariff rates on coal exports. Coal-fired powerplants, especially in the coastal Provinces, increased the use of imported coal to reduce their production costs. The price of electricity was set by the Government, and electricity producers faced financial loss for more than a decade. In 2012, demand for coal was low, and the country faced a surplus of coal. At yearend 2012, coal producers stockpiled about 85 Mt of coal in their warehouses, which was an increase of 58% from the beginning of the year. Major coal ports stockpiled about 43 Mt, and major coal-fired powerplants had about 81 Mt on their sites. In 2012, the price per metric ton of 5,500-kilocalorie coal decreased by 170 yuan ($27) to 640 yuan ($103) at Qinhuangdao Port, which was the country's major transshipment port (China Metal Bulletin, 2013a; General Administration of Customs of the People's Republic of China, 2013).

The Government identified 1,256 coal mines that would be shut down in 2013, which would eliminate 64.2 Mt of production capacity. In China, there was two-tier system of coal prices—contract and market. The coal contract price was negotiated between coal producers and coal-fired powerplant producers at prices under conditions established by the Government. In 2012, the Government decided not to participate in the yearend coal contract meeting. The cancellation of this key coal contract meeting indicated the removal of Government intervention. Coal producers thus had more freedom to sell their coal at higher prices; however, coal-fired powerplant producers could face increased financial losses if the Government does not adjust the power supply rate. Alternatively, coal-fired powerplant producers could import more coal from overseas. The growth rate in the demand for coal in China was expected to be lower during the next several years. Imports accounted for about 7% of the country's coal consumption; however, coal imports accounted for almost 30% of the seaborne coal trade in the Asia and the Pacific region. The country's coal output capacity and production were targeted by the Government to be 4.1 Gt and 3.9 Gt, respectively, at yearend 2015. In 2012, the country consumed about 3.7 Gt of coal. China's coal-fired powerplant producers were sensitive to overseas coal prices. Coal imports would increase if coal prices on the international market were cheaper than in the domestic market. The volume of coal imports also depended on the railway transportation system. Major coal-producing provinces were located in the northern and northwestern parts of the country, and coal consumers were located in the southern and in the coastal Provinces. The Government planned to develop a more reliable coal transportation system to release the constraint during the next few years (Citigroup Global Market Inc., 2013c).

Outlook

China's economy is expected to continue to grow in the near future. The country has replaced Japan as the second largest economy in the world behind the United States, and the Government has set the economic growth rate target at 7% for the next 5 years. The Government recognizes that the country cannot depend solely on exports to sustain its economic growth and that the country needs to increase domestic consumption and to have a more transparent financial and legal system. The expected continuation of China's economic growth implies that a strong demand for mineral commodities is likely to continue. China has shortages of supply of most major minerals, including bauxite, chromium, copper, iron, lead, manganese, nickel, oil, and potash, and it relies on imports to meet its demand. This trend is expected to continue. The Government, therefore, encourages enterprises to invest in such mineral-rich countries as Australia, Brazil, Burma, Chile, Indonesia, and Mongolia to secure minerals for domestic economic development and growth. The Government is expected to continue its effort to protect the country's resources of minerals, such as antimony, coal, molybdenum, rare earths, tin, and tungsten, and to avoid overexploitation.

China's imports of raw minerals have been increasing in recent years. As a result, the Chinese Government has been promoting reduced dependence on mineral commodity imports and encouraging the production of high-value-added and high-quality downstream products. The Government also promotes the secondary nonferrous metals industry to reduce energy consumption. The Government has not yet achieved great success in meeting these goals. As progress is made toward these goals, the country's dependence on most major mineral commodities could decline; however, China will likely continue to play an important role in the world's metal and mineral markets. Also, China's overseas investments will probably continue until the transition to resource independence takes place. China's overall outward investment is expected to continue to increase and may soon exceed inward foreign direct investment.

The environmental, health, safety, and social performance of the mining and metal enterprises are of concern to the Government. The Government has set guidelines for the development of these enterprises in an attempt to improve protection of the environment, but progress has been slow. The Government plans to continue its effort to address the sustainable development of the mining and metal sectors through air and water pollution prevention and treatment, land protection, mine safety, and reclamation of mine sites.

References Cited

Alumina and Aluminum Monthly, 2013a, 2012 alumina market review and outlook for 2013: Beijing Antaike Information Development Co. Ltd., February, 7 p.

Alumina and Aluminum Monthly, 2013b, 2012 aluminum market review and outlook for 2013: Beijing Antaike Information Development Co. Ltd., February, 25 p.

Alumina and Aluminum Monthly, 2013c, Chinese import of alumina by country in Jan Dec 2012: Beijing Antaike Information Development Co. Ltd., February, p. 14.

Alumina and Aluminum Monthly, 2013d, Imports and exports of Al products in December 2012: Beijing Antaike Information Development Co. Ltd., February, p. 25–26.

Alumina and Aluminum Monthly, 2013e, Shanxi Liulin Senze makes success in extracting alumina from coal gangue: Beijing Antaike Information Development Co. Ltd., March, p. 13.

Asian Development Bank, 2013, Asian Economic integration Monitor: Manila, Philippines, Asian Development Bank, 65 p.

Chen Xinliang, 2013, 2012 newly install capacity of blast furnaces and converters: China Steel, no. 3, p. 28–29.

China Metal Bulletin, 2012a, Duobaoshan copper (molybdenum) mine completed: China Metal Bulletin, issue 36, p. 7.

China Metal Bulletin, 2012b, Fuchunjiang and Jiangxi Copper signed a copper project: China Metal Bulletin, issue 28, p. 8.

China Metal Bulletin, 2012c, Fujian Rare Earth invested into Xiamen Tungsten: China Metal Bulletin, issue 44, p. 8.

China Metal Bulletin, 2012d, Policy attracted investment: China Metal Bulletin, issue 22, p. 34.

China Metal Bulletin, 2013a, 2012 surplus of coal in domestic market: China Metal Bulletin, issue 5, p. 19.

China Metal Bulletin, 2013b, 2013 estimation of China rare-earth consumption will be 90,000 t: China Metal Bulletin, issue 15, p. 18.

China Metal Bulletin, 2013c, Aluminum sector preview: China Metal Bulletin, issue 1, p. 24.

China Metal Bulletin, 2013d, Brief discussion of germanium market: China Metal Bulletin, issue 1, p. 32.

China Metal Bulletin, 2013e, Difficult to maintain oversupply of aluminum: China Metal Bulletin, issue 5, p. 28.

China Metal Bulletin, 2013f, Ganzhou Rare Earth Group started operating: China Metal Bulletin, issue 11, p. 20.

China Metal Bulletin, 2013g, Manufacturing industry adjusted: China Metal Bulletin, issue 8, p. 11.

China Metal Bulletin, 2013h, Overview of China lithium market: China Metal Bulletin, issue 3, p. 30.

China Metal Bulletin, 2013i, Reorganization of Tianqi and controlling market price: China Metal Bulletin, issue 9, p. 36.

China Metals, 2012a, Baosteel to have 66 mtpa crude steel capacity by 2015: China Metals, v. 18, no. 391, p. 19.

China Metals, 2012b, China likely to halve tax burdens on IO companies: China Metals, v. 18, no. 398, p. 18.

China Metals, 2012c, IO prices rise on falling inventories: China Metals, v. 18, no. 400, p. 15.

China Metals, 2012d, SRB starts stockpiling metals: China Metals, v. 18, no. 398, p. 1.

China Metals, 2013a, Ansteel's overseas JV begins operations: China Metals, v. 19, no. 406, p. 16.

China Metals, 2013b, Baotou Rare Earth to integrate 12 RE firms in Inner Mongolia: China Metals, v. 19, no. 401, p. 15.

China Metals, 2013c, China's net imports of refined copper hit all-time high in 2012: China Metals, v. 19, no. 402, p. 3.

China Metals, 2013d, Heavy task ahead for steel sector upgrading: China Metals, v. 19, no. 404, p. 5.

China Nonferrous Metals Monthly, 2012a, China's recycle copper industry holds broad prospects: China Nonferrous Metals Monthly, April, p. 1.

China Nonferrous Metals Monthly, 2012b, Favorable policies for copper imported processing trade formulated: China Nonferrous Metals Monthly, November, p. 4.

China Nonferrous Metals Monthly, 2012c, Guangdong clarified rare earth integration schedule: China Nonferrous Metals Monthly, August, p. 1.

China Nonferrous Metals Monthly, 2012d, Hunan Rare Earth Group approved: China Nonferrous Metals Monthly, November, p. 6.

China Nonferrous Metals Monthly, 2012e, Refined aluminum industry suffers from deficit and western investment accelerates: China Nonferrous Metals Monthly, September, p. 1.

China Nonferrous Metals Monthly, 2013, Xinjiang gave full support to expanding and strengthening aluminum industry: China Nonferrous Metals Monthly, January, p. 4.

Citigroup Global Market Inc., 2012, China economics weekly—New leadership new action: New York, New York, Citigroup Global Market Inc., November 8, 14 p.

Citigroup Global Market Inc., 2013a, China economics weekly—Normalized growth and policy adjustment: New York, New York, Citigroup Global Market Inc., January 13, 17 p.

Citigroup Global Market Inc., 2013b, China macro view—Fiscal vista—Expect conservative policy and aggressive reform: New York, New York, Citigroup Global Market Inc., May 10, 20 p.

Citigroup Global Market Inc., 2013c, PRC coal: New York, New York, Citigroup Global Market Inc., January 2, 9 p.

Copper Monthly, 2012, Tiandilong Copper's 150ktpy refined copper project is to be put into operation by the end 2012: Beijing Antaike Information Development Co. Ltd., August, p. 10.

Copper Monthly, 2013a, Chinese imports and exports of Cu products in December 2012: Beijing Antaike Information Development Co. Ltd., February, p. 10.

Copper Monthly, 2013b, Shangrao Hefeng Copper's 50ktpa copper cathode project comes onstream: Beijing Antaike Information Development Co. Ltd., April, p. 9.

General Administration of Customs of the People's Republic of China, 2013, China Customs Statistics 2012:, Beijing, China, General Administration of Customs of the People's Republic of China, annual, CD-ROM.

Hu Ling, 2013, Imports and exports of the major steel products in 2012: China Steel, no. 2, p. 37–44.

Li Xinchuang, 2012, How to protect domestic iron ore resources: China Steel, no. 12, p. 5–10.

Ma Gao-feng, Lei Ning, Guo Jin-liang, Wang Zi-chuan, Xie Ya-ning, Sun Bao-lian, Bai Hong-bin, and Feng Bao-qi, 2012, Recycling rhenium from flue dust in molybdenum concentrate roasting: China Molybdenum Industry, v. 36, no. 2, p. 4–9.

Ma Rongzhang, 2012, Current situation and outlook of China rare earth industry: Association of China Rare Earth Industry, 10 p.

Ministry of Commerce, 2012a, 2013 export products monitoring list: Beijing, China, Ministry of Commerce, issue 97, December 31, 29 p.

Ministry of Commerce, 2012b, 2013 export quota of agriculture and industrial products: Beijing, China, Ministry of Commerce, issue 74, October 31, 1 p.

Ministry of Commerce, 2012c, The first batch rare-earth export quota: Beijing, China, Ministry of Commerce, issue 1158, December 27, 2 p.

Ministry of Commerce, 2012d, The second batch rare-earth export quota: Beijing, China, Ministry of Commerce, issue 627, August 16, 3 p.

Ministry of Finance, 2012, Investment encouraged and disbanded investment tax: Ministry of Finance. (Accessed November 20, 2012, at (http://syz.mof.gov.cn/zhengwuxinxi/zhengcejiedu/201211/t20121119_699228.html.)

Ministry of Industry and Information Technology, 2012a, 2012 amended iron and steel requirement: Beijing, China, Ministry of Industry and Information Technology, announcement 35, 6 p.

Ministry of Industry and Information Technology, 2012b, 2012 list of retired enterprises in the industrial sector: Beijing, China, Ministry of Industry and Information Technology, announcement 39, 21 p.

Ministry of Industry and Information Technology, 2012c, Graphite industrial policy guidelines: Beijing, China, Ministry of Industry and Information Technology, announcement 60, 7 p.

Ministry of Industry and Information Technology, 2013a, Announcement of industry merging guidelines: Beijing, China, Ministry of Industry and Information Technology joint announcement 2013–15, 6 p.

Ministry of Industry and Information Technology, 2013b, Guidelines for copper smelting sector—Solicit opinion: Beijing, China, Ministry of Industry and Information Technology, May 15, 8 p.

Ministry of Land and Resources, 2012, Announcement of rare-earth exploration and mining licenses: Beijing, China, Ministry of Land and Resources, issue 21, September 4, 5 p.

Ministry of Land and Resources, 2013, Zhongguo Guotu Ziyuan Gongbao 2012: Beijing, China, Ministry of Land and Resources, April, 46 p.

Minor Metals Monthly, 2013a, Antimony market overview for 2012: Beijing Antaike Information Development Co. Ltd., February, p. 2.

Minor Metals Monthly, 2013b, Imports and exports of antimony products in Dec. 2012: Beijing Antaike Information Development Co. Ltd., February, p. 12.

National Bureau of Statistics of China, 2012, China statistical yearbook: Beijing, China, National Bureau of Statistics of China, 1,069 p.

National Bureau of Statistics of China, 2013, 2012 national economic and social development statistics: Beijing, China, National Bureau of Statistics of China, February 22, 26 p.

Recycling Resources, 2012, China copper industry: Recycling Resources, no. 12, p. 12.

Shandong Provincial Government, 2012, Shandong iron and steel capacity development guideline: Jinan, Shandong, China, Shandong Provincial Government, issue 2012–37, 2 p.

Sparton Resources Inc., 2013, Sparton announces sale of germanium assets: Toronto, Ontario, Canada, Sparton Resources Inc. press release, January 28, 1 p.

State Council, The, 2012, Situation and policies of China's rare earth industry: Beijing, China, The State Council, June, 20 p.

State Council, The, 2013a, Announcement of structural and function changes at the State Council: Beijing, China, The State Council, March 10, 8 p.

State Council, The, 2013b, List of items approval decentralization decision: Beijing, China, The State Council, issue no. 2013–19, May 15, 1 p.

Wang Tingting, 2013, 2012 iron and steel production and 2013 consumption analysis: China Steel, no. 3, p. 26–27.

Yang Yumei, Wu Weidong, and Xu Qing, 2012, Present situation and future development trend of strontium carbonate industry in China: Inorganic Chemical Industry, v. 44, no. 7, p. 1–5.

Zhang Ming-yan and Xu Sheng-ping, 2012, The distribution and utilization plan on Ge in coal bed: China Mining Magazine, v. 21, no. 9, p. 54–56.

Zhonghua Renmin Gongheguo Guowuyuan Gongbao, 2012a, Plan for development of the national strategic emerging industries during the period of the twelfth five-year plan: Beijing, China, Zhonghua Renmin Gongheguo Guowuyuan Gongbao, issue no. 21, p. 11.

Zhonghua Renmin Gongheguo Guowuyuan Gongbao, 2012b, Plan for energy conservation and emissions reduction during the period of the twelfth five-year plan: Beijing, China, Zhonghua Renmin Gongheguo Guowuyuan Gongbao, issue no. 25, p. 5.

TABLE 1
CHINA: ESTIMATED PRODUCTION OF MINERAL COMMODITIES[1, 2]

(Metric tons unless otherwise specified)

Commodity[3]		2008	2009	2010	2011	2012
METALS						
Aluminum:						
Bauxite, gross weight	thousand metric tons	35,000	40,000	44,000	45,000	47,000
Alumina	do.	22,800	23,800	29,000	34,100	37,700
Metal, refined:						
Primary	do.	13,200	12,900	16,200	18,100	20,300
Secondary	do.	2,700	3,100	4,000	4,100	4,200
Total	do.	15,900	16,000	20,200	22,200	24,500
Antimony:						
Mine, Sb content		166,000	140,000	150,000	150,000	145,000
Metal		158,000	168,000	193,000	200,000	230,000
Bismuth:						
Mine output, Bi content		5,000	6,000	6,500	7,000 [r]	6,000
Metal		13,100	12,300	14,000	15,000 [r]	14,000
Cadmium, smelter		6,960	7,050	7,360	6,670 [r]	7,000
Chromite, gross weight	thousand metric tons	200	200	200	200	200
Cobalt:						
Mine output, Co content		6,630	6,000	6,380	6,800	6,800
Metal		6,700	6,000	4,120	5,430 [r]	5,500
Copper:						
Mine output, Cu content		1,070,000	1,040,000	1,160,000	1,270,000	1,550,000
Metal:						
Smelter, primary	thousand metric tons	2,500	2,700	2,900	3,030 [r]	3,200
Refined:						
Primary	do.	2,700	2,750	2,950	3,390	3,930
Secondary	do.	1,200	1,400	1,700	1,850 [r]	1,950
Total	do.	3,900	4,150	4,650	5,240 [r]	5,880
Germanium		100	95	100	110	105
Gold, mine output, Au content		285	320	345	362	403
Indium, primary and secondary		340	340	330	380	405
Iron and steel:						
Iron ore, gross weight	thousand metric tons	824,000	880,000	1,070,000	1,330,000	1,310,000
Pig iron[4]	do.	470,670	552,830	597,330	640,510	663,500
Ferroalloys	do.	18,300	22,100	24,300	28,400	31,300
Steel, crude[4]	do.	500,490	572,180	637,230	685,280	723,880
Steel, rolled[4]	do.	584,770	694,050	802,760	886,190	955,780
Lead:						
Mine output, Pb content	do.	1,550	1,600	1,980	2,400	2,800
Metal:						
Smelter, primary	do.	2,430	2,630	2,800	3,110 [r]	3,200
Refined:						
Primary	do.	2,350	2,630	2,800	3,200	3,300
Secondary	do.	850	1,150	1,360	1,400	1,400
Total	do.	3,200	3,780	4,160	4,600	4,700
Magnesium, metal and alloy		559,000	501,000	654,000	675,000	698,000
Manganese:						
Ore, Mn content	thousand metric tons	2,200	2,400	2,600	2,800	2,900
Metal		1,130,000	1,310,000	1,370,000	1,480,000 [r]	1,500,000
Mercury, mine output, Hg content		1,300	1,430	1,600	1,500	1,350
Molybdenum, mine output, Mo content		81,000	93,500	96,600	103,000	105,000
Nickel:						
Mine output, Ni content		79,500	84,800	80,000	90,000 [r]	93,300
Matte		114,000	143,000	139,000	166,000 [r]	170,000
Smelter		129,000	165,000	159,000	175,000	229,000

See footnotes at end of table.

TABLE 1—Continued
CHINA: ESTIMATED PRODUCTION OF MINERAL COMMODITIES[1,2]

(Metric tons unless otherwise specified)

Commodity[3]		2008	2009	2010	2011	2012
METALS—Continued						
Niobium and tantalum, mine output:						
Nb_2O_5 content		300	30	32	25 [r]	20
Ta_2O_5 content		900	90	86	75 [r]	70
Rhenium, Re content in NH_4ReO_4[5]	kilograms	1,900	1,900	2,000	2,100	2,200
Silicon, metal	thousand metric tons	1,100	993	1,140	1,350 [r]	1,300
Silver, mine output, Ag content		2,800	2,900	3,500	3,700	3,900
Tin:						
Mine output, Sn content		110,000	97,200	115,000	120,000	110,000
Metal		140,000	140,000	149,000	156,000	148,000
Titanium:						
Ilmenite, TiO_2 equivalent		550,000	550,000	700,000	850,000	800,000
Sponge		57,000	45,800	57,000	68,000	76,800
Tungsten, mine output, W content		50,000	51,000	59,000 [r]	61,800	64,000
Vanadium, V_2O_5 in vanadiferous slag product		46,000	52,000	58,000	65,000	70,000
Zinc:						
Mine output, Zn content	thousand metric tons	3,340	3,330	3,840	4,050	4,900
Refined:						
Primary	do.	4,000	4,200	5,030	5,040	4,720
Secondary	do.	37	90	175	173	170
Total	do.	4,040	4,290	5,210	5,210	4,890
INDUSTRIAL MINERALS						
Asbestos		380,000	440,000	400,000	440,000	420,000
Barite	thousand metric tons	4,600	3,400	4,000	4,100	4,200
Bentonite	do.	3,300	3,400	3,400	3,500	3,500
Boron, mine, B_2O_3 equivalent		140,000	145,000	150,000	150,000	160,000
Bromine		135,000	93,000	100,000	100,000	105,000
Cement, hydraulic[4]	million metric tons	1,400	1,644	1,882	2,099	2,210
Diatomite		440,000	440,000	400,000	440,000	420,000
Dolomite	thousand metric tons	8,000	8,100	8,200	8,200	8,300
Feldspar	do.	2,000	2,000	2,000	2,100	2,100
Fluorspar	do.	4,200	3,800	4,600	4,200	4,600
Graphite		650,000	450,000	700,000	800,000	820,000
Gypsum	thousand metric tons	4,600	4,500	4,700	4,800 [r]	4,900
Kaolin	do.	3,200	3,000	3,260	3,200	3,300
Lime	do.	180,000	185,000	190,000	200,000	220,000
Lithium, Li content, all types		4,500	5,500	6,000	7,200	9,500
Magnesite	thousand metric tons	15,600	13,000	14,000	19,000 [r]	16,000
Mica		750,000	700,000	750,000	760,000	770,000
Nitrogen, N content of ammonia[4]	thousand metric tons	41,140	42,290	40,870	43,250	45,520
Phosphate rock, P_2O_5 equivalent	do.	15,200	18,000	20,400	24,000	28,500
Potash, marketable, K_2O equivalent	do.	2,750	3,200	3,600	3,800	4,100
Rare earths, rare-earth oxide equivalent		125,000	129,000	120,000	105,000	100,000
Salt[4]	thousand metric tons	66,640	66,630	70,380	67,420	69,120
Sodium compounds:						
Mirabilite	do.	6,600	6,000	6,500	6,000	5,500
Soda ash, natural and synthetic[4]	do.	18,540	19,450	20,350	22,940	24,010
Strontium carbonate		335,000	159,000	150,000 [r]	145,000 [r]	140,000
Sulfur:						
Native	thousand metric tons	960	1,000	1,100	1,100	1,200
Content of pyrite	do.	4,300	4,370	4,400	5,300 [r]	5,400
Byproduct, all sources	do.	3,350	4,000	4,100	3,300 [r]	3,300
Total	do.	8,610	9,370	9,600	9,700	9,900
Talc and related materials	do.	2,200	2,300	2,000	2,200	2,200

See footnotes at end of table.

TABLE 1—Continued
CHINA: ESTIMATED PRODUCTION OF MINERAL COMMODITIES[1,2]

(Metric tons unless otherwise specified)

Commodity[3]		2008	2009	2010	2011	2012
MINERAL FUELS AND RELATED MATERIALS						
Coal:						
Anthracite	thousand metric tons	447,000	426,000	500,000 [r]	450,000	500,000
Bituminous	do.	2,110,000	2,320,000	2,420,000 [r]	2,800,000	2,830,000
Lignite	do.	196,000	256,000	320,000 [r]	270,000	330,000
Total	do.	2,750,000	3,000,000	3,240,000	3,520,000	3,660,000
Coke, all types[4]	do.	323,590	345,020	388,640	432,710	447,790
Gas, natural:						
Gross	billion cubic meters	80	85	95	102	107
Marketed	do.	68	73	83	90	95
Petroleum:						
Crude, including crude from oil shale	million 42-gallon barrels	1,380	1,370	1,480	1,480	1,510
Refinery products	do.	3,700	3,750	4,220	4,470	4,640

[r]Revised. do. Ditto.

[1]Estimated data are rounded to no more than three significant digits; may not add to totals shown.

[2]Table includes data available through July 29, 2013.

[3]In addition to the commodities listed, China also produces beryllium, diamond, gallium, iodine, platinum-group metals, selenium, stone, tellurium, uranium, and zirconium, but available information is inadequate to make reliable estimates of output.

[4]Reported by China's National Statistical Bureau.

[5]Includes rhenium from imported copper and molybdenum concentrates.

TABLE 2
CHINA: STRUCTURE OF THE MINERAL INDUSTRY IN 2012

(Thousand metric tons unless otherwise specified)

Commodity	Facilities, major operating companies, and major equity owners[1]	Location of main facilities[2]	Annual capacity[e]
Aluminum:			
Alumina	Chongqing Aluminum Co. [Aluminum Corporation of China (Chinalco)]	Chongqing	800
Do.	Chongqing Dingtai Tuoyuan Alumina Co.	do.	150
Do.	Nanchuan Pioneer Alumina Co.	do.	200
Do.	Guangxi Huayin Aluminum Co. Ltd.	Guangxi, Bose	2,000
Do.	Pingguo Aluminum Co. [Aluminum Corporation of China (Chinalco)]	Guangxi, Pingguo	1,200
Do.	Guizhou Aluminum Plant [Aluminum Corporation of China (Chinalco)]	Guizhou, Guiyang	1,200
Do.	Chalco Zunyi Aluminum Co. Ltd. [Aluminum Corporation of China (Chinalco)]	Guizhou, Zunyi	1,000
Do.	Luoyang Wanji Xiangjiang Aluminum Co. Ltd.	Henan, Luoyang	800
Do.	Sanmenxia Yixiang Aluminum Co. Ltd. (Henan Yima Coal Group)	Henan, Mianchi	600
Do.	Pingdingshan Huiyuan Chemical Co.	Henan, Pingdingshan	300
Do.	Yangquan Coalmine Aluminum (Sanmenxia) Co. Ltd.	Henan, Sanmenxia	1,200
Do.	Orient Hope (Sanmenxia) Aluminum Co. Ltd.	do.	1,200
Do.	Zhengzhou Aluminum Plant [Aluminum Corporation of China (Chinalco)]	Henan, Zhengzhou	2,200
Do.	Zhongzhou Aluminum Plant [Aluminum Corporation of China (Chinalco)]	Hunan, Zhongzhou	2,800
Do.	Shandong Huayu Alumina Co. Ltd. (Shandong Chiping Xinfa Aluminum and Electricity Group)	Shandong, Chiping	1,800
Do.	Longhou Donghai Alumina Co. Ltd. (Nanshan Group)	Shandong, Nanshan, Longkou	1,600
Do.	Shandong Aluminum Plant [Aluminum Corporation of China (Chinalco)]	Shandong, Zibo	2,000
Do.	Bingzhou Weiqiao Aluminum Co.	Shandong, Zouping	1,600
Do.	Shanxi Aluminum Plant [Aluminum Corporation of China (Chinalco)]	Shanxi, Hejin	2,700
Do.	Liulin Senze Group	Shanxi, Liulin	600
Do.	Coalmine Aluminum (Sanmenxia) Co. Ltd.	Shanxi, Sanmenxia	1,200
Do.	Shanxi Luneng Jinbei Aluminum Co. Ltd.	Shanxi, Yuanping	2,000
Do.	Wenshan Aluminum Co. Ltd. (Yunnan Aluminum Co.)	Yunnan, Wenshan	800
Metal	Baiyin Aluminum Plant	Gansu, Baiyin	150
Do.	Lanzhou Aluminum Plant	Gansu, Lanzhou	210
Do.	Liancheng Aluminum Plant	do.	235
Do.	Gansu Dongxi Aluminum Co. Ltd. (formerly Gansu Longxi Aluminum Plant)	Gansu, Longxi	360
Do.	Yinhai Aluminum Co. Ltd.	Guangxi, Laibin	125
Do.	Pingguo Aluminum Co. [Aluminum Corporation of China (Chinalco)]	Guangxi, Pingguo	380
Do.	Guizhou Aluminum Plant [Aluminum Corporation of China (Chinalco)]	Guizhou, Guiyang	400
Do.	Chalco Zunyi Aluminum Co. Ltd. [Aluminum Corporation of China (Chinalco)]	Guizhou, Zunyi	250
Do.	Henan Zhongfu Industry Co. Ltd.	Henan, Gongyi	180
Do.	Jiaozuo Wanfang Aluminum Co. Ltd.	Henan, Jiaozuo	420
Do.	Henan Wanji Aluminum Co. Ltd.	Henan, Luoyang	125
Do.	Henan Zhongmai Mianchi Aluminum Plant	Henan, Mianchi	400
Do.	Shangqiu Aluminum Smelter	Henan, Shangqiu	180
Do.	Yichuan Yugang Longquan Aluminum Co.	Henan, Yichuan	600
Do.	Shangqiu Shenhuo Foguang Aluminum Co. Ltd.	Henan, Yongcheng	280
Do.	Hanjiang Danjiangkou Aluminum Co. Ltd.	Hubei, Danjiangkou	110
Do.	Hunan Chuanquan Aluminum Co. Ltd.	Hunan, Taoyuan	210
Do.	Fushun Aluminum Plant [Aluminum Corporation of China (Chinalco)]	Liaoning, Fushun	340
Do.	Baotou Aluminum Plant	Nei Mongol, Baotou	250
Do.	Orient (East Hope) Aluminum Plant (Orient Group)	do.	800
Do.	Nei Mongol HMHJ Aluminum Electricity Co. Ltd.	Nei Mongol, Holin Gol	400
Do.	Qingtongxia Aluminum Plant (China Power Investment Corp. and Ningxia Qingtongxia Energy Group Co. Ltd.)	Ningxia, Qingtongxia	1,150
Do.	Qiaotou Aluminum Co. Electrolysis Branch	Qinghai, Datong	750
Do.	Qinghai Aluminum Smelter [Aluminum Corporation of China (Chinalco)]	Qinghai, Xining	560
Do.	Qinghai West Mining Baihe Aluminum Co. Ltd.	do.	112
Do.	Tongchuan Xingguang Aluminum Co. Ltd.	Shaanxi, Tongchuan	250
Do.	Shandong Chiping Xinfa Aluminum and Power Group	Shandong, Chiping	360
Do.	Taishan Aluminum-Power Co. Ltd.	Shandong, Fecheng	125

See footnotes at end of table.

TABLE 2—Continued
CHINA: STRUCTURE OF THE MINERAL INDUSTRY IN 2012

(Thousand metric tons unless otherwise specified)

Commodity		Facilities, major operating companies, and major equity owners[1]	Location of main facilities[2]	Annual capacity[e]
Aluminum—Continued:				
Metal—Continued		Shandong Nanshan Aluminum Co. Ltd. (Nanshan Group)	Shandong, Nanshan, Longkou	156
Do.		Shandong Aluminum Plant [Aluminum Corporation of China (Chinalco)]	Shandong, Zibo	120
Do.		Bingzhou Weiqiao Aluminum Co.	Shandong, Zouping	250
Do.		Zouping Aluminum Co. Ltd.	do.	150
Do.		Huaze Aluminum and Power Co. Ltd.	Shanxi, Hejin	400
Do.		New Orient Aluminum Co. Ltd.	Shanxi, Taiyuan	75
Do.		Chalco Shanxi Huasheng Aluminum Co. Ltd. [Aluminum Corporation of China (Chinalco)]	Shanxi, Yongji	220
Do.		Shanxi Guanlv Aluminum Co. Ltd.	Shanxi, Yuncheng	210
Do.		Qient (East Hope) Aluminum Plant (Orient Group)	Xinjiang, Changji Prefecture	540
Do.		Xinjiang Qiya Aluminum Co. Ltd.	do.	450
Do.		Xinjiang Nongliushi Aluminum Co. Ltd.	Xinjiang, Wujiaqu	1,200
Do.		Yunnan Aluminum Plant	Yunnan, Kunming	500
Antimony		Huaxi (China Tin) Group Industrial Co.	Guangxi, Hechi	25
Do.		Hunan Chenzhou Mining Group Co. Ltd.	Hunan, Yuanling	20
Do.		Hsikuangshan Twinkling Star Antimony Co. Ltd. (China Minmetals Group)	Hunan, Lengshuijiang	40
Asbestos		China National Nonmetallic Industry Corp.	Nei Mongol, Baotou; Shanxi, Lai Yuan, and Lu Liang	130
Barite		do.	Guizhou, Xiangshou	NA
Bismuth	metric tons	Guangzhou Smelter	Guangdong, Guangzhou	300
Do.	do.	Hunan Bismuth Industry Co. Ltd.	Hunan, Chouzhou	3,500
Do.	do.	Shizhuyuan Nonferrous Metals Co. Ltd.	Hunan, Shizhuyuan	1,200
Do.	do.	Zhuzhou Smelter (Zhuye Torch Metals Co. Ltd.)	Hunan, Zhuzhou	350
Do.	do.	Yunnan Copper Group Co. Ltd.	Nei Mongol, Chifeng	300
Do.	do.	Yunnan Chihong Zinc and Germanium Co. Ltd.	Yunnan, Qujing	300
Cadmium	do.	Zhuzhou Smelter (Zhuye Torch Metals Co. Ltd.)	Hunan, Zhuzhou	1,000
Do.	do.	Yunnan Chihong Zinc and Germanium Co. Ltd.	Yunnan, Qujing	800
Coal		Hebei Provincial Government	Hebei	70,000
Do.		Heilongjiang Provincial Government	Heilongjiang	100,000
Do.		Henan Provincial Government	Henan	100,000
Do.		Liaoning Provincial Government	Liaoning	70,000
Do.		Nei Mongol Provincial Government	Nei Mongol	90,000
Do.		Shandong Provincial Government	Shandong	60,000
Do.		Shanxi Provincial Government	Shanxi	400,000
Do.		Sichuan Provincial Government	Sichuan	80,000
Do.		Shenhua Coal Corp.	Nei Mongol, Ningxia, and Shaanxi	150,000
Cobalt	metric tons	Jinchuan Nonferrous Metals Corp.	Gansu, Jinchang	10,000
Do.	do.	Huayou Cobalt Co. Ltd.	Zhejiang, Tongxiang	3,000
Copper, refined		Jinchang Smelter (Tongling Nonferrous Metals Group Holding Co. Ltd.)	Anhui, Tongling	170
Do.		Jinlong Smelter (Tongling Nonferrous Metals Group Holding Co. Ltd.)	do.	400
Do.		Wuhu Smelter (Hengxin Copper Industry Group Co.)	Anhui, Wuhu	60
Do.		Zijin Copper Co. Ltd.	Fujian, Shanghang	200
Do.		Baiyin Nonferrous Metals Group Co. Ltd.	Gansu, Baiyin	100
Do.		Jinchuan Nonferrous Metals Corp.	Gansu, Jinchuan	600
Do.		Luoyang Copper Processing Factory	Henan, Luoyang	50
Do.		Daye Nonferrous Metals Co.	Hubei, Daye	400
Do.		Zhangjiagang United Copper Co. (Tongling Nonferrous Metals Group Holding Co. Ltd.)	Jiangsu, Zhangjiagang	200
Do.		Guixi Smelter (Jiangxi Copper Co. Ltd.)	Jiangxi, Guixi	1,200
Do.		Dongfang Copper Co. (Huludao Nonferrous Metals Group)	Liaoning, Huludao	100

See footnotes at end of table.

TABLE 2—Continued
CHINA: STRUCTURE OF THE MINERAL INDUSTRY IN 2012

(Thousand metric tons unless otherwise specified)

Commodity		Facilities, major operating companies, and major equity owners[1]	Location of main facilities[2]	Annual capacity[e]
Copper, refined—Continued		Chifeng Fubang Copper Co. Ltd.	Nei Mongol, Chifeng	100
Do.		Chifeng Jingeng Copper Co. Ltd.	Nei Mongol, Chifeng, Harqin Banner	100
Do.		Shandong Dongying Fangyuan Nonferrous Metals Co. Ltd.	Shandong, Dongying	400
Do.		Shandong Jinsheng Nonferrous Metals Corp.	Shandong, Linyi	100
Do.		Yanggu Xiangguang Copper Co. Ltd. (Shandong Fengxiang Group)	Shandong, Liaocheng, Yanggu	600
Do.		Yantai Penghui Copper Industry Co. Ltd.	Shandong, Yantai	200
Do.		Taiyuan Copper Industry Co.	Shanxi, Taiyuan	100
Do.		Zhongtiaoshan Nonferrous Metals Co.	Shanxi, Yuangu	100
Do.		Huili Kunpeng Co. Ltd.	Sichuan, Huili	100
Do.		Tianjin Datong Copper Co. Ltd. (formerly Tianjin Copper Electrolysis Factory)	Tianjin	200
Do.		Yunnan Smelter (Chinalco Yunnan Copper Group Co. Ltd.)	Yunnan, Kunming	250
Do.		Hangzhou Fuchunjiang Smelting Co. Ltd.	Zhejiang, Fuchunjiang	100
Gallium	metric tons	Chalco Zunyi Aluminum Co. Ltd. [Aluminum Corporation of China (Chinalco)]	Guizhou, Zunyi	40
Do.	do.	Shandong Aluminum Plant	Shandong, Zibo	20
Gas, natural	billion cubic meters	China National Petroleum Corp.	Sichuan	10
Germanium	metric tons	Shaoguan Smelter (Shenzhen Nonfemet Co.)	Guangdong, Shaoquan	30
Do.	do.	Nanjing Germanium Co. Ltd.	Jiangsu, Nanjing	30
Do.	do.	Nei Mongol Xilingol Tongtai Germanium Refining Co. Ltd.	Nei Mongol, Xilinhot	20
Do.	do.	Shanghai Lontai Copper Co. Ltd.	Shanghai	10
Do.	do.	Yunnan Lincang Xinyuan Germanium Industrial Co. Ltd.	Yunnan, Lincang	50
Do.	do.	Yunnan Chihong Zinc and Germanium Industrial Co. Ltd.	Yunnan, Qujing	50
Gold, refined	do.	Zijin Copper Co. Ltd.	Fujian, Shanghang	5
Do.	do.	China National Gold Corp.	Henan, Lingbao	10
Do.	do.	Zhongyan Gold Smelter (Zhongjin Gold Co. Ltd.)	Henan, Sanmenxia	30
Do.	do.	Jiangxi Copper Co. Ltd.	Jiangxi, Guixi	20
Do.	do.	Laizhou Gold Co.	Shandong, Laizhou	15
Do.	do.	Yanggu Xiangguang Copper Co. Ltd. (Shandong Fengxiang Group)	Shandong, Liaocheng, Yanggu	20
Do.	do.	Shandong Yanggu Xiangguang Co. Ltd.	Shandong, Yanggu	20
Do.	do.	Yantai Penghui Copper Industry Co. Ltd.	Shandong, Yantai	5
Do.	do.	Zhaoyuan Gold Co.	Shandong, Zhaoyuan	15
Do.	do.	Great Wall Gold Silver Refinery	Sichuan, Chengdu	100
Do.	do.	Yunnan Chihong Zinc and Germanium Co. Ltd.	Yunnan, Qujing	130
Graphite		Jixi Aoyu Graphite Co. Ltd.	Heilongjiang, Jixi and Luo	60
Do.		Nei Mongol Xinghe Jingxin Graphite Co. Ltd.	Nei Mongol, Xinghe	10
Indium	metric tons	Shaoguan Smelter (Shenzhen Nonfemet Co.)	Guangdong, Shaoquan	25
Do.	do.	Guangxi Tanghan Zinc & Indium Co. Ltd.	Guangxi, Hechi	30
Do.	do.	Laibin Smelter [Liuzhou Huaxi (China Tin) Group Co.]	Guangxi, Laibin	50
Do.	do.	Guangxi Debang Technology Co. Ltd.	Guangxi, Liuzhou	85
Do.	do.	Liuzhou Zinc Products Co.	do.	20
Do.	do.	Yintai Technology Co. Ltd.	do.	40
Do.	do.	Yuguang Gold-Lead Co. Ltd.	Henan, Jiyuan	10
Do.	do.	Hsikuangshan Twinkling Star Antimony Co. Ltd. (China Minmetals Group)	Hunan, Lengshuijiang	7
Do.	do.	Xiangtan Zhengtan Nonferrous Metal Co. Ltd.	Hunan, Xiangtan	75
Do.	do.	Zhuzhou Smelter	Hunan, Zhuzhou	60
Do.	do.	Nanjing Germanium Co. Ltd.	Jiangsu, Nanjing	150
Do.	do.	Nanjing Sanyou Electronic Material Co. Ltd.	do.	50
Do.	do.	Huludao Nonferrous Metals Group Co.	Liaoning, Huludao	50
Do.	do.	Yunnan Chengfeng Nonferrous Metals Co. Ltd.	Yunnan, Gejiu	10
Do.	do.	Yunnan Mengzi Mining and Smelting Co. Ltd.	Yunnan, Honghe	30

See footnotes at end of table.

TABLE 2—Continued
CHINA: STRUCTURE OF THE MINERAL INDUSTRY IN 2012

(Thousand metric tons unless otherwise specified)

Commodity	Facilities, major operating companies, and major equity owners[1]	Location of main facilities[2]	Annual capacity[e]
Iron and steel:			
Iron ore	Ma'anshan Iron and Steel Co.	Anhui, Maanshan	10,000
Do.	Shoudu (Capital) Mining Co.	Beijing	20,000
Do.	Jiuquan Iron and Steel Co.	Gansu, Jiayuguan	4,000
Do.	Hainan Iron Mine	Hainan, Changjiang	4,600
Do.	Handan Xingtai Metallurgical Bureau (Hebei Iron and Steel Group Co.)	Hebei, Handan	3,800
Do.	Tangshan Iron and Steel Co. (Hebei Iron and Steel Group Co.)	Hebei, Tangshan	3,000
Do.	Wuhan Iron and Steel (Group) Co. (Wugang)	Hubei, Wuhan	5,100
Do.	Meishan Metallurgical Co.	Jiangsu, Nanjing	2,000
Do.	Banshigou Iron Mine Mining Co.	Jilin, Hunjiang	1,400
Do.	Anshan Mining Co.	Liaoning, Anshan	30,000
Do.	Benxi Iron and Steel Co.	Liaoning, Benxi	13,700
Do.	Baotou Iron and Steel and Rare Earth Co.	Nei Mongol, Baotou	10,000
Do.	Taiyuan Iron and Steel Co.	Shanxi, Taiyuan	4,000
Do.	Dabaoshan Mining Co.	Guangdong, Qujiang	1,670
Do.	Panzhihua Mining Co.	Sichuan, Panzhihua	13,000
Do.	Kunming Iron and Steel Co.	Yunnan, Kunming	1,400
Ferroalloys	Shoudu (Capital) Iron and Steel (Group) Co.	Beijing	35
Do.	Qingshan Holding Group Co. Ltd.	Fujian, Fu'an	300
Do.	Desheng Nickel Industry Co. Ltd.	Fujian, Luoyuanwan	920
Do.	Northwest Ferroalloy Co.	Gansu, Yongdeng	60
Do.	Zunyi Ferroalloy Co.	Guizhou, Zunhi	100
Do.	Zhejiang Huaguang Smelting Group	Jiangxi, Hengfeng	50
Do.	Jilin Ferroalloy Co.	Jilin, Jilin	250
Do.	Jinzhou Ferroalloy Co.	Liaoning, Jinzhou	90
Do.	Liaoyang Ferroalloy Co.	Liaoning, Liaoyang	70
Do.	Shanghai Iron and Steel Co. Ltd.	Shanghai	180
Do.	Emei Ferroalloy Co.	Sichuan, Emei	70
Do.	Hengshan Ferroalloy Co.	Zhejiang, Jiande	70
Crude steel	Ma'anshan Iron and Steel Co.	Anhui, Maanshan	10,000
Do.	Liuzhou Iron and Steel Group	Guangxi, Liuzhou	6,000
Do.	Shougang-Tangshan Iron and Steel Group Co. Ltd.	Hebei, Caofeidian	10,000
Do.	Handan Iron and Steel General Work (Hebei Iron and Steel Group Co.)	Hebei, Handan	12,000
Do.	Shougang Qianan Iron and Steel Co. Ltd. (Shougang)	Hebei, Qianan	7,800
Do.	Tangshan Iron and Steel Co. (Taigang) (Hebei Iron and Steel Group Co.)	Hebei, Tangshan	15,000
Do.	Wuhan Iron and Steel (Group) Co. (Wugang)	Hubei, Wuhan	12,000
Do.	Shagang Group Co. Ltd.	Jiangsu, Zhangjiagang	30,000
Do.	Anshan Iron and Steel (Group) Co. (Angang) (Anben Iron and Steel Group)	Liaoning, Anshan	16,000
Do.	Benxi Iron and Steel Co. (Bengang) (Anben Iron and Steel Group)	Liaoning, Benxi	6,000
Do.	Anshan Iron and Steel (Group) Co. (Angang) (Anben Iron and Steel Group)	Liaoning, Yingkou, Bayuquan	6,500
Do.	Baotou Iron and Steel and Rare Earth Co. (Baogang Group)	Nei Mongol, Baotou	10,000
Do.	Baoshan Iron and Steel (Group) Corp. (Baosteel) [Baogang Group]	Shanghai	19,000
Do.	Shanghai Iron and Steel Co. Ltd.	do.	6,000
Do.	Shandong Jinan Iron and Steel Group Co. (Shandong Iron and Steel Group)	Shandong, Jinan	10,000
Do.	Shandong Laiwu Iron and Steel Group Co. (Shandong Iron and Steel Group)	Shandong, Laiwu	10,000
Do.	Taiyuan Iron and Steel Co. (Taigang)	Shanxi, Taiyuan	5,000
Do.	Panzhihua Iron and Steel (Group) Co. (Pangang)	Sichuan, Panzhihua	6,000
Do.	Xinjiang Biyi Iron and Steel Group (Baogang Group)	Xinjiang, Urumqi	6,000

See footnotes at end of table.

TABLE 2—Continued
CHINA: STRUCTURE OF THE MINERAL INDUSTRY IN 2012

(Thousand metric tons unless otherwise specified)

Commodity	Facilities, major operating companies, and major equity owners[1]	Location of main facilities[2]	Annual capacity[e]
Lead	Jiuhua Smelter (Tongling Nonferrous Metals Group Holding Co. Ltd.)	Anhui, Chizhou	80
Do.	Baiyin Nonferrous Metals Co. Ltd.	Gansu, Baiyin	80
Do.	Shaoguan Smelter (Shenzhen Nonfemet Co.)	Guangdong, Shaoquan	100
Do.	Laibin Smelter [Huaxi (China Tin) Group Co.]	Guangxi, Laibin	100
Do.	Hechi Nanfang Nonferrous Metals Smelting Co. Ltd.	Guangxi, Hechi	80
Do.	Anyang Smelter (Yubei Metal Co.)	Henan, Anyang	160
Do.	Jiyuan Wangyang Smelter (Jiquan Wangyang Smeltery Group Co. Ltd.)	Henan, Jiaozuo	160
Do.	Jinli Smelter (Jiyuan Jinli Smelting Co.)	Henan, Jiyuan	300
Do.	Jiyuan Smelter (Yuguang Gold-Lead Co. Ltd.)	do.	350
Do.	Henan Lingye Co. Ltd.	Henan, Lingbao	100
Do.	Hanjiang Smelter	Hubei, Luhekou	50
Do.	Shuikoushan Nonferrous Metals Co. Ltd.	Hunan, Hengyang	100
Do.	Zhuzhou Smelter (Zhuye Torch Metals Co. Ltd.)	Hunan, Zhuzhou	100
Do.	Xuzhou Chunxing Alloy Co. Ltd.	Jiangsu, Xuzhou	150
Do.	Jiangxi Jinde Lead Co. Ltd.	Jiangxi, Shangrao	80
Do.	Huludao Nonferrous Metals Group Co. Ltd.	Liaoning, Huludao	30
Do.	Shaanxi Dongling Group	Shaaxi, Baoji	100
Do.	Yunnan Tin Co. Ltd. (Yunnan Tin Corp.)	Yunnan, Gejiu	100
Do.	Kunming Smelter	Yunnan, Kunming	100
Do.	Yunnan Chihong Zinc and Germanium Co. Ltd.	Yunnan, Qujing	100
Lithium, $LiCO_3$	Tibet Mineral Development Co. Ltd.	Gansu, Baiyin	5
Do.	Jiangxi Ganfeng Lithium Co. Ltd.	Jiangxi, Xinyu	3
Do.	Sichuan Ni/Co Guorun New Material Co. Ltd.	Sichuan, Pengshan	2
Do.	Sichuan Shehong Lithium Co. Ltd.	Sichuan, Shehong	2
Do.	Sichuan Tianqi Lithium Industry Co. Ltd. (Chengdu Tianqi Group Co. Ltd.)	Sichuan, Suining	10
Do.	Sichuan Aba Guangsheng Lithium Industrial Co. Ltd.	Sichuan, Wenchuan	2
Do.	Qinghai Yanhu Industry Group Co. Ltd.	Qinghai, Golmud	10
Do.	Qinghai CITIC Guoan Technology Development Co. Ltd.	do.	20
Do.	Qinghai Lithium Industry Co. Ltd.	Qinghai, Xining	20
Do.	Xinjiang Haoxin Lithium Salt Development Co. Ltd. (former Xinjiang Lithium Co.)	Xinjiang, Urumqi	5
Magnesium	Zunyi Titanium Co. Ltd.	Guizhou, Zunyi	24
Do.	Ningxia Huayuan Magnesium Group	Ningxia, Yinchuan	15
Do.	Huayu Enterprises (Group) Ltd.	Shanxi, Jishan	35
Do.	Taiyuan Tongxiang Magnesium Metal Co. Ltd.	Shanxi, Taiyuan	45
Do.	Taiyuan Yiwei Magnesium Co. Ltd.	do.	21
Do.	Wenxi Biyun Magnesium Co. Ltd.	Shanxi, Wenxi	30
Do.	Wenxi Yinguang Magnesium Group	do.	40
Manganese, metal	Chongqing Tycoon Manganese Co. Ltd.	Chongqing	23
Do.	Guangxi Dameng Manganese Industry Co. Ltd.	Guangxi, Nanning	70
Molybdenum, concentrate	Luoyang Luanchuan Molybdenum Industry Group Co., Ltd.	Henan, Luanchuan	30
Do.	Jinduicheng Molybdenum Industry Group Co. Ltd.	Shaanxi, Huaxian	30

See footnotes at end of table.

TABLE 2—Continued
CHINA: STRUCTURE OF THE MINERAL INDUSTRY IN 2012

(Thousand metric tons unless otherwise specified)

Commodity	Facilities, major operating companies, and major equity owners[1]	Location of main facilities[2]	Annual capacity[e]
Nickel, refined	Jinchuan Nonferrous Metals Corp.	Gansu, Jinchuan	130
Do.	Guangxi Yinyi Science and Technic Mine	Guangxi, Yulin, Bohai	10
Do.	Guangxi Yulin Weinie Co. Ltd.	Guangxi, Bobai	18
Do.	Jiangxi Jiangli Science and Technology Co. Ltd.	Jiangxi, Fenyi	50
Do.	Jilin Jien Nickel Industry Co. Ltd.	Jilin, Panshi	10
Do.	Inco New Nickel Materials (Dalian) Co. Ltd.	Liaoning, Dalian	32
Do.	Schaanxi Huaze Nickel and Cobalt Metal Co. Ltd.	Shaanxi, Xian	5
Do.	Chengdu Electro-Metallurgy Factory	Sichuan, Chengdu	5
Do.	Huili Kunpeng Co. Ltd.	Sichuan, Huili	10
Do.	Sichuan Ni/Co Guorun New Material Co. Ltd.	Sichuan, Pengshan	10
Do.	Xinjiang Fukang Smelter	Xinjiang, Fukang	15
Do.	Xinjiang Xinxin Mining Co. Ltd.	Xinjiang, Fuyun	7
Do.	Yuanjiang Nickel Industry Co. Ltd.	Yunnan, Yuxi	5
Palladium and platinum kilograms	Jinchuan Nonferrous Metals Corp.	Gansu, Jinchang	3,500
Petroleum, crude	Shengli Bureau	Hebei, Shengli	33,500
Do.	Daqing Bureau	Heilongjiang, Daqing	55,000
Do.	Liaohe Bureau	Liaoning, Liaohe	15,000
Do.	Bohai Offshore Oil Corp.	Bohai	4,000
Do.	Nanhai East Corp.	Nanhai	5,000
Potash	Qinghai Yanhu Industry Group Co. Ltd.	Qinghai, Charhan	2,000
Do.	Xinjiang Lop Nur Potassic Salt Scientific and Technology Development Co.	Xinjiang, Ruoqiang	1,200
Rare earths	Fujian Changting Jinlong Rare Earth Co. Ltd.	Fujian, Changting	4
Do.	Gansu Rare Earths Co.	Gansu, Baiyin	32
Do.	Zhujiang Smelter	Guangdong, Guangzhou	5
Do.	Jiangyin Jiahua Advanced Material Resources Co. Ltd. (Neo Material Technologies)	Jiangsu, Jiangyin	3
Do.	Liyang Rhodia Rare Earth New Material Co. Ltd. (Rhodia Group)	Jiangsu, Liyang	12
Do.	Jiangsu Guosheng Rare Earth Co. Ltd.	Jiangsu, Taixing	5
Do.	Yixing Xinwei Leeshing Rare Earth Co. Ltd. (China Rare Earth Holdings Ltd.)	Jiangsu, Yixing	6
Do.	Dingnan Nanfang Rare Earth Co. Ltd.	Jiangxi, Ganzhou, Dingnan	4
Do.	Longnan Guangdong Rising Rare Earth Smelting Co. Ltd.	Jiangxi, Ganzhou, Longnan	4
Do.	Baotou Iron and Steel and Rare Earths Corp. (Baogang Group)	Nei Mongol, Baotou	55
Do.	Leshan Primet (Puruimei) New Materials Co. Ltd. (US Primet LLC)	Sichuan, Leshan	8
Do.	Sichuan Jiangxi Copper Rare Earth Co. Ltd. (Jiangxi Copper Co. Ltd.)	Sichuan, Mianning	18
Rhenium, rhenate kilograms	Guixi Smelter (Jiangxi Copper Co. Ltd.)	Jiangxi, Guixi	3,000
Do. do.	Western Xinxing Metal Materials Co. Ltd.	Shaanxi, Luonan	200
Salt	Shandong Haihua Group Co. Ltd.	Shandong, Weifang	1,400
Do.	Zigong Zhangjiaba Salt Chemical Plant	Sichuan, Zigong	250
Selenium metric tons	Jinchuan Nonferrous Metals Corp.	Gansu, Jinchang	50
Do. do.	Guixi Smelter (Jiangxi Copper Co. Ltd.)	Jiangxi, Guixi	300

See footnotes at end of table.

TABLE 2—Continued
CHINA: STRUCTURE OF THE MINERAL INDUSTRY IN 2012

(Thousand metric tons unless otherwise specified)

Commodity		Facilities, major operating companies, and major equity owners[1]	Location of main facilities[2]	Annual capacity[e]
Silver	metric tons	Zijin Copper Co. Ltd.	Fujian, Shanghang	125
Do.	do.	Jinchuan Nonferrous Metals Corp.	Gansu, Jinchang	150
Do.	do.	Laibin Smelter [Huaxi (China Tin) Group Co.]	Guangxi, Laibin	80
Do.	do.	Daye Nonferrous Metals Co.	Hubei, Daye	300
Do.	do.	Jiyuan Smelter (Yuguang Gold-Lead Co. Ltd.)	Henan, Jiyuan	730
Do.	do.	Jiangxi Copper Co. Ltd.	Jiangxi, Guixi	430
Do.	do.	Huludao Nonferrous Metals Group Co. Ltd.	Liaoning, Huludao	80
Do.	do.	Yanggu Xiangguang Copper Co. Ltd. (Shandong Fengxiang Group)	Shandong, Liaocheng, Yanggu	600
Do.	do.	Yantai Penghui Copper Industry Co. Ltd.	Shandong, Yantai	80
Do.	do.	Great Wall Gold Silver Refinery	Sichuan, Chengdu	300
Do.	do.	Yunnan Chengfeng Nonferrous Metals Co. Ltd.	Yunnan, Gejiu	150
Do.	do.	Yunnan Tin Co. Ltd. (Yunnan Tin Corp.)	do.	160
Do.	do.	Yunnan Smelter (Yunnan Copper Group Co. Ltd.)	Yunnan, Kunming	450
Do.	do.	Yunnan Chihong Zinc and Germanium Co. Ltd.	Yunnan, Qujing	150
Strontium, carbonate		Chongqing Chonglong Strontium Co. Ltd.	Chongqing	20
Do.		Chongqing Tongliang Redbutterfly Strontium Co.	do.	40
Do.		Shijiazhuang Zhengding Xian Jinshi Chemical Co. Ltd	Hebei, Shijiazhuang	3
Do.		Hebei Xinji Chemical Group	Hebei, Xinji	2
Do.		Nanjing Jinyan Strontium Co. Ltd.	Jiangsu, Lishui	2
Talc		China National Nonmetallic Industry Corp.	Guangxi, Longshen	130
Do.		do.	Liaoning, Haicheng	50
Do.		do.	Shandong, Qixia	5
Tellurium, concentrate	metric tons	Jiangxi Copper Co. Ltd.	Jiangxi, Guixi	50
Tin, smelter		Guihuacheng Smelter (Guangxi Pinggui PGMA Co. Ltd.	Guangxi, Hezhou	8
Do.		Laibin Smelter (Guangxi China Tin Group Co. Ltd.)	Guangxi, Laibin	25
Do.		Chenzhou Smelter (Yunnan Tin Co. Ltd.)	Hunan, Chenzhou	20
Do.		Nanshan Tin Co. Ltd.	Jiangxi, Nankang	10
Do.		Yunnan Chengfeng Nonferrous Metals Co. Ltd.	Yunnan, Gejiu	20
Do.		Yunnan Tin Co. Ltd. (Yunnan Tin Corp.)	do.	70
Do.		Yunnan Gejiu Zili Metallurgy Co. Ltd.	Yunnan, Huogudu	20
Titanium, sponge		Jinchuan Nonferrous Metals Corp.	Gansu, Jinchuan	15
Do.		Guizhou Southwest Titanium Co. Ltd.	Guizhou, Guiyang	3
Do.		Zunbao Titanium Co. Ltd.	Guizhou, Tongzi	10
Do.		Zunyi Titanium Co. Ltd.	Guizhou, Zunyi	20
Do.		Tangshan Tianhe Titanium Co. Ltd.	Hebei, Tangshan	10
Do.		Luoyang Sun Rui Wanji Titanium Industry Co. Ltd.	Henan, Xinan	10
Do.		Chaoyang Baisheng Zirconium Co. Ltd.	Liaoning, Chaoyang	8
Do.		Chaoyang Jintai Titanium Co. Ltd.	do.	7
Do.		Fushun Titanium Co. Ltd.	Liaoning, Fushun	5
Do.		Jinzhou Huashen Nonferrous Metals Plant	Liaoning, Jinzhou	10
Do.		Baoti Titanium Industry Co. Ltd.	Shaanxi, Baoji	10
Do.		Gangqi Xinyu Titanium Co. Ltd.	Sichuan, Panzhihua	5
Do.		Hengwei Titanium Co. Ltd.	do.	5
Do.		Panzhihua Iron and Steel (Group) Co. (Pangang)	do.	15
Do.		Yunnan Metallurgical Group	Yunnan, Lufeng	10
Tungsten, concentrate		Ninghua Hangluoken Tungsten Mine (Amoi Tungsten Co. Ltd.)	Fujian, Ninghua	4
Do.		Shizhuyuan Nonferrous Metals Co.	Hunan, Chenzhou	5
Do.		Yaogangxian Tungsten Mine	Hunan, Yizhang	3
Do.		Jiangxi Tungsten and Rare Earth Co. Ltd.	Jiangxi, Gangzhou	15

See footnotes at end of table.

TABLE 2—Continued
CHINA: STRUCTURE OF THE MINERAL INDUSTRY IN 2012

(Thousand metric tons unless otherwise specified)

Commodity	Facilities, major operating companies, and major equity owners[1]	Location of main facilities[2]	Annual capacity[e]
Zinc	Northwest China Lead-Zinc Smelter (Baiyin Nonferrous Metals Co. Ltd.)	Gansu, Baiyin	150
Do.	Shaoguan Smelter (Shenzhen Nonfemet Co.)	Guangdong, Shaoquan	270
Do.	Hechi Nanfang Nonferrous Metal Smelting Co. Ltd.	Guangxi, Hechi	200
Do.	Liuzhou Nonferrous Metal Smelting Co. Ltd. (former Liuzhou Zinc Products Factory)	Guangxi, Liuzhou	100
Do.	Yugang Gold-Lead Co. Ltd.	Henan, Jiyuan	300
Do.	Shuikoushan Nonferrous Metals Co. Ltd.	Hunan, Hengyan	60
Do.	Hsikuangshan Twinkling Star Antimony Co. Ltd. (China Minmetals Group)	Hunan, Lengshuijiang	40
Do.	Zhuzhou Smelter (Zhuye Torch Metals Co. Ltd.)	Hunan, Zhuzhou	500
Do.	Huludao Zinc Smelting Co. (Huludao Nonferrous Metals Group. Co. Ltd.)	Liaoning, Huludao	390
Do.	Zijin Bayannur Co. Ltd. (Zijin Mining Group)	Nei Mongol, Bayannar League	220
Do.	Chifeng NFC Kumba Hongye Zinc Co. Ltd. (China Nonferrous Metals Mining Group Co. Ltd.)	Nei Mongol, Chifeng	230
Do.	Xingan Copper and Zinc Smelter	Nei Mongol, Xilinuole	100
Do.	Dongling Zinc Industry Co. Ltd. (Dongling Group)	Shaanxi, Baoji	250
Do.	Laibin Smelter (Guangxi China Tin Group Co. Ltd.)	Yunnan, Laibin	60
Do.	Yunnan Jinding Zinc Co. Ltd. (Sichuan Hongda Group)	Yunnan, Lanping	120
Do.	Yunnan Chihong Zinc and Germanium Co. Ltd.	Yunnan, Qujing	280

[e]Estimated; estimated data are rounded to no more than three significant digits. Do., do. Ditto. NA Not available.
[1]Most companies are owned by either the central Government or a Provincial government.
[2]Listed by Province or Autonomous Region, followed by locality.

TABLE 3
CHINA: EXPORTS OF SELECTED MINERAL COMMODITIES IN 2012

Commodity	Quantity (metric tons)	Value (thousands)
METALS		
Aluminum:		
Alumina	43,293	$39,414
Metal and alloys:		
Unwrought	631,214	1,467,688
Semimanufactures	2,830,000	9,798,654
Antimony:		
Metal, unwrought	9,583	118,324
Oxide	40,598	430,144
Barium sulfate	2,890,000	241,911
Copper, metal and alloys:		
Unwrought	274,180	2,284,400
Semimanufactures	493,049	4,223,854
Iron and steel:		
Pig iron and cast iron	300,000	145,053
Steel:		
Bars and rods	11,760,000	8,193,287
Shapes and sections	3,480,000	2,364,937
Sheets and plates	26,970,000	22,339,222
Tube and pipe	1,560,000	4,195,360
Wire of steel or iron	1,800,000	2,071,884
Ferroalloys	640,000	2,067,012
Scrap	923	551
Manganese, unwrought	42,746	123,912
Molybdenum, ores and concentrates	11,499	205,521
Tin, metal and alloys, unwrought	1,738	43,610
Tungsten, tungstates	2,533	87,723
Zinc:		
Metal and alloys, unwrought	7,937	19,560
Oxide and peroxide	11,567	21,114
INDUSTRIAL MINERALS		
Barite	2,940,000	358,073
Cement	12,000,000	683,627
Fluorspar	430,000	156,744
Granite	7,450,000	3,165,931
Graphite, natural	260,000	287,568
Magnesia, fused	2,130,000	611,823
Rare-earth products	16,265	905,999
Talc	750,000	189,484
MINERAL FUELS AND RELATED MATERIALS		
Coal	9,260,000	1,585,995
Coke, semicoke	1,020,000	445,059
Petroleum:		
Crude oil	2,430,000	2,226,025
Refinery products	24,290,000	21,328,891

Source: General Administration of Customs of the People's Republic of China, 2012, China monthly exports and imports, no. 12.

TABLE 4
CHINA: IMPORTS OF SELECTED MINERAL COMMODITIES IN 2012

(Metric tons unless otherwise specified)

Commodity		Quantity	Value (thousands)
METALS			
Aluminum:			
Bauxite		39,637,831	$1,887,890
Alumina		5,020,000	1,818,301
Metal and alloys, unwrought		639,812	1,400,385
Semimanufactures		531,133	3,389,576
Scrap		2,590,000	4,127,913
Chromium, chromite		9,290,000	2,033,813
Cobalt:			
Ore and concentrates		166,491	347,499
Unwrought and powder		9,979	128,909
Copper:			
Ore and concentrates		7,830,000	16,908,979
Metal and alloys, unwrought		3,979,074	31,843,552
Semimanufactures		668,552	6,743,245
Scrap		4,860,000	14,862,098
Iron and steel:			
Iron ore		743,550,000	95,605,352
Steel:			
Bars and rods		890,000	1,519,660
Seamless pipe		430,000	1,735,102
Shapes and sections		370,000	411,772
Sheets and plates		11,660,000	12,854,279
Scrap		4,970,000	3,090,323
Manganese ore		12,370,000	2,185,360
Nickel:			
Ore and concentrates		64,999,859	5,261,148
Metal, refined greater than 99.95% Ni		10,888	194,627
Metal, other refined		146,695	2,596,966
Titanium dioxide		163,440	545,191
INDUSTRIAL MINERALS			
Diamond	kilograms	3,274	5,756,262
Nitrogen, phosphorus, and potassium fertilizers:			
Compound fertilizers		1,320,000	753,927
Diammonium phosphate		160,000	102,950
Potassium chloride		6,340,000	2,918,533
Potassium sulfate		170,000	85,593
Urea		170,978	71,349
MINERAL FUELS AND RELATED MATERIALS			
Coal		288,510,000	28,706,582
Liquefied natural gas		14,680,000	8,222,584
Petroleum:			
Crude oil		271,020,000	220,665,916
Refinery products		39,820,000	32,992,995

Source: General Administration of Customs of the People's Republic of China, 2012, China monthly exports and imports, no. 12.

The Mineral Industry of Fiji

By Lin Shi

In 2012, gold continued to be the mainstay of the mineral industry of Fiji, although gold production was only of minor significance to the country's economy, accounting for about 1% of the country's gross domestic product (GDP). Fiji's GDP in 2012 was valued at about $3.9 billion, and the average annual inflation rate as measured by the consumer price index was about 4.3% (World Bank, The, 2013).

Fiji's mineral industry also included a cement plant and several quarries that produced coral and river sands, limestone, and stone and crushed gravel. Although gold had been produced in Fiji since the first gold discovery in the 1930s, the only significant mine was the Vatukoula Mine, which is located in the Nakauvadra Mountains on the main island of Viti Levu about 100 kilometers (km) northwest of the capital, Suva, and 8 km inland from the coast (table 2).

Government Policies and Programs

The Mineral Resources Department (MRD) is Fiji's national geological survey and mining organization. The MRD develops mining policies, provides geologic information, assists mining investors, and facilitates the exploration and development of mineral and petroleum resources in the country (Mineral Resources Department, 2012).

Production

In 2012, gold production in Fiji increased by 1.9% to 1,653 kilograms (kg) (reported as 53,152 troy ounces) from 1,622 kg (reported as 52,157 troy ounces) in 2011. This increase was a result of continued investment in the Vatukoula Mine's infrastructure, which allowed the mine's owner, Vatukoula Gold Mines plc., to access the high-grade ore bodies, although lower ore tonnage was delivered to the mill during the year (Vatukoula Gold Mines plc., 2013).

Structure of the Mineral Industry

Fiji had one British-owned gold and silver mining company, one locally operated cement company, and several small locally operated industrial mineral quarries that produced construction materials, such as limestone, sand and gravel, and other construction aggregates. Fiji was a regional gold producer, and it hosts several undeveloped major porphyry copper-gold and epithermal gold deposits.

Mineral Trade

In 2012 (the latest year for which data were available), the leading export partners of Fiji were the United States, which received 13.3% of Fiji's exports; Australia, 12%; Japan, 6.3%; Samoa, 5.8%; and Tonga, 5.1%. Fiji's major mineral commodity export was gold. The leading import partners of Fiji were Singapore, which supplied 32.6% of Fiji's imports; Australia, 15.4%; New Zealand, 14.4%; and China, 10.7%. Fiji's major mineral commodity import category was petroleum products. The country increased its imports of petroleum products in 2012, which included mineral fuels, mineral oils, and distillations. Imports of mineral products were expected to increase steadily in 2013 and beyond (Fiji Islands Bureau of Statistics, 2013; U.S. Central Intelligence Agency, 2013).

Commodity Review

Metals

Bauxite and Alumina.—Shangdong Xinfa Group of China built a bauxite mine in Fiji, and the mine went into production in November 2011. By yearend 2012, the mine had supplied about 300,000 metric tons (t) of bauxite to the company's alumina refinery in China. China imported in total about 500,000 t of bauxite from Fiji in 2012 (Alumina and Aluminum Monthly, 2013).

Gold.—Vatukoula Gold Mines acquired 100% of Fiji's primary gold metal producing mine, the Vatukoula gold mine, in 2008. In 2012, the company processed 480,000 t of ore and sold 1,636 kg (52,616 troy ounces) of gold at an average cost of $1,627 per troy ounce. The company's production was limited by the flooding in February and April 2012, which hampered access to some of the mine's ore bodies. Operating costs on a per ton basis increased to an average of $179 owing to an increase in diesel prices during the year. Although the "mineral resource base" remained at 130,000 kg (4.2 million troy ounces), Vatukoula Gold Mines decreased its mineral reserves by 1,600 kg (50,000 troy ounces) to 22,000 kg (700,000 troy ounces) owing to the mine's depletion (Vatukoula Gold Mines plc., 2013, p. 4, 7, 10).

The Namosi Joint Venture project was a large undeveloped copper-gold deposit with reserves estimated to be about 1.3 billion metric tons of ore at a grade of 0.12 gram per metric ton (g/t) gold. Newcrest Mining Ltd. of Australia owned and managed 69.94% of the project (Newcrest Mining Ltd., 2013, p. 5).

In August 2011, Nautilus Minerals Inc. of Canada acquired 14 special prospecting licenses to explore the potential mineral deposits offshore Fiji. The licenses were for an exploration area that covers 60,000 square kilometers of ocean floor. Each of the licenses had an initial 2-year term. Although it retained an interest in Nautilus as a significant shareholder, Teck Resources Ltd. of Canada elected not to participate in exploration activities in Fijian waters (Nautilus Minerals Inc., 2011, 2012).

Industrial Minerals

Cement.—Fiji Industries Ltd.'s cement mill broke down in November 2012, which caused a cement supply shortage countrywide. Fiji imported cement from New Zealand during this time to help meet domestic demand (ICR Newsroom, 2012).

Outlook

Vatukoula Gold Mines is likely to open some high-grade areas in its mine to increase gold production. It is also likely to continue with its reserve and resource drilling program to increase the company's mineral resource base (Vatukoula Gold Mines plc., 2013).

References Cited

Alumina and Aluminum Monthly, 2013, Shangdong Xinfa actively develops overseas bauxite resource: eijing Antaike Information Development Co. Ltd., February, p. 19.

Fiji Bureau of Statistics, 2013, Economic statistics—Table 2, imports by HS—International merchandise trade statistics—Annual 2012: Fiji Bureau of Statistics release no. 24, 2013. (Accessed July 9, 2013, at http://www.statsfiji.gov.fj/index.php/economic/45-economic-statistics/national-accounts/103-summary-of-merchandise-trade-statistics-fjd000.)

ICR Newsroom, 2012, Fiji Industries Ltd begins cement production: Cemnet.com, December 7. (Accessed July 9, 2013, at http://www.cemnet.com/News/story/151194/fiji-industries-ltd-begins-cement-production.html.)

Mineral Resources Department [Fiji], 2012, Mining and geology in Fiji: Mineral Resources Department. (Accessed July 27, 2012, at http://www.mrd.gov.fj/gfiji/Index.html.)

Nautilus Minerals Inc., 2011, Nautilus granted exploration tenements in Fijian waters: Nautilus Minerals Inc. news release, August 11. (Accessed July 9, 2012, at http://www.nautilusminerals.com/s/Media-NewsReleases.asp?ReportID=471055&_Title=Nautilus-granted-exploration-tenements-in-Fijian-waters.)

Nautilus Minerals Inc., 2012, Fiji update on partnering arrangement: Nautilus Minerals Inc. news release, February 22. (Accessed July 9, 2012, at http://www.nautilusminerals.com/s/Media-NewsReleases.asp?ReportID=508707&_Type=News-Releases&_Title=Fiji-Update-on-Partnering-Arrangement.)

Newcrest Mining Ltd., 2013, Market release—December 2012 resources and reserves statement: Melbourne, Victoria, Australia, Newcrest Mining Ltd., February 8, 6 p.

U.S. Central Intelligence Agency, 2013, Fiji, in The world factbook: U.S. Central Intelligence Agency. (Accessed July 24, 2013, at https://www.cia.gov/library/publications/the-world-factbook/geos/fj.html.)

Vatukoula Gold Mines plc., 2013, Vatukoula Gold Mines annual report 2012: Vatukoula Gold Mines plc., 85 p. (Accessed July 9, 2013, at http://www.vgmplc.com/downloads/vgm-annual-report-2012-final-lr.pdf.)

World Bank, The, 2013, Data, indicators: The World Bank. (Accessed July 24, 2013, at http://data.worldbank.org/indicator/FP.CPI.TOTL.ZG.)

TABLE 1
FIJI: ESTIMATED PRODUCTION OF MINERAL COMMODITIES[1, 2]

(Thousand metric tons unless otherwise specified)

Commodity[3]		2008	2009	2010	2011	2012
Bauxite		--	--	--	50	500
Cement, hydraulic		143	110	120	120	110
Gold, mine output, Au content	kilograms	871	1,040	1,856 [4]	1,622 [r, 4]	1,653 [4]
Limestone		215	215	50	50	50
Silver, mine output, Ag content	kilograms	265 [4]	293	500	400	300

[r]Revised. do. Ditto. -- Zero.

[1]Estimated data are rounded to no more than three significant digits.

[2]Table includes data available through July 9, 2013.

[3]In addition to the commodities listed, bauxite, crushed and dimension stone, coral sand, marble, and other construction materials are produced, but data are insufficient to make reliable estimates of output.

[4]Reported figure.

TABLE 2
FIJI: STRUCTURE OF THE MINERAL INDUSTRY IN 2012

(Thousand metric tons unless otherwise specified)

Commodity		Major operating companies and major equity owners	Location of main facilities	Annual capacity[e]
Bauxite		Shangdong Xinfa Group	NA	500
Cement		Fiji Industries Ltd. (FIL) (an affiliate company of Holcim Group)	Lami, Suva, Fiji	150
Gold	kilograms	Vatukoula Gold Mines plc., 100%	Vatukoula, island of Viti Levu	1,900
Silver	do.	do.	do.	500

[e]Estimated. do. Ditto. NA Not available.

THE MINERAL INDUSTRY OF INDIA

By Chin S. Kuo

India has significant resources of metallic and industrial minerals. The country's reserves and resources of barite, bauxite, chromium, coal, iron ore, limestone, and manganese ore were all among the 10 largest in the world. The country produced 10 metals, 47 industrial minerals, 23 minor minerals, 4 mineral fuels, and 3 atomic minerals. In terms of output, India was ranked second in the production of barite, graphite, and talc, and third in the production of chromite, coal, rare earths, and zinc (slab) in the world. The country was also among the world's eight leading producers of aluminum, bauxite, iron ore, kyanite, manganese ore, mica (sheet), and steel (Ministry of Mines, 2013, p. 15, 191–194).

For the most part, the State governments own the minerals within their territorial jurisdiction and the national Government owns the minerals in the offshore areas. In India, the per capita consumption of most metals, including aluminum, copper, lead, zinc, and related products, was one of the lowest in the world. In the next 15 years, however, the demand for metals and minerals is likely to increase by four to five times owing to the improvement in living standards. Because these raw materials are vital for infrastructure, capital goods, and basic industries, development of a sustainable mining industry was the Government's top priority.

Minerals in the National Economy

In 2012, the mineral sector's contribution to the gross domestic product was 2.6%, and this percentage was likely to increase to between 3% and 4% in the near future owing to increasing domestic demand for, and therefore production of, metals and minerals. Overall mineral production in terms of tonnage decreased by 5.1% in 2012, and the total value of mineral production decreased by 0.1%. Mineral fuels accounted for 66.9% of the total value; metals, 18.5%; and industrial minerals, 14.7%. In 2011 (the latest year for which data were available), the value of mineral exports increased by 29.14% and that of mineral imports increased by 27.47% compared with those of 2010 (Ministry of Mines, 2013, p. 11–12, 185–186).

Government Policies and Programs

The Government was considering lifting its 25-year ban on issuing new asbestos mine licenses if the mining companies met the country's health and safety standards. The State of Rajasthan held a large deposit of asbestos, but its mines were idle. Only three private asbestos mines were active in the State of Andhra Pradesh, and together they produced a few hundred metric tons per year (t/yr) of chrysotile. Almost all the asbestos used in India was imported from Brazil, Canada, Kazakhstan, Russia, and Zimbabwe. About 90% of asbestos was used in roofing sheet, 5% was used for drinking water pipes, and the rest was used for vehicle brake linings and other products (Verma, 2012).

The Government imposed a 2% tax on imports of cut and polished diamond in 2012. The prices of these items subsequently went up by 2%. There was no effect on exports of cut and polished diamond, as most of the diamond was cut and polished in India. In addition, gold and platinum imports were each subject to a 2% tax. The import duty on silver was set at 6% of its import value (Diamonds.net, 2012).

Production

In 2012, production of such mineral commodities as barite, cement, bituminous coal, gem diamond, pig iron, and silver increased by more than 10% whereas output of cobalt, fire-refined copper, gold, iron ore, and manganese ore and concentrate (and their Mn content) decreased by more than 10%. The decrease (by 38%) in the production of cobalt metal was owing to one of the producers (Rubamin) operating below its capacity. Reduction in iron ore production by 19% was owing to the increase in mining taxes and the court-imposed partial ban and restriction on iron ore mining in the States of Karnataka and Odisha, respectively. In the State of Goa, iron ore production was suspended. Production of silver increased by 62% whereas that of gold was estimated to have decreased by 22%. For the industrial minerals, output levels remained about the same, except for gem diamond, which increased by 125% owing to strong demand in the domestic jewelry industry (table 1).

Structure of the Mineral Industry

India's mineral sector includes mining and mineral processing industries, which are the backbone of the country's industrial production. The mineral sector provides the basic raw materials to the manufacturing sector. India's mining industry was characterized by a large number of small operating mines. Small mines in the private sector continued to be operated either as proprietary or partnership ventures. Public sector undertakings under the Ministry of Mines were Hindustan Copper Ltd. (HCL), Mineral Exploration Corp. Ltd., and National Aluminium Co. Ltd. (Nalco). The public sector companies accounted for 67.7% of the total value of mineral production. The number of mines that reported mineral production was 3,108 in 2012 and included 1,976 industrial mines, 559 metal mines, and 573 coal mines. Production from opencast mines accounted for 80% of the total mine output. The number of underground operations decreased to 82. Total employment in the mineral industry was estimated to be more than 500,000 (table 2; Ministry of Mines, 2013, p. 13).

Mineral Trade

The total value of exports of ores and minerals was about $27.7 billion in 2011 (the latest year for which data were available). Diamond (mostly cut and polished) was the principal item of export, accounting for 76.4% of all mineral and ore exports; iron ore, 12.7%; granite, 3.6%; and alumina, 0.1%. The total value of imports of fuels, minerals, and ores was about $160 billion. Crude petroleum was the main component of these imports, accounting for 68.2%; diamond (rough), 14.0%; coal, 8.4%; natural gas, 3.5%; and copper ore and concentrate, 2.8% (Ministry of Mines, 2013, p. 187–190).

The country continued to be largely self-sufficient in such mineral commodities as bauxite, chromite, ilmenite, iron ore, manganese ore, and rutile, among the metals; and barite, dolomite, feldspar, limestone, silica minerals, sillimanite, and talc, among the industrial minerals (Ministry of Mines, 2013, p. 16).

Commodity Review

Metals

Bauxite and Alumina.—Vedanta Aluminium temporarily shut down its 1-million-metric-ton-per-year (Mt/yr) alumina refinery at Lanjigarh in the State of Odisha in December owing to the unavailability of bauxite ore. To run the refinery at full capacity, the company needed 300,000 metric tons (t) of high-grade bauxite each month and depended on externally sourced bauxite from other States. The Odisha government had committed to providing a total of 150 million metric tons (Mt) of bauxite for the refinery. The closure of the refinery could affect 2,500 workers directly and 4,500 others indirectly. The company's first greenfield aluminum smelter at Jharsuguda in Odisha had a capacity of 500,000 t/yr and produced aluminum using alumina processed at Lanjigarh (Seth, 2012b).

Gold.—Kolar Gold Ltd. announced an update of the gold resource at its Mallappakonda deposit in the Kolar gold belt to 6,070 kilograms (kg) from 1,910 kg (195,000 troy ounces from 61,500 troy ounces) of contained gold. The gold belt is an Archaean greenstone belt located in southern India in the States of Andhra Pradesh, Karnataka, and Tamil Nadu. The gold resource amounted to 3.46 Mt at an average grade of 1.76 grams per metric ton (g/t) gold. The cutoff grade used for resource reporting was 1 g/t gold. The associated silver was at slightly higher values than the gold. Resource drilling was to continue as part of the company's new 160-hole drill program (London Stock Exchange, 2012).

Iron and Steel.—The State government of Goa temporarily suspended all iron ore mining activities owing to some illegalities and irregularities. The mining operations of Sesa Goa Ltd. came to a standstill. The company had a production capacity of 14.5 Mt/yr in Goa. The State of Goa was India's second-ranked producer of iron ore and the leading exporter. The State produced 50 Mt/yr and exported almost all of it. In 2011, the State of Karnataka banned iron ore mining, but later the Supreme Court allowed partial mining. The State of Odisha restricted the iron ore output of 10 mining companies, allowing them to produce only the amount needed for their internal use, which resulted in a production decrease in the State of 27 Mt for the year (Mineweb.com, 2012).

NSL Consolidated Ltd. of Australia produced the first salable beneficiated iron ore from its Kurnool dry separation plant in the State of Andhra Pradesh. The beneficiated ore had a grade of 58% iron. The plant would source its feed from the nearby Mangal iron ore mine. The company expected to produce 200,000 t/yr of beneficiated ore in the first phase of operations. The company also was scheduled to complete construction of a wet beneficiation plant that could produce grades of between 58% and 62% iron in 2013 (Feary, 2012).

Under a modernization and expansion program, the production capacity of the Bhilai steel plant, which was a unit of Steel Authority of India Ltd. (SAIL), would increase to 7 Mt/yr of crude steel from 4.9 Mt/yr. The existing Dalli Rajhara iron ore mine had a remaining reserve of 80 Mt with increasing silica content in the ore. To sustain the expanded level of production, new mines at Raoghat were being developed with associated facilities and a 95-kilometer (km) railroad was being constructed by Indian Railways from Rajhara to Raoghat (Hindu, The, 2012).

Platinum-Group Metals.—The Geological Survey of India discovered platinum-group metals (PGMs) and chromite reserves in the Dhenkanal District in the State of Odisha. Deposits of chromite were found in numerous small patches of ultramafic and basic rocks occurring within the gneissic rocks. Deposits of PGMs were found within a 1-meter (m)-thick band in the Chander region and in the Bangur region in the Keonjhar District in the Baula-Nuasahi ultramafic belt where a resource of 14. 2 Mt was estimated. Four prospects for PGMs were also found in the States of Karnataka (1), Odisha (1), and Tamil Nadu (2) (Das, 2012a).

Industrial Minerals

Barite.—Andhra Pradesh Mineral Development Corp., which was a state-owned mineral exporter, issued a 36,000-t/yr tender for barite powder for 2 years. The material would come from its Mangampet barite project in Andhra Pradesh. In India, the company owned most of the country's barite reserves and produced 77% of the Nation's barite output. The company's production capacity was 1.6 Mt/yr of barite. It also had two granite projects, two limestone quarries, and a ball clay extraction facility (Sharma, 2012).

Cement.—Dalmia Cement (Bharat) Ltd. acquired a 50% stake in Calcom Cement India Ltd. for $47 million. Calcom Cement operated a 1-Mt/yr clinker plant in the State of Assam. Calcom Cement was in the process of expanding its production capacity to 2.1 Mt/yr and had a presence in markets in northeastern India. Dalmia Cement had the capacity to produce 9 Mt/yr of cement and also held about a 45% interest in OCL India Ltd., which had a production capacity of 5.5 Mt/yr. Dalmia Cement planned to set up two greenfield cement plants in the State of Karnataka with a capacity of 2.5 Mt/yr each (Global Cement News, 2012b).

India Cements planned a brownfield expansion totaling 3 Mt/yr at two of its cement plants (Dalavoi and Sankaridurg) in the State of Tamil Nadu. At Dalavoi, a new line would be added to increase the cement plant's existing 2.16-Mt/yr capacity by 2.55 Mt/yr, and a 40-megawatt (MW)-capacity powerplant (incorporating two 20-MW units) would also be set up. At Sankaridurg, the production capacity of the 700,000-t/yr-capacity cement plant would be doubled to 1.4 Mt/yr of cement (Aggregate Research, 2012c).

Burnpur Cement Ltd. planned to set up an 800-metric-ton-per-day (t/d) integrated cement plant at Patratu in the State of Jharkhand. The State government had allotted the nearby limestone mines at Benti Bagda to the company. The first phase of construction was expected to be completed in March 2013, and production would begin in April. The company had a cement plant at Asansol with a production capacity of 1,000 t/d (Aggregate Research, 2012a).

OCL India Ltd. planned to raise $40 million from the International Finance Corp. to help invest in the company's new $102 million 1.35-Mt/yr grinding plant at Medinipur in the State of West Bengal. The plant was expected to be operational by January 2014 and would use clinker from the company's existing plant at Rajgangpur in the State of Odisha (International Cement Research, 2012b).

Associated Cement Cos. Ltd. planned to increase its production capacity to 35 Mt/yr from 10 Mt/yr of cement with an investment of $650 million. The company planned to set up a 4-Mt/yr cement unit and a 2.8-Mt/yr clinker unit at Jamul in the State of Chhattisgarh. Grinding units were also planned at Sindri in the State of Jharkhand and at Kharagpur in the State of West Bengal (Global Cement News, 2012a).

Lafarge S.A. of France planned to increase its production capacity in India to 8 Mt/yr from 6.5 Mt/yr of cement. The company had four cement plants in the country—at Arasmeta and Sonadih in Chhattisgarh, at Jojobera in Jharkhand, and at Mejia in West Bengal. Production lines at Jojobera and Mejia were commissioned in 2012. The company also had four greenfield cement projects in the States of Himachal Pradesh, Karnataka, Meghalaya, and Rajasthan, respectively (Aggregate Research, 2012d).

JK Lakshmi Cements planned to revive operations at its Udaipur cement plant to increase the production capacity by 1.4 Mt/yr in the next 2 years. The revival of the plant was expected to be completed in September 2014. The company's expansion plan was to increase clinker capacity at its unit in the State of Rajasthan to 4.3 Mt/yr by 0.5 Mt/yr and to construct a 2.7-Mt/yr greenfield cement plant at Durg in the State of Chhattisgarh. The company's total capacity would increase to nearly 10 Mt/yr from the current 5.3 Mt/yr (International Cement Research, 2012a).

Clay and Shale.—A leading Indian kaolin producer, 20 Microns Ltd., expanded production of calcined kaolin at its Bhuj plant to 65,000 t/yr from 40,000 t/yr. The company exported 35% of its output to the Asia and the Pacific region, to the Middle East, and to Western Europe. English India Clays Ltd., which was the leading single producer of kaolin in Asia, had a capacity of 240,000 t/yr from its Veli plant in the State of Kerala and produced 180,000 t/yr of hydrous kaolin and 60,000 t/yr of calcined kaolin. Ashapura Group had mine reserves in Kerala of between 2.5 Mt and 3 Mt at grades of 96% kaolinite and operated a 180,000-t/yr plant that produced both hydrous and calcined kaolin (Ollett, 2012).

Diamond.—Rio Tinto plc of the United Kingdom planned to develop India's second diamond mine at Bundelkhand in the State of Madhya Pradesh. The project at prefeasibility stage covered 954 hectares (ha), and it would take 8 to 10 years to develop the deposit into an operational mine. Data indicated a field of eight kimberlitic rock pipes containing an estimated 27.4 million carats of diamond. The prefeasibility study of the project was expected to be completed in December 2012. The company planned to mine the Atri deposit first, which was the largest of the kimberlitic pipes. India's first diamond mine was at Majhgawan, Panna District, in Madhya Pradesh (Economic Times, The, 2012).

Feldspar.—Producers of ceramic tiles and sanitaryware asked the Government to ban exports of feldspar and quartz. The industries consumed about 84% of domestic feldspar production. Strong foreign demand for these two minerals resulted in increased exports, which led to a local shortage and price increases. Shortage of these minerals would force domestic industries to rely heavily on imports, which could result in higher costs and more expensive end products (Feytis, 2012a).

Graphite.—India was the second-ranked producer of graphite in the world after China. Its reserves of 11 Mt ranked India as the country with the second largest reserves after China. Graphite deposits of economic importance were found in the States of Andhra Pradesh, Arunachal Pradesh, and Chhattisgarh. Graphite was produced in the States of Jharkhand, Odisha, and Tamil Nadu. Three of the major graphite producers were Agrawal Graphite Industries, Tamil Nadu Minerals Ltd., and Tirupati Carbons and Graphite Pvt. Ltd. Tirupati's new project, which would include an enhanced processing plant, would produce 800 metric tons per month of high-quality flake graphite. Chotanagpur Graphite Industries, in a joint venture with Mega Graphite, was expected to increase the existing production in Jharkhand by 5,000 to 9,000 t/yr of high-quality graphite. Agrawal Graphite Industries was building a new high-purity graphite processing plant with an installed capacity of 2,000 t/yr. The main processing plant at Belpara in Odisha had an installed capacity of 12,000 t/yr of processed graphite (Salwan, 2012).

Rare Earths.—India was the world's third-ranked producer and was home to large deposits of light rare-earth minerals, which are found in the sands of the country's coastlines. The country had reserves of about 3.1 Mt of rare-earth minerals, the most after China, the United States, and the Commonwealth of Independent States. State-owned Indian Rare Earths Ltd. was building a $25 million rare-earth processing plant with a capacity of 11,000 t/yr in Puri District in the State of Odisha and was expecting operations to begin in September 2012. The plant's production would account for 4% of global rare-earth elements output. The refined product would be separated at the company's Aluva facility (Cordier, 2012; Mukherji and Wright, 2012).

Toyota Tsusho and its Indian subsidiary Toyotsu Rare Earths Orissa Pvt. Ltd. were constructing a processing plant for rare-earth oxides (REOs) in Odisha. Elements to be produced included cerium, lanthanum, and neodymium. Production at the plant would reach 4,000 t/yr of REOs, all for export to Japan, which would meet 14% of Japan's annual consumption (Currie, 2012).

Soda Ash.—Gujarat Heavy Chemicals Ltd., which was one of India's leading soda ash producers, was closed by the Gujarat Pollution Control Board for causing water and air pollution. The company's 850,000-t/yr synthetic soda ash plant was located at Sutrapada in the State of Gujarat (Lismore, 2012).

Sulfur.—India relied on imports of 1.8 Mt/yr of sulfur for its fertilizer industry. The country consumed 3.8 Mt/yr of sulfur, of which 2 Mt/yr was used for fertilizers, and was expected to increase its sulfur consumption by 18% to 4.5 Mt/yr by 2016, of which 2 Mt/yr would be imported. India produced 2.4 Mt/yr of sulfur, including 1.4 Mt/yr from oil refineries and 1 Mt/yr from copper and zinc smelters. The country exported 400,000 t/yr of its domestic sulfur (Feytis, 2012b).

Mineral Fuels and Related Materials

Coal.—Coal India Ltd. (CIL) objected to the supply of discounted coal to domestic aluminum, cement, and steel sectors that sold their end products at free market rates not regulated by the Government. The objection was a sign that raw material costs could rise for the infrastructure sector. Such a move would affect such companies as Grasim Industries Ltd., Hindalco Industries Ltd., and Lafarge India Pvt. Ltd. CIL supplied 312 Mt of coal to the power sector at negotiated prices through long-term agreements and fuel supply pacts; these prices were about one-half the price of coal sold on the international market (Aggregate Research, 2012b).

SAIL was looking for mining contractors to develop its Tasra coking coal deposit in the State of Jharkhand, which had a reserve of 117 Mt. It planned to increase production to 4 Mt/yr with an investment of $370 million; the expansion would include a coal washery. The company was considering offering washed coal rejects to contract miners in exchange for setting up a powerplant near the mine site. Coking coal accounted for 30% of SAIL's steel production cost. SAIL consumed 14 Mt/yr of coking coal to produce 16 Mt/yr of steel. The company had domestic supplies of 4 Mt/yr of coking coal from its captive mines and from CIL; the balance of the coking coal used was imported (Das, 2012b).

Petroleum.—Essar Oil Ltd. completed a $1.81 billion expansion of its Vadinar oil refinery and a 7.5-Mt/yr isomerization unit in the State of Gujarat in 2011. The refinery was India's second largest refinery in terms of capacity. The expansion increased the refinery's capacity to 375,000 barrels per day (bbl/d) from the 300,000 bbl/d. Plans were underway in September 2012 to further increase the refinery's capacity to 405,000 bbl/d (Oil & Gas Journal, 2012).

Uranium.—The Governments of Canada and India finalized a deal that would allow a nuclear cooperation agreement to be implemented. The agreement would allow Canadian companies to export and import controlled nuclear equipment, materials, and technology to and from India's facilities for civilian and peaceful applications. Canada had large and high-quality reserves of uranium and produced uranium in the Province of Saskatchewan; it could become a significant supplier to India. In India, uranium reserves were found in the States of Andhra Pradesh and Jharkhand, and Uranium Corp. of India Ltd. (UCIL) mined small amounts of uranium in Andhra Pradesh (World Nuclear News, 2012).

UCIL commissioned a $208 million uranium processing plant at its Tummalapalle underground mine in Andhra Pradesh, which was estimated to have one of the world's largest uranium reserves (150,000 t U). A second uranium processing plant was also planned. The plant would treat dolomite/limestone-based uraniferous ore using an alkali leaching processing method. In the first phase, it would process 3,000 t/d of ore, which would be increased to 4,500 t/d in the second phase. India's first uranium mine was located at Jaduguda in the State of Jharkhand (Seth, 2012a).

Reserves and Resources

The country's mineral resources include large deposits of barite, bauxite, chromium, coal, iron ore, limestone, manganese, and uranium. Barite deposits occur in the State of Andhra Pradesh. Bauxite resources are found in the States of Andhra Pradesh, Chhattisgarh, Gujarat, and Odisha. Iron ore deposits in the form of hematite and magnetite occur in the States of Bihar, Karnataka, Madhya Pradesh, Odisha, and Tamil Nadu. India's coal resources amounted to 277 billion metric tons (Gt) and are located in the States of Andhra Pradesh, Chhattisgarh, Jharkhand, Madhya Pradesh, Maharashtra, Odisha, and West Bengal. Its lignite reserves totaled 39 Gt and are found in the States of Gujarat, Jammu and Kashmir, Kerala, Rajasthan, and Tamil Nadu (table 3).

Outlook

India is expected to continue to be largely self-sufficient in the minerals and metals that constitute the primary raw materials for its various industries. The Government's lifting of the ban on new asbestos mining is expected to increase asbestos production. Production of alumina and aluminum is expected to increase slightly when Vedanta Aluminium secures the sources for bauxite ore from Odisha and other States in 2013. Owing to the suspension, ban, and restriction of iron ore mining in several States, India is expected to produce less iron ore and to reduce iron ore exports in the near future. Cement consumption in India is expected to increase and so is the country's total production capacity. Several cement companies are expected to expand their production capacities by a total of 20 Mt/yr by 2014. Production of rare earths in India is expected to increase because of the addition of two rare-earth processing plants. The country is, by and large, self-sufficient in coal and lignite. By 2015, India's consumption of natural gas is expected to exceed its domestic production. Imports of liquefied natural gas (LNG) are expected to increase gradually as development of the LNG terminals and the pipeline infrastructure progresses.

References Cited

Aggregate Research, 2012a, Burnpur Cement to build unit in Patratu: Aggregate Research, August 29. (Accessed September 5, 2012, at http://www.aggregateresearch.com/article.aspx?id=25807.)

Aggregate Research, 2012b, Coal India board objects to cheap coal for cement, metals: Aggregate Research, November 28. (Accessed November 29, 2012, at http://www.aggregateresearch.com/print.aspx?id=26753.)

Aggregate Research, 2012c, India Cements starts spadework on brownfield expansion plans: Aggregate Research, October 25. (Accessed October 31, 2012, at http://www.aggregateresearch.com/article.aspx?id=26286.)

Aggregate Research, 2012d, Lafarge plans expansion in India: Aggregate Research, March 21. (Accessed March 29, 2012, at http://www.aggregateresearch.com/print.aspx?id=24744.)

Cordier, D.J., 2012, Rare earths: U.S. Geological Survey Mineral Commodity Summaries 2012, p. 128–129.

Currie, Adam, 2012, India revisits rare earth potential: Rare Earth Investing News, September 17. (Accessed September 17, 2012, at http://rareearthinvestingnews.com/7975/?utm_source=resource+investing+news&utm_campaign=f8b600e.)

Das, Ajoy, 2012a, New platinum reserves discovered in India: Mining Weekly, April 19. (Accessed April 20, 2012, at http://www.miningweekly.com/print-version/new-platinum-reserves-discovered-in-india-2012-04-19.)

Das, Ajoy, 2012b, SAIL looking for contract miners in Jharkhand: Mining Weekly, February 3. (Accessed February 3, 2012, at http://www.miningweekly.com/print-version/sail-looking-for-contract-miners-in-jharkhand-2012-02-03.)

Diamonds.net, 2012, India imposes 2% tax on polished imports: Diamonds.net, January 17. (Accessed July 24, 2012, at http://www.diamonds.net/mediacenter/tradealert.aspx?articleid=38679.)

Economic Times, The, 2012, Rio Tinto is gearing up to develop India's second diamond mine in Bundelkhand: The Economic Times, March 18. (Accessed May 23, 2012, at http://articles.economictimes.indiatimes.com/2012-03-18/news/31207441.)

Feary, Christine, 2012, NSL Consolidated delivers milestone first saleable iron ore from Kurnool plant: Proactiveinvestors.com, April 19. (Accessed April 19, 2012, at http://www.proactiveinvestors.com.au/companies/news/27841.)

Feytis, Alexandra, 2012a, Feldspar's future in flux: Industrial Minerals, no. 538, July, p. 42.

Feytis, Alexandra, 2012b, Indian sulfur consumption to reach 4.5m tpa by 2016-17: Industrial Minerals Online News, May 1, 1 p.

Global Cement News, 2012a, ACC to upgrade and consolidate: Global Cement News, April 4. (Accessed April 4, 2012, at http://www.globalcement.com/news.)

Global Cement News, 2012b, Dalmia Bharat picks up 50% stake in Calcom: Global Cement News, January 18. (Accessed January 18, 2012, at http://www.propubs.com/resources/global-cement/news/itemlist/tag/GCW32?utm.)

Hindu, The, 2012, Bhilai steel plant bets big on Raoghat mines: The Hindu, December 2. (Accessed December 3, 2012, at http://www.thehindu.com/business/industry/article4154170.ece.)

International Cement Research, 2012a, India plant revival: International Cement Research, December, p. 18.

International Cement Research, 2012b, OCL to raise funds for new plant: International Cement Research, December, p. 11.

Lismore, Siobhan, 2012, India's Gujarat Heavy Chemicals served soda ash closure notice: Industrial Minerals, no. 533, February, p. 12.

London Stock Exchange, 2012, Maiden resource statement—Mallappakonda: London Stock Exchange, June 13. (Accessed June 20, 2012, at http://www.londonstockexchange.com/exchange/news/market-news/market-news-detail.htm.)

Mineweb.com, 2012, Goa suspends all mining activities temporarily: Mineweb.com, September 11. (Accessed September 11, 2012, at http://www.mineweb.com/mineweb/view/mineweb/en/page504?oid=158430&sn=detail&pid=92730.)

Ministry of Mines, 2013, Annual report 2012–13: New Delhi, India, Ministry of Mines, August, 212 p.

Mukherji, Biman, and Wright, Tom, 2012, India bets on rare-earth minerals: The Wall Street Journal, August 14, p. A16.

Oil & Gas Journal, 2012, Essar Energy completes Vadinar refinery expansion: Oil & Gas Journal, March 30. (Accessed April 16, 2012, at http://www.ogj.com/articles/2012/03/essar-energy-completes-vadinar-refinery-expansion.html?cmpid=EnlRefiningApril162012.)

Ollett, John, 2012, Kaolin bounces back: Industrial Minerals, no. 533, February, p. 26.

Salwan, Shruti, 2012, The threat of Indian graphite: Industrial Minerals, no. 540, September, p. 27.

Seth, Shivom, 2012a, India's new uranium operation could host one of world's largest reserves: Mineweb.com, April 30. (Accessed April 30, 2012, at http://www.mineweb.com/mineweb/view/mineweb/en/page72103?oid=150382&sn=detail&pid=102055.)

Seth, Shivom, 2012b, Vedanta to shut down alumina refinery tomorrow: Mineweb.com, December 4. (Accessed December 5, 2012, at http://www.mineweb.com/mineweb/content/en/mineweb-base-metals?oid=165148&sn=detail.)

Sharma, Gaurav, 2012, APMDC issues new barites tender: Industrial Minerals Online News, June 27, 1 p.

Verma, Raghavendra, 2012, India considers lifting 25-year ban on new asbestos mining: Industrial Minerals, February 23. (Accessed February 23, 2012, at http://www.indmin.com/print.aspx?articleid=2983850.)

World Nuclear News, 2012, Clearing the way for Canada-India U trade: World Nuclear News, November 7. (Accessed November 14, 2012, at http://www.world-nuclear-news.org/NP-Clearing_the_way_for_Canada-India_U_trade-0.)

TABLE 1
INDIA: ESTIMATED PRODUCTION OF MINERAL COMMODITIES[1,2]

(Metric tons unless otherwise specified)

Commodity[3]		2008	2009	2010	2011	2012
METALS						
Aluminum:						
Bauxite, gross weight	thousand metric tons	21,210 [r,4]	16,000 [r]	18,000 [r]	19,000 [r]	19,000
Alumina, Al_2O_3 equivalent	do.	3,820	3,900	3,640	3,880	3,900
Metal, primary	do.	1,402 [4]	1,598 [4]	1,607 [4]	1,667 [4]	1,700
Cadmium metal		599 [4]	627 [4]	632 [4]	616 [4]	620
Chromium, chromite, gross weight	thousand metric tons	3,900 [4]	3,760 [4]	3,800	3,850	3,900
Cobalt metal[4]		858	1,001	1,187 [r]	1,299 [r]	800
Copper:						
Mine output, Cu content[4]		30,600	29,500	35,500	37,700	34,000
Metal, primary:						
Smelter		651,000 [4]	705,100 [4]	748,800 [4]	670,000 [r,4]	680,000
Refinery:						
Electrolytic, cathode[4]		654,200	705,100	654,900	671,100	670,000
Fire refined		15,000	10,000	9,000	2,000	1,000
Total		669,000	715,000	664,000	673,000	671,000
Gold metal, smelter	kilograms	2,700 [4]	2,800	2,700	2,300	1,800
Iron and steel:						
Iron ore and concentrate:						
Gross weight	thousand metric tons	213,000 [r]	217,000 [r]	210,000 [r]	177,000 [r]	144,000
Fe content	do.	136,000 [r]	139,000 [r]	134,000 [r]	113,000 [r]	92,000
Metal:						
Pig iron[4]	do.	29,000	34,000	38,685	43,600 [r]	48,000
Direct-reduced iron[4]	do.	20,900 [r]	23,400 [r]	24,800 [r]	21,300 [r]	19,700
Ferroalloys:						
Ferrochromium, including charge chrome		750,000 [4]	873,385 [4]	850,000	830,000	800,000
Ferrochromiumsilicon		10,000	10,000	10,000	11,000 [r]	11,000
Ferromanganese[4]		386,200 [r]	399,100 [r]	440,000 [r]	440,000 [r]	463,900
Ferrosilicon		92,000	101,337 [4]	101,000	105,000	108,000
Silicomanganese[4]		848,700 [r]	875,500 [r]	1,000,000 [r]	1,433,600 [r]	1,522,600
Other		9,000	10,000 [r]	9,000	10,000 [r]	9,000
Steel, crude[4]	thousand metric tons	57,800	63,500	68,300	73,500 [r]	77,600
Semimanufactures[5]	do.	49,000	50,000	51,000	53,000	52,000
Lead:						
Mine output, Pb content		87,300 [4]	92,000	97,000	115,000	118,000
Metal, refined:[4]						
Primary		62,000	61,700	75,000	120,000	122,000
Secondary		232,000	275,000	305,000	306,000	310,000
Total		294,000	336,700	380,000	426,000	432,000
Manganese:						
Ore and concentrate, gross weight[4]	thousand metric tons	2,293	2,374	2,858	2,542	2,225
Mn content[4]	do.	826 [r]	845 [r]	1,013 [r]	895 [r]	800
Selenium	kilograms	14,000	15,000	15,000	16,000	16,000
Silver, mine and smelter output[4]	do.	96,000	138,100	165,100	203,500	330,000
Titanium mineral concentrates, gross weight:						
Ilmenite		610,000	700,000	540,000	550,000	560,000
Rutile		21,000	21,000	24,000	25,000	26,000
Zinc:						
Mine output, concentrate:						
Gross weight		1,120,000	1,260,000	1,400,000	1,350,000	1,400,000
Zn content[4]		613,600	695,000	740,000	710,000	750,000

See footnotes at end of table.

TABLE 1—Continued
INDIA: ESTIMATED PRODUCTION OF MINERAL COMMODITIES[1, 2]

(Metric tons unless otherwise specified)

Commodity[3]		2008	2009	2010	2011	2012
METALS—Continued						
Zinc—Continued:						
Metal:						
Primary		545,800 [4]	584,100 [4]	663,300 [4]	750,000	755,000
Secondary		22,000	30,000	37,000	45,000	45,000
Total		568,000	614,000	700,000	795,000	800,000
Zirconium concentrate, zircon, gross weight		30,000	37,000	38,000	39,000	40,000
INDUSTRIAL MINERALS						
Abrasives, natural, n.e.s.:[6]						
Garnet		125,000	127,000	130,000	133,000	135,000
Jasper		8,900	8,700	8,800	8,900	8,800
Asbestos[4]		304 [r]	261 [r]	254 [r]	250 [r]	245
Barite	thousand metric tons	1,100	1,200	1,300	1,350	1,700
Bromine, elemental		1,500	1,500	1,600 [r]	1,600 [r]	1,700
Cement, hydraulic	thousand metric tons	185,000	205,000	220,000	240,000 [r]	270,000
Chalk		125,000	125,000	130,000	130,000	132,000
Clays:						
Ball		430,000	440,000	440,000	450,000	460,000
Fire		390,000	395,000	400,000	410,000	415,000
Kaolin:						
Salable crude	thousand metric tons	570	580	580	600	600
Processed	do.	210	210	220	220	230
Total	do.	780	790	800	820	830
Other	do.	85	85	90	90	90
Diamond:						
Gem	thousand carats	15	14	13	12	27
Industrial	do.	38	38	37	36	35
Total	do.	53	52	50	48	62
Feldspar		385,436 [4]	390,000	400,000	420,000	430,000
Fluorspar:						
Concentrates, metallurgical grade		6,814 [r, 4]	8,786 [r, 4]	8,400	8,500	8,600
Other fluorspar materials, acid grade		3,176 [r, 4]	4,996 [r, 4]	4,600	4,800	5,000
Gemstones, excluding diamond:						
Agate, including chalcedony pebble		160	160	150	150	140
Graphite[7]		140,000	130,000	140,000	150,000	160,000
Gypsum	thousand metric tons	3,760 [r]	3,500 [r]	4,530 [r]	3,620 [r]	3,600
Kyanite and related materials:						
Kyanite		4,982 [r, 4]	4,839 [r, 4]	5,513 [r, 4]	5,500 [r]	5,600
Sillimanite		38,469 [r, 4]	33,699 [r, 4]	37,183 [r, 4]	37,000 [r]	38,000
Lime	thousand metric tons	13,000	13,000	14,000	15,000	15,000
Magnesite		350,000	340,000	345,000	350,000	355,000
Mica:						
Crude		2,241 [r, 4]	1,161 [r, 4]	1,265 [r, 4]	1,689 [r, 4]	1,700
Scrap and waste		5,140 [r, 4]	7,495 [r, 4]	7,508 [r, 4]	12,095 [r, 4]	12,000
Total		7,381 [r, 4]	8,656 [r, 4]	8,773 [r, 4]	13,784 [r, 4]	13,700
Nitrogen, N content of ammonia	thousand metric tons	11,100	11,200	11,500	11,800	12,000
Phosphate rock, including apatite	do.	1,220	1,230	1,240	1,250	1,260
Pigments, mineral, natural, ocher		1,117 [r, 4]	890 [r, 4]	1,237 [r, 4]	1,100 [r]	1,200
Pyrites, gross weight		120,000	115,000	115,000	110,000	110,000

See footnotes at end of table.

TABLE 1—Continued
INDIA: ESTIMATED PRODUCTION OF MINERAL COMMODITIES[1,2]

(Metric tons unless otherwise specified)

Commodity[3]		2008	2009	2010	2011	2012
INDUSTRIAL MINERALS—Continued						
Rare-earth metals, monazite concentrate, gross weight		5,000	5,000	5,200	5,200	5,400
Salt:						
Rock	thousand metric tons	3	3	3	2	2
Other	do.	16,000	16,500	17,000	16,000	17,000
Total	do.	16,000	16,500	17,000	16,000	17,000
Sand:						
Calcareous	do.	275	280	285	290	295
Silica	do.	1,700	1,700	1,800	1,800	1,900
Other	do.	3,200	3,300	3,400	3,500	3,600
Slate		13,500	14,000	14,500	15,000	15,500
Soda ash	thousand metric tons	1,500	1,400 [r]	1,500	1,400 [r]	1,500
Stone, sand and gravel:						
Calcite		55,000	55,000	56,000	56,000	57,000
Dolomite	thousand metric tons	3,100	3,100	3,200	3,200	3,300
Limestone	do.	127,000	130,000	132,000	135,000	138,000
Quartz and quartzite	do.	280	280	290	290	300
Sulfur:						
Byproduct from metallurgy	do.	1,000	1,000	1,000	1,000	1,000
Byproduct from petroleum	do.	1,400	1,400	1,400	1,400	1,400
Talc and related materials:						
Pyrophyllite		87,000	88,000	90,000	90,000	92,000
Steatite, soapstone		560,000	550,000	550,000	560,000	570,000
Vermiculite		11,742 [4]	12,000	12,000	13,000	13,000
Wollastonite		125,000	135,000	145,000	150,000	155,000
MINERAL FUELS AND RELATED MATERIALS						
Coal:						
Bituminous	thousand metric tons	420,000	450,000	480,000	500,000	550,000
Lignite	do.	26,000	25,000	27,000	28,000	30,000
Total	do.	446,000	475,000	507,000	528,000	580,000
Gas, natural:						
Gross	million cubic meters	33,061 [4]	35,000	37,000	38,000	40,000
Marketable	do.	27,457 [4]	30,000	32,000	33,000	35,000
Petroleum:						
Crude	thousand 42-gallon barrels	253,000 [4]	255,000	260,000	265,000	270,000
Refinery products:						
Liquefied petroleum gas	do.	53,000 [4]	55,000	58,000	60,000	62,000
Gasoline	do.	119,000 [4]	125,000	130,000	135,000	137,000
Kerosene and jet fuel	do.	122,000 [4]	120,000	125,000	130,000	133,000
Distillate fuel oil	do.	468,000 [4]	480,000	490,000	500,000	520,000
Residual fuel oil	do.	130,000 [4]	140,000	145,000	150,000	152,000
Other	do.	319,000 [4]	315,000	310,000	300,000	280,000
Total	do.	1,211,000 [4]	1,240,000	1,260,000	1,280,000 [r]	1,280,000

[r]Revised. do. Ditto.

[1]Estimated data are rounded to no more than three significant digits; may not add to totals shown.

[2]Table includes data available through October 17, 2013.

[3]In addition to the commodities listed, boron, corundum, diaspore, other gemstones (aquamarine, emerald, garnet, ruby, and spinel), and uranium are produced, but output is not reported, and available information is inadequate to make reliable estimates of output.

[4]Reported figure.

[5]Excludes production from steel miniplants.

[6]Not elsewhere specified.

[7]India's marketable production is 10% to 20% of mine production.

TABLE 2
INDIA: STRUCTURE OF THE MINERAL INDUSTRY IN 2012

(Thousand metric tons unless otherwise specified)

Commodity	Major operating companies and major equity owners	Location of main facilities	Annual capacity[e]
Alumina	Indian Aluminium Co. Ltd. (Indian interests, 60.4%, and Alcan Aluminium Ltd., 39.6%)	Belgaum refinery, Karnataka	280
Do.	National Aluminium Co. Ltd. (Government, 100%)	Dhamanjodi refinery, Odisha	1,580
Do.	Bharat Aluminium Co. Ltd. [Government, 49%, and Sterlite Industries (India) Ltd., 51%]	Korba refinery, Chhattisgarh	200
Do.	Utkal Alumina International Ltd. (Hindalco Industries Ltd., 100%)	Koraput refinery, Odisha	1,500 [1]
Do.	Madras Aluminium Co. Ltd. [Sterlite Industries (India) Ltd., 80%, and others, 20%]	Mettur refinery, Tamil Nadu	80
Do.	Indian Aluminium Co. Ltd. (Indian interests, 60.4%, and Alcan Aluminium Ltd., 39.6%)	Muri refinery, Jharkhand	88
Do.	Hindalco Industries Ltd. (Birla Group, 33%; foreign investors, 26%; private Indian investors, 23%; financial institutions, 18%)	Renukoot refinery, Uttar Pradesh	450
Aluminum	Indian Aluminium Co. Ltd. (Indian interests, 60.4%, and Alcan Aluminium Ltd., 39.6%)	Alupuram smelter, Kerala	20
Do.	Vedanta Aluminium Ltd. (Vedanta Resources plc, 100%)	Jharsuguda, Odisha	500
Do.	National Aluminium Co. Ltd. (Government, 100%)	Angul smelter, Odisha	345
Do.	Indian Aluminium Co. Ltd. (Indian interests, 60.4%, and Alcan Aluminium Ltd., 39.6%)	Belgaum smelter, Karnataka	70
Do.	Hindalco Industries Ltd. (Birla Group, 33%; foreign investors, 26%; private Indian investors, 23%; financial institutions, 18%)	Hirakud smelter, Odisha	100
Do.	Bharat Aluminium Co. Ltd. [Government, 49%, and Sterlite Industries (India) Ltd., 51%]	Korba smelter, Chhattisgarh	350
Do.	Madras Aluminium Co. Ltd. [Sterlite Industries (India) Ltd., 80%, and others, 20%]	Mettur smelter, Tamil Nadu	40
Do.	Hindalco Industries Ltd. (Birla Group, 33%; foreign investors, 26%; private Indian investors, 23%; financial institutions, 18%)	Renukoot smelter, Uttar Pradesh	275
Barite	Andhra Pradesh Mineral Development Corp. Ltd. (Andhra Pradesh State government, 100%)	Cuddapah District mines, Andhra Pradesh	1,600
Do.	ICL Ltd.	do.	300
Do.	Associated Mineral Corp.	do.	75
Do.	Pragathi Minerals	do.	50
Do.	Vijayalaxmi Minerals Trading Co.	do.	50
Bauxite	Bharat Aluminium Co. Ltd. [Government, 49%, and Sterlite Industries (India) Ltd., 51%]	Amarkantak Mine, Madhya Pradesh	200
Do.	Indian Aluminium Co. Ltd. (Indian interests, 60.4%, and Alcan Aluminium Ltd., 39.6%)	Kolhapur District mines, Maharashtra	600
Do.	Gujarat Mineral Development Corp. (Gujarat State government, 100%)	Kutch and Saurashtra Mines, Gujarat	500
Do.	Hindalco Aluminium Co. Ltd. (Birla Group, 33%; foreign investors, 26%; private Indian investors, 23%; financial institutions, 18%)	Mines in Lohardaga District, Jharkhand	750
Do.	Indian Aluminium Co. Ltd. (Indian interests, 60.4%, and Alcan Aluminium Ltd., 39.6%)	do.	200
Do.	National Aluminium Co. Ltd. (Government, 100%)	Mines in Panchpatmali Hills, Koraput District, Odisha	4,800
Do.	Minerals & Minerals Ltd. (Government, 100%)	Mines in Richuguta, Palamau District, Jharkhand	200
Boron	Borax Morarji Ltd.	Ambernath, Maharashtra	17
Cement	Ultratech Cement Ltd.	11 integrated plants and 15 grinding units	45,000
Do.	Century Cement [Century Textiles and Industries Ltd. (a subsidiary of the Birla Group, 100%)]	Baikunth Plant, Madhya Pradesh	1,120
Do.	Ambuja Cements Ltd. (Holcim Group, 14.8%)	Plants in 7 States	25,000

See footnotes at end of table.

TABLE 2—Continued
INDIA: STRUCTURE OF THE MINERAL INDUSTRY IN 2012

(Thousand metric tons unless otherwise specified)

Commodity	Major operating companies and major equity owners	Location of main facilities	Annual capacity[e]
Cement—Continued	Coromandel Fertilizers Ltd. [Chevron Chemical Co., 23.55%; International Minerals and Chemical Co., 20.89%; Parry and Co., 10.64%; E.I.D. Parry (India) Ltd., 6.65%; others, 38.27%]	Chilamkur plant, Andhra Pradesh	1,000
Do.	Dalmia Cement (Bharat) Ltd.	Dalmiapuram and Ariyalur, Tamil Nadu; and Kadapa, Andhra Pradesh	9,000
Do.	Associated Cement Cos. Ltd. (ACC) (Government, 34.86%; Holcim Ltd., 46%; and private shareholders, 19.14%)	Gagal plant, Himachal Pradesh	1,830
Do.	do.	Chanda plant, Maharashtra	2,520
Do.	Raymond Cement Works (a division of Raymond Woolen Mills Ltd., JK Singhania, principal shareholder)	Gopalnagar plant, West Bengal	1,250
Do.	Shree Cement Ltd.	Haridwar plant, Uttarakhand	1,800
Do.	OCL India Ltd.	Kapilas and Rajgangpur, Odisha	5,500
Do.	Rajashree Cement (a division of Indian Rayon and Industries Ltd., 100%)	Khor plant, Karnataka	1,020
Do.	Associated Cement Cos. Ltd. (ACC) (Government, 34.86%; Holcim Ltd., 50.1%; and private shareholders, 15.04%)	Kymore plant, Madhya Pradesh	1,500
Do.	My Home Industries Ltd. (joint venture of My Home Group and CRH plc)	Mellacheruvu and Visakhapatnam in Andhra Pradesh	4,600
Do.	HeidelbergCement India Ltd.	Narasingarh plant, Haryana	1,090
Do.	Cement Corp. of India Ltd. (Government, 100%)	Nayagaon plant, Madhya Pradesh	1,330
Do.	JK Cement Works (a division of JK Synthetics Ltd.), 100%	Nimbahera plant, Rajasthan	1,460
Do.	India Cements Co. Ltd. (Government, 26%; Life Insurance Corp. of India, 24%; others, 50%)	Sankarnagar plant and 2 plants, Tamil Nadu; 4 plants, Andhra Pradesh; Mahi plant, Rajasthan	16,000
Do.	Prism Cement Ltd.	Satna plant, Madhya Pradesh	3,000
Do.	Jaiprakash Associates Ltd.	Sewagram, Gujarat	2,400
Do.	Shree Digvijay Cement Co. Ltd.	Shreeniwas plant, Maharashtra	1,070
Do.	JK Lakshmi Cement Ltd. (a division of Straw Products Ltd., JK Singhania, principal shareholder)	Sirohi plant, Rajasthan and Ahmadabad, Gujarat	4,700
Do.	Lafarge S.A.	Arasmeta and Sonadih, Chhattisgarh; Jojobera, Jharkhand; and Mejia, West Bengal	1,400
Do.	Manikgarth Cement [Century Textiles and Industries Ltd. (a subsidiary of the Birla Group, 100%)]	Tehsil Rajura plant, Maharashtra	1,000
Do.	Vikram Cement [Grasim Industries Ltd. (a subsidiary of the Birla Group, 100%)]	Vikram plant, Madhya Pradesh	1,000
Do.	Raasi Cement Ltd. (Andhra Pradesh State government, 50%, and Development Co. Ltd., 50%)	Vishnupuram plant, Andhra Pradesh	1,000
Do.	Associated Cement Cos. Ltd. (ACC) (Government, 34.86%; Holcim Ltd., 46%; and private shareholders, 19.14%)	Wadi plant, Karnataka	6,680
Chromium	Ferro Alloys Corp. Ltd.	Bhadrak, Cuttack District, Odisha	120
Do.	Orissa Mining Corp. Ltd. (Orissa Industries Ltd., 100%)	do.	300
Do.	Tata Steel Ltd.	do.	100
Do.	Ferro Alloys Corp. Ltd.	Dhenkanal and Kendujhar District, Odisha	150
Do.	Orissa Mining Corp. Ltd. (Orissa Industries Ltd., 100%)	do.	200
Do.	Mysore Minerals Ltd.	Hassan District, Karnataka	125
Do.	Orissa Mining Corp. Ltd. (Orissa Industries Ltd., 100%)	do.	100
Do.	Ferro Alloys Corp. Ltd.	Khammam District, Andhra Pradesh	100
Coal, bituminous million metric tons	Bharat Coking Coal Ltd. [a subsidiary of Coal India Ltd. (Government, 100%)]	Bihar and West Bengal	26
Do. do.	Central Coalfields Ltd. [a subsidiary of Coal India Ltd. (Government, 100%)]	Bihar	27

See footnotes at end of table.

TABLE 2—Continued
INDIA: STRUCTURE OF THE MINERAL INDUSTRY IN 2012

(Thousand metric tons unless otherwise specified)

Commodity		Major operating companies and major equity owners	Location of main facilities	Annual capacity[e]
Coal, bituminous—Continued	million metric tons	Eastern Coalfields Ltd. [a subsidiary of Coal India Ltd. (Government, 100%)]	Bihar and West Bengal	21
Do.	do.	Mahanadi Coalfields Ltd. [a subsidiary of Coal India Ltd. (Government, 100%)]	Odisha	21
Do.	do.	North Eastern Coalfields Ltd. [a subsidiary of Coal India Ltd. (Government, 100%)]	Assam	640
Do.	do.	Northern Coalfields Ltd. [a subsidiary of Coal India Ltd. (Government, 100%)]	Madhya Pradesh and Uttar Pradesh	24
Do.	do.	Singareni Collieries Co. Ltd. (Andhra Pradesh State government, 50%, and Government, 50%)	Andhra Pradesh and Maharashtra	18
Do.	do.	South Eastern Coalfields Ltd. [a subsidiary of Coal India Ltd. (Government, 100%)]	Chhattisgarh	36
Do.	do.	Western Coalfields Ltd. [a subsidiary of Coal India Ltd. (Government, 100%)]	Madhya Pradesh and Maharashtra	18
Coal, lignite	do.	Neyveli Lignite Corp. Ltd. (NLC) (Government, 100%)	Tamil Nadu	17
Copper, mine		Hindustan Copper Ltd. (HCL) (Government, 100%)	Indian Copper Complex Mines, Ghatsila District, Jharkhand	31
Do.		do.	Khetri Copper Complex Mines, Khetrinagar Rajasthan	15
Do.		do.	Malanjkhand Copper Complex Mines, Balaghar District, Madhya Pradesh	2,000
Copper, metal		Hindalco Industries Ltd. (Birla Group, 33%; foreign investors, 26%; private Indian investors, 23%; financial institutions, 18%)	Birla Copper Complex smelter, Dahej, Gujarat	70
Do.		Hindustan Copper Ltd. (HCL) (Government, 100%)	Indian Copper Complex smelter-refinery, Ghatsila District, Jharkhand	20
Do.		do.	Khetri Copper Complex smelter-refinery, Khetrinagar District, Rajasthan	45
Do.		Sterlite Industries (India) Ltd.	Tuticorin smelter, Tamil Nadu	400
Do.		do.	Silvassa refinery, Gujarat	300
Do.		Jhagadis Copper Ltd.	Jhagadia, Gujarat	50
Diamond	carats	National Mineral Development Corp. Ltd. (NMDC) (Government, 100%)	Mahjgawan Mine	25,000
Gold	kilograms	Hutti Gold Mines Co.	Hutti Mine, Karnataka	3,000
Graphite		Agrawal Graphite Industries Ltd.	Belpara District, Odisha	12
Do.		Tamil Nadu Minerals Ltd.	Sivaganga District, Tamil Nadu	NA
Iron and steel, crude steel		Visvesvaraya Iron and Steel Ltd. (Karnataka State government, 60%, and Government-owned Steel Authority of India Ltd., 40%)	Bhadravati steel plant, Karnataka	180
Do.		Steel Authority of India Ltd. (Government, 100%)	Bhilai steel plant, Jharkhand	4,930
Do.		do.	Bokaro steel plant, Jharkhand	4,600
Do.		Indian Iron and Steel Co. Ltd. (a wholly owned subsidiary of Government-owned Steel Authority of India Ltd.), 100%	Burnpur steel plant, West Bengal	1,500
Do.		Ispat Industries Ltd.	Dolvi, Maharashtra	3,000
Do.		Steel Authority of India Ltd. (Government, 100%)	Durgapur steel plant, West Bengal	1,600
Do.		Tata Steel Ltd.	Jamshedpur steel plant, Jharkhand	6,800
Do.		do.	Jagdalpur, Chhattisgarh	2,000
Do.		do.	Duburi, Odisha	3,000
Do.		Steel Authority of India Ltd. (Government, 100%)	Rourkela steel plant, Odisha	1,800
Do.		Rashtriya Ispat Nigam Ltd.	Visakhapatnam steel plant, Andhra Pradesh	3,200
Do.		JSW Steel Co. Ltd.	Vijayanagar, Karnataka	7,800
Do.		Ministeel plants (privately owned)	180 plants located throughout India	4,700
Do.		Essar Steel Co. Ltd.	Hazira, Gujarat	3,000
Do.		Lloyds Steel Industries Ltd.	Wardha, Maharashtra	500
Do.		MSP Steel and Power Ltd.	Raipur, Chhattisgarh	750

See footnotes at end of table.

TABLE 2—Continued
INDIA: STRUCTURE OF THE MINERAL INDUSTRY IN 2012

(Thousand metric tons unless otherwise specified)

Commodity	Major operating companies and major equity owners	Location of main facilities	Annual capacity[e]
Iron ore	National Mineral Development Corp. Ltd. (NMDC) (Government, 100%)	Bailadila, Chhattisgarh	9,000
Do.	Steel Authority of India Ltd. (Government, 100%)	Bastar and Durg District, Chhattisgarh; Bolani, Odisha; and Chiria, Jharkhand	7,000
Do.	Kudremukh Iron Ore Co. Ltd. (Government, 100%)	Kudremukh, Chikmagalur District, Karnataka	10,300
Do.	National Mineral Development Corp. Ltd. (NMDC) (Government, 100%)	Donimalai, Karnataka	9,000
Do.	Chowgule and Co. Ltd.	Goa	2,500
Do.	Dempo Mining Corp. Ltd.	do.	2,500
Do.	V.M. Salgaocar & Bros. Pvt. Ltd.	do.	2,500
Do.	Sesa Goa Ltd. (Vedanta Resources plc, 51%)	Codli and Sonshi, Goa	NA
Do.	Steel Authority of India Ltd. (Government, 100%)	Kendujhar District, Odisha	3,000
Do.	Tata Steel Ltd.	do.	2,000
Do.	NSL Consolidated Ltd. (China Metallurgical Group Corp., 10%)	Mangal, Andhra Pradesh	200
Do.	Indian Iron and Steel Co. Ltd. (a wholly owned subsidiary of Government-owned Steel Authority of India Ltd.), 100%	Singhbhum District, Bihar	2,500
Do.	Steel Authority of India Ltd. (Government, 100%)	do.	3,500
Do.	Tata Steel Ltd.	do.	3,500
Kaolin	20 Microns Ltd.	Bhuj, Gujarat	65
Do.	English India Clays Ltd.	Veli, Kerala	240
Kyanite	Associated Mining Co.	Bhandara District, Maharashtra	10
Do.	Maharashtra Mineral Corp. Ltd.	do.	10
Do.	Bihar State Mineral Development Corp. Ltd. (Bihar State government, 100%)	Singhbhum District, Bihar	10
Do.	Hindustan Copper Ltd. (HCL) (Government, 100%)	do.	22
Lead:			
Primary	Hindustan Zinc Ltd. (Sterlite Opportunities and Ventures Ltd., 64.9%, and Government, 29.5%)	Chanderiya smelters, Rajasthan	85
Do.	do.	Tundoo smelter, Bihar	8
Secondary	Indian Lead Co.	Thane refinery, Mumbai, Maharashtra	25
Do.	do.	Wada, Mumbai, Maharashtra	40
Lead ore	Hindustan Zinc Ltd. (Sterlite Opportunities and Ventures Ltd., 64.9%, and Government, 29.5%)	Agnigundala Mine, Andhra Pradesh	72
Do.	do.	Sargipalli Mine, Odisha	150
Lead-zinc ore	do.	Rampura-Agucha Mine, Rajasthan	1,300
Do.	do.	Zawar Mine group, Rajasthan	1,200
Magnesite	Steel Authority of India Ltd. (Government, 100%)	Salem, Tamil Nadu	150
Do.	Dalmia Magnesite Corp.	do.	72
Do.	Tamil Nadu Magnesite Ltd. (Tamil Nadu State government, 100%)	do.	150
Manganese ore[2]	Manganese Ore India Ltd. (Government, 100%)	Adilabad, Andhra Pradesh	NA
Do.	Falechand Marsingdas	Andhra Pradesh	NA
Do.	Manganese Ore India Ltd. (Government, 100%)	Balaghat, Madhya Pradesh	NA
Do.	J.A. Trivedi Bros.	do.	NA
Do.	Sandur Manganese and Iron Ores Ltd.	Bellary, Karnataka	NA
Do.	Manganese Ore India Ltd. (Government, 100%)	Bhandara, Maharashtra	NA
Do.	Eastern Mining Co.	North Kanara, Karnataka	NA
Do.	Mysore Minerals Ltd.	do.	NA
Do.	Manganese Ore India Ltd. (Government, 100%)	Keonjhar, Odisha	NA
Do.	Mangilah, Rungta (Pvt.) Ltd.	do.	NA
Do.	Orissa Mining Corp. Ltd.	do.	NA
Do.	Rungta Mines (Pvt.) Ltd.	do.	NA
Do.	S. Lall & Co.	do.	NA

See footnotes at end of table.

TABLE 2—Continued
INDIA: STRUCTURE OF THE MINERAL INDUSTRY IN 2012

(Thousand metric tons unless otherwise specified)

Commodity		Major operating companies and major equity owners	Location of main facilities	Annual capacity[e]
Manganese ore[2]—Continued		Tata Steel Ltd.	Keonjhar, Odisha	NA
Do.		Orissa Mineral Development Co. Ltd.	Koraput, Odisha	NA
Do.		Orissa Mining Corp. Ltd.	do.	NA
Do.		Mysore Minerals Ltd.	Shimoga, Karnataka	NA
Do.		Aryan Mining & Trading Corp.	Sundargarh, Odisha	NA
Do.		Orissa Manganese & Minerals (Pvt.) Ltd.	do.	NA
Do.		Tata Steel	do.	NA
Do.		R.B.S. Shreeram Durga Prasad and Falechand Marsingdas	Vizianagaram, Andhra Pradesh	NA
Mica	metric tons	Micafab India Pvt. Ltd.	Sydapuram Mandal, Andhra Pradesh	4,500
Do.	do.	Premier Mica Co.	Rjupalem, Andhra Pradesh	200
Petroleum, refined products	thousand 42-gallon barrels per day	Cochin Refineries Ltd. (Oil and Natural Gas Corp., 55%, and private interests, 45%)	Ambalamugal refinery, Kerala	93
Do.	do.	Indian Oil Corp. (Oil and Natural Gas Corp., 91%, and private interests, 9%)	Barauni refinery, Bihar	66
Do.	do.	Bongaigaon Refinery and Petrochemicals Ltd. (a subsidiary of Government-owned Oil and Natural Gas Corp.), 100%	Bongaigaon refinery, Assam	27
Do.	do.	Indian Oil Corp. (Oil and Natural Gas Corp., 91%, and private interests, 9%)	Digboi refinery, Assam	12
Do.	do.	do.	Guwahati refinery, Assam	20
Do.	do.	do.	Haldia refinery, West Bengal	61
Do.	do.	Reliance Industries Ltd.	Jamnagar refinery, Gujarat	540
Do.	do.	do.	Koyali refinery, Gujarat	185
Do.	do.	Madras Refineries Ltd. (Oil and Natural Gas Corp., 52%, and private interests, 48%)	Madras refinery, Tamil Nadu	131
Do.	do.	Bharat Petroleum Corp. Ltd. (Oil and Natural Gas Corp., 67%, and private interests, 33%)	Mahul refinery, Mumbai, Maharashtra	135
Do.	do.	Hindustan Petroleum Corp. Ltd. (Oil and Natural Gas Corp., 51%, and private interests, 49%)	do.	110
Do.	do.	Essar Oil Ltd.	Vadinar refinery, Gujarat	375
Do.	do.	do.	Visakhapatnam refinery, Andhra Pradesh	90
Do.	do.	Indian Oil Corp. (Oil and Natural Gas Corp., 91%, and private interests, 9%)	Mathura refinery, Uttar Pradesh	156
Do.	do.	do.	Panipat refinery, Haryana	240
Phosphate rock[3]		Rajasthan State Mineral Development Corp. Ltd. (Rajasthan State government, 100%)	Badgaon, Dakankotra, Kanpur, Kharbaria-ka-Guda, and Sallopat Mines, Rajasthan	NA
Do.		Pyrites Phosphates and Chemicals Ltd.	Durmala and Maldeota underground mines, Uttar Pradesh	NA
Do.		Madhya Pradesh State Mining Corp. Ltd. (Madhya Pradesh State government, 100%)	Hirapur and Khatamba Mines, Jharkhand	NA
Do.		Rajasthan State Mines and Minerals Ltd. (Rajasthan State government, 100%)	Jhamarkotra Mine, Rajasthan	NA
Do.		Hindustan Zinc Ltd. (HZL) (Sterlite Opportunities and Ventures Ltd., 64.9%, and Government, 29.5%)	Maton Mine, Rajasthan	NA
Titanium, ilmenite-rutile ore		Kerala Minerals and Metals Ltd. (Kerala State government, 100%)	Chavara, Kerala	100
Do.		Indian Rare Earths Ltd. (IREL) (Government, 100%)	do.	250
Do.		do.	Ganjam, Odisha	220
Do.		do.	Manavalakurichi, Tamil Nadu	65
Do.		Trimex Group	Chennai, Andhra Pradesh	200
Do.		VV Mineral Ltd.	Kanyakumari, Tamil Nadu	450

See footnotes at end of table.

TABLE 2—Continued
INDIA: STRUCTURE OF THE MINERAL INDUSTRY IN 2012

(Thousand metric tons unless otherwise specified)

Commodity		Major operating companies and major equity owners	Location of main facilities	Annual capacity[e]
Uranium ore	metric tons per day	Uranium Corp. of India	Jaduguda, Jharkhand	2,190
Do.	do.	do.	Tummalapalle, Andhra Pradesh	3,000
Zinc		Binani Zinc Ltd.	Binanipuram smelter, Kerala	38
Do.		Hindustan Zinc Ltd. (HZL) (Sterlite Opportunities and Ventures Ltd., 64.9%, and Government, 29.5%)	Chanderiya smelter, Rajasthan	340
Do.		do.	Debari smelter, Rajasthan	78
Do.		do.	Visakhapatnam (Vizag) smelter, Andhra Pradesh	54

[e]Estimated. Do., do. Ditto. NA Not available.
[1]Scheduled startup was delayed to 2013.
[2]Capacity of clusters of surface mines varies extremely, depending on demand. Estimated total capacity is 3.0 million metric tons per year (Mt/yr).
[3]Estimated total phosphate rock capacity is 1.5 Mt/yr.

TABLE 3
INDIA: ESTIMATED RESERVES OF MAJOR MINERAL COMMODITIES IN 2012

(Thousand metric tons unless otherwise specified)

Commodity		Reserves
Barite		34,000
Bauxite		539,000
Chromite ore		54,000
Coal:		
Bituminous		110,000,000
Lignite		39,000,000
Copper ore		394,000
Gold, in metal	kilograms	67,000
Graphite		11,000
Ilmenite and rutile		193,000
Iron ore		8,120,000
Kyanite and sillimanite		1,380
Lead and zinc ore		63,000
Limestone		7,500,000
Magnesite		70,000
Manganese ore		49,000
Phosphate rock		34,800
Talc and pyrophyllite		74,600
Uranium		150
Zircon		1,350

Source: Indian Minerals Yearbook 2010, Indian Bureau of Mines.

THE MINERAL INDUSTRY OF INDONESIA

By Chin S. Kuo

Indonesia is rich in reserves of copper and gold. In addition, the country has abundant mineral resources, such as coal, natural gas, nickel, tin, and, in smaller amounts, bauxite, petroleum, and silver. Indonesia's tin output was ranked second in the world after China (Carlin, 2013). The country was ranked among the world's 5 leading producers of copper and nickel and among the world's top 10 producers of gold and natural gas. Indonesia was one of the world's leading exporters of bituminous coal, liquefied natural gas (LNG), and refined tin (Ministry of Industry, 2013, p. 9).

Minerals in the National Economy

Indonesia's real gross domestic product (GDP) growth was 6.2% in 2012. The country's industrial output accounted for 23.9% of its total GDP. The value of mineral commodity production accounted for 11.8% of the GDP. The country's mineral industry—primarily the cement, metal mining, and oil and gas industries—contributed mostly to domestic industrial production. The industrial sector grew by 5.7% in 2012. The fertilizer industry and the mining and quarrying industries grew by 10.3% and 1.5%, respectively, during the year. The cement and iron and steel industries increased by 7.9% and 6.5%, respectively, whereas the oil and gas industry registered negative growth of 2.7% (Ministry of Industry, 2013, p. 9, 50–51).

Government Policies and Programs

In 2012, the Government established a team to evaluate the adjustment of contracts of work (COW) and coal contracts of work (CCOW), as required by the 2009 Law on Mineral and Coal Mining. In addition, the evaluation team was to determine the Government's position on mining work areas and Government revenue and to enforce COW and CCOW holders' obligations regarding processing and refining of minerals and coal. Funding of the team would come from the Ministry of Energy and Mineral Resources (Surowidjojo, 2012a).

Regulation No. 7 of 2012 on increasing the added value of minerals through processing and refining was passed on February 6, with the aim of developing the country's domestic mineral processing industry and deriving more revenue from its mineral sector. Value-added minerals affected by the regulation include metals, nonmetallic minerals, coal, and stone. The regulation sets out the minimum levels of processing that the minerals must be subjected to prior to export and prohibits the export of unprocessed minerals (minerals in raw form). This ban on unprocessed mineral exports was to be imposed gradually, beginning in May 2012, with full implementation in 2014. The regulation provides for cooperation among the holders of mining permits and other parties with respect to the sale and purchase of ores or concentrates, to activities to undertake processing and (or) refining, and to the joint development of processing and (or) refining facilities or infrastructure (Surowidjojo, 2012b).

The regulation to restrict ore exports went into effect on May 1, 2012, beginning with bauxite and certain other unprocessed metal ores. Tin ore exports had been banned in 2010, and PT Timah had built smelters and exported refined tin. It was estimated that the export ban that began in May could result in a 75% decrease in the exports of bauxite and nickel ore (together) for the year (Yieh Corp., 2012).

On May 6, the Government also imposed a 20% duty on 14 mineral ore exports that were not yet subject to the export ban, including copper, gold, and nickel. Later in the year, the list was extended to include 21 other mineral commodities. In total, 65 specific types of mineral ores and concentrates were subject to the duty. The export duty does not apply to coal. The duty is designed to increase revenues from the mining sector and is part of the Government's effort to push mining companies to process raw ore domestically and export higher-value finished metals (Mining Weekly, 2012).

Under additional new rules that went into effect in 2012, the Government requires foreign companies to reduce their stakes in mines by the 10th year of production so that domestic ownership is at least 51%. The move is part of a global trend toward increased resource nationalization but is likely to hinder new investments in mining (Thaher and Chatterjee, 2012). Meanwhile, the Government was renegotiating existing contracts with Freeport McMoRan Copper & Gold Inc. and Newmont Mining Corp., both of the United States. Freeport McMoRan owned 90.64% of the joint venture that operated the Grasberg copper and gold mine and the Government owned 9.36%; the company agreed to divest a 9.36% interest to a potential acquirer, such as the Province of Papua. Rio Tinto plc of the United Kingdom was expected to retain the rights to 40% of production from the Grasberg Mine from 2021. Newmont Mining had already divested some of its interest in PT Newmont Nusa Tenggara in 2009 and owned only a minority stake in 2012. Newcrest Mining Ltd. of Australia, which owned an 82.5% stake in the Gosowong Mine [state-owned PT Antam Tbk (Antam) owned 17.5%], would not be affected until its existing COW runs out in 2029. Kingsrose Mining Ltd. of Australia under its existing COW was supposed to start selling down its 85% stake in PT Natarang Mining to 49% beginning in 2012. Intrepid Mines Ltd., also of Australia, which had a mining concession for the Tujuh Bukit copper-gold-silver mine in East Java Province, might also be affected (Australia's Paydirt, 2012a).

Production

In 2012, production of bauxite decreased by an estimated 27.5% owing to Government's restriction on bauxite exports, which led to a cutback in output. Production of mined copper decreased by 33.7% owing to the lower grade of the ore mined at Grasberg. As a result, the output of smelted copper also decreased by 27.6% compared with that of 2011. The output of gold and silver—both byproducts of copper mining—decreased by 38.8% and 19.5%, respectively. Because of strong domestic demand (fueled in part by the Government's support for housing and infrastructure development), production of cement increased by 13.3%. Output of phosphate rock was estimated to have increased by 33% after several years of steady production owing to increasing demand for phosphorus fertilizers (table 1).

Structure of the Mineral Industry

State-owned PT Antam Tbk (Antam) produced bauxite, gold, nickel, and silver. Other state-owned companies—PT Krakatau Steel, PT Pertamina, PT Tambang Batubara Bukit Asam, and PT Tambang Timah Tbk—were engaged in the production of steel, oil, coal, and tin, respectively. Privately owned PT Indocement Tunggal Prakarsa Tbk was the leading cement producer in the country. International companies were active in Indonesia's metals mining and processing industries. Partially foreign-owned PT Freeport Indonesia Co. and PT Newmont Nusa Tenggara were engaged in the mining of copper and gold. PT International Nickel Indonesia Tbk produced nickel ore and matte, and PT Koba Tin produced tin ore and tin metal (table 2).

Mineral Trade

In 2012, Indonesia's total exports were valued at $190.0 billion; mineral commodity exports included minerals and metals (bauxite, copper, nickel, and tin) and mineral fuels (coal, LNG, natural gas, and petroleum). The major export partners were, in descending order of export value, China, Japan, the United States, India, Singapore, Malaysia, and the Republic of Korea. Total imports were valued at $191.7 billion, and mineral commodity imports included crude petroleum, iron and steel, and petroleum products. The major import partners were, in descending order of the value of the imports received, China, Japan, the United States, Singapore, Thailand, the Republic of Korea, and Malaysia (Ministry of Industry, 2013, p. 9, 54–55).

Commodity Review

Metals

Bauxite and Alumina.—Indonesia was the sixth-ranked bauxite producing country in the world. Outotec Oyj of Finland agreed with Antam to carry out a feasibility study for a smelter-grade alumina refinery to be built in North Sumatra Province. The feasibility study would include comprehensive mineralogical investigations, laboratory-scale hydrometallurgical tests, and basic engineering of the alumina refinery using Outotec® alumina refining technologies. The country had an aluminum smelter in North Sumatra Province that imported alumina from Australia (Outotec Oyj, 2012).

The Government awarded mineral export permits to 22 companies, including two permits for bauxite producers. Bauxite ore was exported principally to China and Japan. Indonesian exports accounted for 49% of global bauxite exports in 2011. The export restrictions on unprocessed minerals that Indonesia imposed in 2012 caused Japan to consider taking action against Indonesia through the World Trade Organization. China reduced its domestic alumina and aluminum production capacities in June and might increase its bauxite imports from Australia (Syrett, 2012).

Copper.—PT Emas Mineral Murni's Beutong copper project is located in Aceh Province, 180 kilometers (km) southeast of Banda Aceh. The tenement covered an area of 100 square kilometers and was held under an exploration license. Tigers Realm Minerals Pty. Ltd. held a 40% equity interest in the project and had an agreement with Emas Mineral Murni to earn up to 80% equity interest, depending on the results of exploration. The project was a large porphyry copper (with gold and molybdenum) deposit; skarn-hosted copper-gold mineralization was located approximately 300 meters (m) north of the porphyry system. Tigers Realm completed 4,150 m of drilling focused on infill and extension of the higher grade central core of the porphyry deposit and on defining and expanding the skarn-hosted high-grade copper-gold mineralization. The goal of the exploration program was to develop a resource of 100 million metric tons (Mt). A feasibility study and open pit mining were planned pending development of the resource (Tigers Realm Group, 2012).

Gold and Silver.—G-Resources Group Ltd. of Hong Kong revised its schedule for first commercial production of gold at its Martabe mining project in North Sumatra Province to July 2012. Initial ore mining began at the Purnama pit, and the project was about 80% complete by mid-February. The mine had a resource of 245 metric tons (t) (7.86 million troy ounces) and 2,290 t (73.48 million troy ounces) of contained gold and silver, respectively. Production in 2012 called for 7.8 metric tons per year (t/yr) (250,000 troy ounces per year) of gold and 60 to 90 t/yr (2 million to 3 million troy ounces per year) of silver (G-Resources Group Ltd., 2012).

Straits Resources Ltd. controlled and operated the Mt. Muro gold mine west of Balikpapan, Central Kalimantan Province, through its 100%-owned subsidiary PT Indo Muro Kencana. The company upgraded the mine's resource by 56% and extended the mine life to 6 years. The mine was operated under a third-generation COW, which covered 47,940 hectares (ha). Production at Mt. Muro was expected to ramp up to targeted output of 3.11 t (100,000 troy ounces) of gold equivalent in 2012 (Straits Resources Ltd., 2012).

PT Agis Resources signed an agreement to acquire a 60% stake in a gold mining company in West Sumatra Province. The gold mining company held a mining license on a 2,500-ha area. Agis Resources was a joint venture between PT Agis (51%) and Fujian Xinjifu Enterprises Group Co. Ltd. of China (49%). PT China Coal Geology Mining would be the contractor for gold exploration and exploitation (Jakarta Post, The, 2012).

Sumatra Copper and Gold plc of Australia brought a relatively small-scale Tembang brownfield gold mining project in West Sumatra Province into production. The project included two small open pits and an underground mine, and the company planned to keep construction and mining costs as low as possible. The expected life of the operation was relatively short, but might be extended as additional exploration in and around the mining area was underway. The Tembang vein system was an intermediate-sulfidation epithermal gold mineralization with gold- and silver-bearing quartz veins hosted by Tertiary volcanics. Proven and probable reserves were estimated to contain 12.5 t (403,000 troy ounces) of gold and 171 t (5.5 million troy ounces) of silver (Mineweb.com, 2012).

Robust Resources Ltd. finished a scoping study of its Lakuwahi polymetallic project on Romang Island in 2012. After completing more than 150 drill holes at the Bata Hitam, Batu Jagung, and Batu Mas prospects, a resource of 36.7 t (1.18 million troy ounces) gold equivalent was estimated. In terms of precious and base metals, there were 18.4 t (592,000 troy ounces) of gold, 862 t (27.7 million troy ounces) of silver, 43,100 t of copper, 316,000 t of lead, and 308,000 t of zinc. The oxide ore would be for heap leaching, and production of gold and silver was slated for 2014. More drilling would require $15 million from the company's funds (Australia's Paydirt, 2012b).

Tin.—An estimated 14 tin smelters in Indonesia halted production after tin prices fell in August. Although other smelters were still operating, they scaled back their output by between 20% and 40%. PT Timah was the country's leading producer of tin but lost market share to the private smelters, which purchased most of the small-scale mine output. PT Koba's COW was due to expire in March 2013, and all operations were expected to be suspended pending the outcome of an application for a 10-year extension. Indonesia was the world's leading exporter of tin, accounting for 40% of global exports. Tin was used in electronics and packaging (Rusmana and Listiyorini, 2012).

Industrial Minerals

Cement.—State-owned PT Semen Gresik Tbk completed the new $304 million Tuban IV cement plant in East Java Province with a production capacity of 2.5 million metric tons per year (Mt/yr) in May. The main equipment was the vertical cement mill, which had the capacity to produce 500 metric tons per hour. The company's total production capacity at Tuban would be increased to 22.5 Mt/yr from 20 Mt/yr in 2011. With the start of operations at the Tonasa V cement plant in South Sulawesi Province in the second quarter of 2012, the company's production capacity at Tunasa increased to 25 Mt/yr by yearend. The company planned to build two new cement plants with a capacity of 3 Mt/yr each in 2013—one at Padang, West Sumatra Province, and the other at Rembang, Central Java Province—for a cost of $756 million. The company also planned to change its name to Semen Indonesia to raise its international image (PT Semen Gresik Tbk, 2012).

PT Indocement Tunggal Prakasa planned to build a $500 million cement plant with a production capacity of 3 Mt/yr in the Regency of Pati, Central Java Province. The process of licensing the plant was expected to be completed in 2012, followed by construction work. The plant was expected to be operational in 2015 (Global Cement News, 2012).

Anhui Conch Cement of China planned to begin construction of a $400 million cement plant at Tanjung, South Kalimantan Province, in 2012. The plant planned to produce 2.5 Mt/yr of cement for domestic use. The plant included a cement-grinding line, a port, a 60-megawatt-capacity powerplant, and other supporting infrastructure. The company was waiting for the completion of the land acquisition process and an operation license (Aggregate Research, 2012).

PT Holcim Indonesia Tbk planned to increase its cement production capacity by 1.7 Mt/yr in 2014 when its new $450 million Tuban cement plant in East Java Province is completed. The company's grinding capacity in 2012 was 8.2 Mt/yr. The plant would serve Java and the inter-island cement market. Coal for the kiln would be sourced from the island of Kalimantan. The company secured a loan of $150 million from BNP Paribas and KWF IPEX-Bank of Germany (International Cement Research, 2012b).

PT Semen Bosowa Maros had a cement plant at Maros in South Sulawesi Province that underwent a phase 2 expansion to reach a capacity of 3.75 Mt/yr. The project involved an investment of $326 million and was scheduled for completion in 2014. The company also planned to construct cement plants at Rembang in Central Java Province and at Banyuwangi in East Java Province and planned to have in place a total capacity of 11 Mt/yr in 2015 (International Cement Research, 2012a).

Mineral Fuels

Coal.—PT Adaro Energy Tbk reported the estimated coal resources of its subsidiary PT Mustika Indah Permai (MIP)'s property at Lahat in South Sumatra Province to be 286.4 Mt, of which 272.6 Mt was the estimated coal reserves. At a production level of 10 Mt/yr, the coal reserves were sufficient to sustain a mine life of 26 years. The concession area contained three main coal seams and two minor coal seams. The seams were suitable for extraction by open pit mining. Coal production was expected to begin in the second half of 2012. Mining and hauling contractor PT Saptaindra Sejati would be used to mine and transport coal. Adaro Energy owned a 75% equity interest in MIP, which was acquired from Elite Rich Investment Ltd. of Hong Kong for $222.5 million. The MIP production license was granted in April 2010 for a period of 20 years and was extendable up to two times. The coal was relatively low in sulfur and low in ash, and would be marketed to China, India, Indonesia, the Philippines, Taiwan, Thailand, and Vietnam (PT Adaro Energy Tbk, 2012).

Realm Resources Ltd. of Australia reported a 29% increase in the coal resources of its Katingan Ria project, which is located northwest of Palangkarya in Central Kalimantan Province, to 102.2 Mt. The concession area covered about 4,250 ha. A feasibility study was undertaken to develop the project by the opencut mining method. The project was permitted to allow mining operations to begin in June 2011. The company also completed its phase 2 exploration drilling program (42 holes for 2,844 m). Realm Resources held a 51% interest in Katingan Ria and had an option to increase its ownership to 75% (Realm Resources Ltd., 2012).

Padang Resources Ltd. of Australia planned to acquire a 70% interest in Paser Pte. Ltd. of Singapore, which, in turn, planned to acquire a 100% joint-venture interest with

PT Gunung Mentari Mining in a coal project at Petangis, Batu Engau District, East Kalimantan Province. The project covered an area of 43 ha, and Padang Resources was in the process of doing reconnaissance and drilling work to validate the economic viability of the project. The company received a due diligence report on the permit holder, who had obtained a mining license for the production, transport, and sales of coal (Boston.com, 2012).

Yinfu Gold Corp. of the United States agreed to acquire a 51% interest in Hitric Resources Pte. Ltd. of Singapore, which owned an 80% interest in a coal mine in Tanah Bumbu, South Kalimantan Province. The exploration license covered 1,116 ha and was upgraded into a production license with an expected period of 15 years. Production was scheduled to start in March 2013 with an initial output of 150,000 t/yr of bituminous coal and was expected to double to 300,000 t/yr in the second year of production (GlobeNewswire, 2012).

Borneo Resource Investments Ltd. began producing bituminous coal from its PT Integra Prima Coal concession in East Kalimantan Province, which covered an area of 1,300 ha and had an estimated reserve of 8 Mt of coal. Analysis of the coal showed a high thermal value of 7,400 British thermal units per metric ton and low ash and sulfur content. The company negotiated a 1-year supply contract with an India-based buyer to provide 50,000 metric tons per month of coal (Yahoo! Finance, 2012).

United Tractors (a unit of Astra International of Indonesia), which was Indonesia's leading heavy equipment distributor, was considering the acquisition of two coal mines in East Kalimantan Province and Central Kalimantan Province, respectively, with funds from its $700 million rights issuance in 2011. The mines would increase the company's coal production capacity to 6.5 Mt/yr from 4.5 Mt/yr in 2011. Coal mining was the company's new business division and contributed 10% of the company's revenue in 2011 (Jakarta Globe, The, 2012).

Cokal Ltd. of Australia, which was a global metallurgical coal group, completed an initial report for the Bumi Barito Mineral (BBM) coal project in Central Kalimantan Province on using barges on the Barito River as the primary transportation mode for coal. The report included options to increase the tonnage per trip and the number of operating days for barging operations. The company reported an inferred resource of 60 Mt of metallurgical coal from the BBM project and continued to define further coal resources (Cokal Ltd., 2012b). Cokal also acquired a 75.2% interest in PT Silangkop Nusa Raya, which held three exploration licenses (about 13,000 ha) in West Kalimantan Province. Local Indonesian partners held the remaining equity interest. Analysis of surface samples of coal from surrounding areas showed good coking coal properties. Initial mapping and preliminary geologic work were expected in the second half of 2012 (Cokal Ltd., 2012a).

Indonesia and Joint Stock Company Russian Railways planned to build a $2.4 billion rail line in East Kalimantan Province that would be used initially to transport coal and was expected to be operational in the first quarter of 2017. The Russian company did not plan to invest directly in the project but would work on its technical and economic feasibility studies. Construction of the rail line was to begin in 2013. The first phase of the project would cost $1.7 billion and would include 185 km of rail line to carry 20 Mt/yr of coal. The line would begin in Balikpapan Port and run through Kutai Barat Regency to the border of Central Kalimantan Province in Murung Raya Regency; an extension of 60 km would be added in the second phase. Funding would come from the private sector and the Russian state development bank Vnesheconombank. Indonesia's bituminous coal was exported mostly to China and India in 2012 (Reuters, 2012).

Natural Gas.— PT Pertamina and state-owned electricity utility Perusahaan Listrik Negara (PLN) might delay their joint development of eight mini-LNG receiving terminals and powerplants because Pertamina had not yet managed to secure the LNG to feed the regasification plants. The eight mini-regasification plants would have a combined capacity of 1 Mt/yr of LNG and had been expected to come into operation during the period from 2012 to 2014. The facilities were intended to supply gas to PLN-operated powerplants at 10 sites (Petroleum Economist, 2012).

Petroleum.—Pertamina awarded a contract to Foster Wheeler AG of Switzerland for management of an upgrade of its 348,000-barrel-per-day (bbl/d) Cilacap oil refinery on Java Island. The project included construction of a 62,000-bbl/d residual fluid catalytic cracking complex, a liquefied petroleum gas (LPG) Merox unit, and propylene recovery and gasoline hydrotreating units. The upgrade would increase production of LPG by 350,000 t/yr and produce 140,000 t/yr of propylene (Oil & Gas Journal, 2012).

Outlook

The new Government regulations that focus on value-added mineral products and ban exports of unprocessed minerals are expected to stimulate domestic and foreign investments in downstream mineral extraction industries. Antam's alumina refinery project is an example of following the regulation on value-added products in the years to come. For the near future, production of tin is expected to decrease because of the continued depressed tin prices. The cement industry is in the process of expanding capacity and is expected to add about 13 Mt/yr in the next 2 to 3 years. The Government is expected to encourage investment in new oil and gas exploration to stem the decline in production.

References Cited

Aggregate Research, 2012, Anhui Conch plans US$400m plant for South Kalimantan: Aggregate Research, March 19. (Accessed March 21, 2012, at http://www.aggregateresearch.com/article.aspx?id=24720.)

Australia's Paydirt, 2012a, Australian miners play down Indonesia law impact: Australia's Paydirt, April, v. 1, issue 193, p. 92.

Australia's Paydirt, 2012b, Robust's maiden monkey off its back: Australia's Paydirt, February, v. 1, issue 191, p. 8.

Boston.com, 2012, Padang Resources Ltd.'s acquisition of Paser Resources Pte. Ltd.: Boston.com, April 16. (Accessed May 8, 2013, at http://finance.boston.com/boston/news/read?guid=21084168.)

Carlin, J.F., Jr., 2013, Tin: U.S. Geological Survey Mineral Commodity Summaries 2013, p. 170–171.

Cokal Ltd., 2012a, Cokal expands met coal potential into West Kalimantan: Brisbane, Australia, Cokal Ltd. media release, April 26, 3 p.

Cokal Ltd., 2012b, Results of barge scoping study for metallurgical coal project in Central Kalimantan, Indonesia: Brisbane, Queensland, Australia, Cokal Ltd. media release, February 16, 2 p.

Global Cement News, 2012, New Java plant for Indocement: Global Cement News, January 31. (Accessed February 13, 2012, at http://www.globalcement.com/news/itemlist/tag/gcw34?utm_source=newsletter&utm_medium=email&utm_campaign=gcw34.)

GlobeNewswire, 2012, Yinfu Gold Corp. acquires interest in Kalimantan coal mine: GlobeNewswire, November 1. (Accessed November 7, 2012, at http://www.globenewswire.com/news-release/2012/11/01/501697/10010701/en/.)

G-Resources Group Ltd., 2012, G-Resources updates schedule for Martabe project: Hong Kong, China, G-Resources Group Ltd. media release, February 22, 2 p.

International Cement Research, 2012a, Bosowa's eastern initiatives: International Cement Research, June, p. 32.

International Cement Research, 2012b, Holcim improves Java operations: International Cement Research, June, p. 30.

Jakarta Globe, The, 2012, United Tractors considers coal deals: The Jakarta Globe, April 20. (Accessed April 23, 2012, at http://www.thejakartaglobe.com/business/united-tractors-considers-coal-deals/512866.)

Jakarta Post, The, 2012, Electronic retailer Agis to acquire gold miner: The Jakarta Post, December 18, p. 13.

Mineweb.com, 2012, New Sumatra low cost gold project eyes peers Way Linggo and Martabe: Mineweb.com, April 25. (Accessed April 26, 2012, at http://www.mineweb.com/mineweb/view/mineweb/en/page103118?oid=150140&sn=detail&pid=72730.)

Mining Weekly, 2012, Indonesia to impose 20% tax on raw metal exports: Mining Weekly, May 3. (Accessed May 4, 2012, at http://www.miningweekly.com/print-version/indonesia-to-impose-20-tax-on-raw-metal-exports-2012-05-03.)

Ministry of Industry, 2013, Industry facts and figures: Jakarta, Indonesia, Ministry of Industry, June, 65 p.

Oil & Gas Journal, 2012, Contract let for Cilacap refinery upgrade: Oil & Gas Journal, March 27. (Accessed April 16, 2012, at http://www.ogj.com/articles/2012/03/contract-let-for-cilacap-refinery-upgrade.html?cmpid=EnlRefiningApril162012.)

Outotec Oyj, 2012, Outotec to conduct feasibility study for the first alumina refinery to be built in Indonesia: Espoo, Finland, Outotec Oyj press release, March 8, 1 p.

Petroleum Economist, 2012, News in brief: Petroleum Economist, v. 79, no. 4, May, p. 49.

PT Adaro Energy Tbk, 2012, Adaro Energy completes JORC report for Mustika Indah Permai coal reserves: Jakarta, Indonesia, PT Adaro Energy Tbk news release, March 22, 5 p.

PT Semen Gresik Tbk, 2012, SMGR's new cement plant of Tuban IV operates: PT Semen Gresik Tbk, April 9. (Accessed August 2, 2012, at http://www.semengresik.com/eng/post/SMGRe28099S.)

Realm Resources Ltd., 2012, Flagship coal project increases resources by 29%: Sydney, New South Wales, Australia, Realm Resources Ltd. media release, March 19, 4 p.

Reuters, 2012, Indonesia, Russian firm plan $2.4 billion rail line: ThomsonReuters, February 7. (Accessed February 9, 2012, at http://www.reuters.com/article/2012/02/07/indonesia-infrastructure-russia-idUSL4E8D72R520120207.)

Rusmana, Yoga, and Listiyorini, Eko, 2012, Indonesian tin smelters halt output amid bear market rout: Bloomberg Businessweek, August 7. (Accessed August 13, 2012, at http://www.businessweek.com/printer/articles/302258?type=bloomberg.)

Straits Resources Ltd., 2012, Operations—Mt Muro gold mine: Straits Resources Ltd. (Accessed May 6, 2013, at http://www.straits.com.au/operations/mt-muro-gold.html.)

Surowidjojo, Lubis, 2012a, Establishment of an evaluation team for contract of work and coal contract of work adjustments: Legal500.com, February. (Accessed March 8, 2012, at http://www.legal500.com/c/indonesia/developments/17230.)

Surowidjojo, Lubis, 2012b, Minister of Energy and Mineral Resources implementing regulation on domestic mineral processing: Legal500.com, February. (Accessed March 8, 2012, at http://www.legal500.com/c/indonesia/developments/17561.)

Syrett, Laura, 2012, Indonesia issues additional bauxite mining permits: Industrial Minerals, July 3. (Accessed February 11, 2015, via http://www.indmin.com/Article/3055190/AluminaBauxite-LatestNews/Indonesia-issues-additional-bauxite-mining-permits.html).

Thaher, Reza, and Chatterjee, Neil, 2012, Indonesia rattles foreign miners with '51% after 10 years' ownership change: Mineweb.com, March 7. (Accessed March 8, 2012, at http://www.mineweb.com/mineweb/view/mineweb/en/page72068?oid=146885&sn=detail&pid=110649.)

Tigers Realm Group, 2012, The Beutong copper project: Tigers Realm Group. (Accessed October 11, 2012, at http://www.tigersrealmgroup.com/beutong-copper-project.php.)

Yahoo! Finance, 2012, Borneo Resource Investments Ltd. commences surface coal harvesting at its PT Integra Prima Coal concession: Yahoo! Finance, June 28. (Accessed June 29, 2012, at http://finance.yahoo.com/news/borneo-investments-ltd-commences-surface-130000772.html.)

Yieh Corp., 2012, Indonesia bans bauxite exports: Yieh Corp. Market News, May 3. (Accessed May 4, 2012, at http://www.yieh.com/2.2.01.01stainlesssteelnews.aspx? no=57277&division=A7.)

TABLE 1
INDONESIA: PRODUCTION OF MINERAL COMMODITIES[1]

(Metric tons unless otherwise specified)

Commodity		2008	2009	2010[e]	2011[e]	2012[e]
METALS						
Aluminum:						
Bauxite, wet basis, gross weight[e]	thousand metric tons	17,000	15,000	27,000	40,000	29,000
Metal, primary		242,500	257,600	253,300 [2]	244,100 [2]	248,000 [2]
Chromite sand, dry basis[e]		1,000	1,000	1,000	1,000	1,200
Cobalt, mine, Co content[e]		1,300	1,200	1,600	1,200 [r]	1,300
Copper:						
Mine, Cu content		632,600	998,530	878,376 [2]	542,700 [2]	360,000
Metal:						
Smelter, primary		253,300	295,900	276,800 [2]	276,200 [2]	200,000
Refinery, primary		254,000	289,200	278,200 [2]	257,000 [2]	272,000
Gold, mine output, Au content[3]	kilograms	64,390	140,488	106,316 [2]	96,100 [2]	58,800 [2]
Iron and steel:						
Iron sand, dry basis[e]		65,000	44,552 [2]	45,610 [2]	46,000	48,000
Metal:						
Ferroalloys:						
Ferronickel		87,800	62,700	93,300 [2]	98,200 [r]	100,000
Ferromanganese[e]		12,000	12,000	12,000	12,000	13,000
Silicomanganese[e]		7,000	7,000	8,000	8,000	9,000
Pig iron, direct-reduced iron	thousand metric tons	1,290	1,230	1,360 [2]	1,200 [r]	1,300
Steel, crude[e]	do.	3,915 [2]	3,500	3,700	3,600 [r]	3,700
Steel, semimanufactured[e]	do.	5,200	5,000	4,900	5,100	5,000
Manganese:						
Ore and concentrate, gross weight[2]		183,000	253,600	207,400	119,100	138,000
Mn content		64,100	88,800	72,600	41,700	39,500 [2]
Nickel:						
Mine output, Ni content[2, 4]		219,300	202,800	235,800	218,200	228,000
Matte, Ni content		73,356	68,228	77,186 [2]	67,800 [r, 2]	70,000
Ferronickel, Ni content		17,566	12,550	18,688 [2]	19,700 [2]	20,000
Silver, mine output, Ag content	kilograms	226,051	359,451	271,534 [2]	310,400 [2]	250,000
Tin:						
Mine output, Sn content		53,228	46,078	43,258 [2]	42,000	41,000
Metal[5]		53,471	51,418	43,832 [2]	43,000	42,000
Titanium mineral concentrates, ilmenite, gross weight		9,000	9,000	60,000	18,000	20,000
Zirconium concentrates, gross weight[e]		65,000	63,000	50,000	130,000	120,000
INDUSTRIAL MINERALS						
Cement, hydraulic	thousand metric tons	38,530 [r]	36,910 [r]	39,500 [r, 2]	45,000 [r, 2]	51,000 [2]
Clays:[e]						
Bentonite		6,000	6,000	6,500	6,500	7,000
Fire clay	thousand metric tons	2,100	2,200	2,200	2,300	2,300
Kaolin powder		150,000	186,010 [2]	170,000	175,000	180,000
Diamond:[e]						
Industrial	thousand carats	28	28	30	30	31
Gem	do.	7	7	7	7	7
Total	do.	35	35	37	37	38
Feldspar[e]		26,000	10,730 [2]	20,000	18,000	19,000
Gypsum[e]		6,000	8,133 [2]	7,000	7,500	8,000
Nitrogen, N content of ammonia[e]	thousand metric tons	4,500	4,600	4,800	5,000	5,100
Phosphate rock[e]		600	600	600	600	800
Salt, all types[e]	thousand metric tons	700	585 [2]	600	650	700

See footnotes at end of table.

TABLE 1—Continued
INDONESIA: PRODUCTION OF MINERAL COMMODITIES[1]

(Metric tons unless otherwise specified)

Commodity		2008	2009	2010[e]	2011[e]	2012[e]
INDUSTRIAL MINERALS—Continued						
Stone:[e]						
Dolomite		3,300	1,885 [2]	2,500	2,400	2,600
Granite	thousand metric tons	4,400	4,500	4,600	4,700	4,800
Limestone	do.	1,800	1,912 [2]	1,900	2,000	2,000
Marble	do.	7,000	7,489 [2]	8,000	7,800	8,000
Quartz sand and silica stone		38,000	32,105 [2]	36,000	37,000	38,000
Sulfur, elemental[e]		500	473 [2]	500	520	540
Zeolite[e]		1,400	1,530 [2]	1,400	1,500	1,600
MINERAL FUELS AND RELATED MATERIALS						
Coal:						
Anthracite[e]	thousand metric tons	54,000	34,348 [2]	118,988 [2]	110,000	100,000
Bituminous	do.	188,717	196,209	137,801 [2]	150,000	140,000
Gas, natural:						
Gross	million cubic meters	81,842	73,587	77,741 [2]	80,000	79,000
Marketed[e]	do.	78,985 [2]	70,000	75,000	76,000	74,000
Petroleum, crude including condensate	thousand 42-gallon barrels	311,000	346,000	341,000 [2]	340,000	342,000

[e]Estimated; estimated data are rounded to no more than three significant digits; may not add to totals shown. [r]Revised. do. Ditto.
[1]Table includes data available through September 11, 2013.
[2]Reported figure.
[3]Includes Au content of copper ore and output by Government-controlled foreign contractor operations. Gold output by operators of so-called people's mines and illegal small-scale mines is not available but may be as much as 20 metric tons per year (t/yr).
[4]Includes a small amount of cobalt that was not recovered separately.
[5]Output by Central Government-controlled foreign contractor operations. Tin output from small tin smelters is not available but may be as much as 40,000 t/yr.

TABLE 2
INDONESIA: STRUCTURE OF THE MINERAL INDUSTRY IN 2012

(Thousand metric tons unless otherwise specified)

Commodity	Major operating companies and major equity owners	Locations of main facilities	Annual capacity[e]
Aluminum:			
Bauxite	PT Antam Tbk (Government, 65%)	Kijang, Bintan Island, Riau	1,300
Metal	PT Indonesia Asahan Aluminum (Nippon Asahan Aluminum Co. Ltd., 59%, and Government, 41%)	Kual Tanjung, North Sumatra	250
Cement	PT Indocement Tunggal Prakarsa Tbk	Cirebon and Citeureup, West Java; Tarjun, South Kalimantan	18,600
Do.	PT Semen Andalas Indonesia (Lafarge S.A., 99%)	Besar, Aceh	1,400
Do.	do.	Lhok, Aceh	1,600
Do.	PT Semen Baturaja	Baturaja-Ogan Komering Ulu, South Sumatra	1,250
Do.	PT Semen Bosowa Maros	Kabupaten Maros, Sulawesi Selatan	1,800
Do.	PT Holcim Tbk	Narogong, East Java	9,700
Do.	PT Semen Gresik Tbk	Gresik and Tuban, East Java	10,700
Do.	PT Semen Padang	West Sumatra	5,440
Do.	PT Semen Tonasa	Pangkep and Tonasa, South Sulawesi	6,000
Coal	PT Adaro Indonesia (New Hope Corp., 50%; PT Asminco Bara Utama, 40%; Mission Energy, 10%)	Paringin and Tutupan, South Kalimantan	35,000
Do.	PT Arutmin Indonesia (PT Bumi Resources Tbk, 80%, and Bakrie Group, 20%)	Mulia, Senakin, and Satui, South Kalimantan, and Asam-Asam, East Kalimantan	20,000
Do.	PT Berau Coal (PT United Tractor, 60%; PT Armadian, 30%; Nissho Iwai, 10%)	Berau, East Kalimantan	13,000
Do.	PT Kaltim Prima Coal Co. (PT Bumi Resources Tbk, 100%)	East Kutai Regency, East Kalimantan	36,000
Do.	PT Kideco Jaya Agung (Samtan Co. Ltd., 100%)	Pasir, East Kalimantan	12,000
Do.	PT Tambang Batubara Bukit Asam (state owned)	Tanjung Enim and Ombilin, South Sumatra	19,000
Do.	United Tractors	Central Kalimantan and East Kalimantan	6,500

See footnotes at end of table.

TABLE 2—Continued
INDONESIA: STRUCTURE OF THE MINERAL INDUSTRY IN 2012

(Thousand metric tons unless otherwise specified)

Commodity		Major operating companies and major equity owners	Locations of main facilities	Annual capacity[e]
Copper:				
Concentrate		PT Freeport Indonesia Co. (Freeport-McMoRan Copper & Gold Inc., 81.28%; Government, 9.36%; others, 9.36%)	Ertsberg and Grasberg, Papua	800
Do.		PT Newmont Nusa Tenggara (Newmont Mining Corp., 45%; Sumitomo Corp., 35%; PT Pukuafu Indah, 20%)	Sumbawa Island, West Nusa Tenggara	300
Metal		PT Smelting Co. (Mitsubishi Materials Corp., 60.5%; PT Freeport Indonesia Co., 25%; others, 14.5%)	Gresik, East Java	270
Gas:				
Natural	million cubic meters per day	ExxonMobil Oil Indonesia	Arun and Aceh, North Sumatra	48
Do.	do.	Roy M. Huffington (subsidiary of HUFFCO Group)	Badak, East Kalimantan	28
Do.	do.	Total Indonesie	Offshore East Kalimantan	59
Liquefied		PT Arun LNG Co. Ltd. (Government, 55%; Mobil Oil Co., 30%; Japan Indonesia LNG Co., 15%)	Balang Lancang amd Aceh, North Sumatra	12,500
Do.		PT Badak LNG Co. Ltd. (Government, 55%; HUFFCO Group, 30%; Japan Indonesia LNG Co., 15%)	Bontang, East Kalimantan	22,500
Coalbed methane		Ephindo Energy Pvt. Ltd. (PT Pertamina, 52%; Dart Energy Ltd., 24%)	Sangatta, East Kalimantan	22,600
Gold	metric tons	Aurora Gold Ltd. (100%)	Balikpapan, Central Kalimantan	60
Do.	do.	Archipelago Resources plc (95%)	Tok Tindung, North Sulawesi	5
Do.	do.	G-Resurces Group Ltd.	Martabe, North Sumatra	8
Do.	do.	PT Antam Tbk (Government, 65%)	Bogor, West Java	3
Do.	do.	PT Freeport Indonesia Co. (Freeport-McMoRan Copper & Gold Inc., 81.28%; Government, 9.36%; others, 9.36%)	Ertsberg and Grasberg, Papua	110
Do.	do.	PT Indo Muro Kencana (Straits Resources Ltd., 100%)	Balikpapan, Central Kalimantan	4
Do.	do.	PT Newmont Nusa Tenggara (Newmont Mining Corp., 45%; Sumitomo Corp., 35%; PT Pukuafu Indah, 20%)	Sumbawa Island, West Nusa Tenggara	16
Do.	do.	PT Nusa Halmahera (PT Aneka Tambang Tbk, 17.5%, and PT Newcrest Mining Ltd., 82.5%)	Halmahera Island, Maluku	24
Do.	do.	PT Prima Lirang Mining (Billiton BV, 90%, and PT Prima Maluku Indah, 10%)	Lerokis, Wetar Island	3
Do.	do.	Sumatra Copper & Gold plc	Tembang, West Sumatra	NA
Nickel:				
Ferronickel	metric tons	PT Antam Tbk (Government, 65%)	Pomalaa, South Sulawesi	100
In ore		do.	Pomalaa, South Sulawesi, and on Gebe Island	80
Do.		PT Vale Indonesia Tbk (Vale Canada Ltd., 59%; Sumitomo Metal Mining Co. Ltd., 20%; others, 21%)	Soroako, South Sulawesi	70
In matte		PT Antam Tbk (Government, 65%)	Pomalaa, South Sulawesi	24
Do.		PT Vale Indonesia Tbk (Vale Canada Ltd., 59%; Sumitomo Metal Mining Co. Ltd., 20%; others, 21%)	Soroako, South Sulawesi	68
Nickel-iron, ore		PT Yiwan Mining (China Nickel Resources Holdings Co. Ltd., 80%)	Mekarsari, West Java	3,000
Nitrogen		PT Asean-Aceh Fertilizer (Government, 60%, and other members of the Association of Southeast Asian Nations, 40%)	Lhokseumawe, North Sumatra	506
Do.		PT Pupuk Iskandar Muda (Government, 100%)	do.	506
Do.		PT Pupuk Kalimantan Timur (Government, 100%)	Bontang, East Kalimantan	1,850
Do.		PT Pupuk Kujang	Cikampek, West Java	330
Do.		PT Pupuk Sriwijawa (Government, 100%)	Palembang, South Sumatra	1,440

See footnotes at end of table.

TABLE 2—Continued
INDONESIA: STRUCTURE OF THE MINERAL INDUSTRY IN 2012

(Thousand metric tons unless otherwise specified)

Commodity		Major operating companies and major equity owners	Locations of main facilities	Annual capacity[e]
Petroleum:				
Crude	thousand barrels per day	BP Indonesia (a subsidiary of BP p.l.c.)	Arjuna and Arimbi, offshore West Java	170
Do.	do.	China National Offshore Oil Co.	Offshore southeastern Sumatra	100
Do.	do.	Maxus Southeast Asia Ltd. (subsidiary of Maxus Energy)	Cinta and Rama, offshore southeast Sumatra	95
Do.	do.	PT Pertamina (Government, 100%)	Jatibarang, West Java, and Bunyu, offshore East Kalimantan	80
Do.	do.	PT Caltex Pacific Indonesia (Texaco Inc., 50%, and Chevron Corp., 50%)	Minas, Duri, and Bangko, central Sumatra	700
Do	do.	Total Indonesie (subsidiary of Total S.A.)	Handi and Bakapai onshore and offshore East Kalimantan	180
Refined	do.	PT Pertamina (Government, 100%)	6 locations	1,047
Silver		PT Antam Tbk (Government, 65%)	Bogor, West Java	25
Do.		PT Freeport Indonesia Co. (Freeport-McMoRan Copper & Gold Inc., 81.28%; Government, 9.36%; others, 9.36%)	Ertsberg and Grasberg, Papua	220
Do.		PT Kelian Equatorial Mining (Rio Tinto Group, 90%, and PT Harita Jaya Raya, 10%)	180 kilometers west of Samarinda	10
Steel, crude		PT Ispat Indo	Sidoarjo, Surabaya	700
Do.		PT Krakatau Steel (Government, 100%)	Cilegon, West Java	2,400
Do.		PT Komatsu Indonesia Tbk	Jakarta	8
Do.		PT Wahana Garuda Lestari	Pulogadung, Jakarta	410
Tin:				
In ore		PT Koba Tin (Malaysia Smelting Corp., 75%, and PT Tambang Timah Tbk, 25%)	Koba, Bangka Island	25
Do.		PT Tambang Timah Tbk (Government, 65%)	Onshore and offshore islands of Bangka, Belitung, and Singkep	60
Metal		Mentok Tin Smelter (PT Tambang Timah Tbk)	Mentok, Bangka Island, South Sumatra	68
Do.		Koba Tin Smelter (PT Koba Tin)	Koba, Bangka Island, South Sumatra	25

[e]Estimated; estimated data are rounded to no more than three significant digits. Do., do. Ditto. NA Not available.

The Mineral Industry of Japan

By Chin S. Kuo

Japan is a resource-poor industrialized country, and its mineral sector was dominated by the metals and metal products industries. Japan remained the world's second-ranked producer of steel after China in 2012, and it was a producer and consumer of nonferrous metals as well. Production of gold, magnesium, and silver partially met its domestic demand, but a large proportion of the raw materials needed to support its manufacturing industries was met through imports. These imports included ores and concentrates to produce copper, lead, nickel, and zinc. The country also imported intermediate products and refined them into metals, such as molybdenum, tin, and tungsten. In terms of value, copper concentrate, iron ore, and rare-earth elements were the most important mineral commodity imports for Japan. The country also imported significant amounts of coke, crude oil, and natural gas, including liquefied natural gas (LNG).

Minerals in the National Economy

Japan was the world's fourth-ranked economy after the United States, China, and India. In 2012, Japan's gross domestic product (GDP) based on purchasing power parity was $4.7 trillion, which was an increase of 2% from that of 2011. Industrial production accounted for 26.3% of the GDP, which was an increase of 2% compared with that of 2011. The mineral processing industry was large and included the processing and production of chemicals, fabricated metal products, industrial mineral products, iron and steel, nonferrous metals, and petroleum products for manufacturing and construction industries. Japan's total exports were valued at $734 billion, of which iron and steel products accounted for 5.5%, and the country's total imports were valued at $831 billion, of which petroleum accounted for 15.5%; LNG, 5.7%; and coal, 3.5% (U.S. Central Intelligence Agency, 2013).

Government Policies and Programs

The focus of Japan's resource development policy was to secure stable overseas sources of oil, gas, and mineral resources. The Government planned to continue to invest in developing Africa's mineral resources, particularly rare (minor) metals. The Government stockpiled gallium and indium in addition to seven other metals (chromium, cobalt, manganese, molybdenum, nickel, tungsten, and vanadium) for the needs of its high-tech industry. Japan Oil, Gas and Metals National Corp. had established the Geologic Remote Sensing Center in Botswana, which trained geological engineers and conducted joint research programs in geologic analysis, exploration, and development. In addition, private companies were also active in the effort to secure access to overseas mineral resources. Sumitomo Corp. joined the Ambatovy nickel project in Madagascar, Itochu Corp. joined the Platreef platinum-group metal (PGM) exploration project in South Africa, and Nippon Steel Corp. joined the Revuboe coking coal project in Mozambique (T. Matsushita, Senior Vice Minister, Ministry of Economy, Trade and Industry, written commun., February 6, 2012).

Production

Japan's mined output of silver and gold decreased by 20% and 17%, respectively, owing to depleted reserves; production of primary and secondary gold also decreased by 22% and 19%, respectively. Production of other nonferrous metals, such as regular-grade aluminum, cobalt metal, refined primary copper, and tin, increased by more than 15%, whereas production of high-purity aluminum, antimony metal, and molybdenum metal decreased by more than 15%. In the iron and steel sector, Japan's production of crude steel and pig iron stayed at about the same level as in 2011, with the exception of ferronickel and ferrochrome output, which increased by 33% and 13%, respectively; nickel content of ferronickel increased by 17%. In the industrial minerals sector, production of cement increased by 6.7%, whereas that of nitrogen decreased by 12.9%. In the energy sector, production of bituminous coal, carbon black, and crude petroleum decreased by 22%, 6.3%, and 4.6%, respectively; the decreases in output of the latter two were owing to depleted reserves (table 1).

Structure of the Mineral Industry

Japan's mineral industry is characterized by small-scale, low-tonnage mining operations and high-value-added mineral and metal processing and manufacturing activities. Its mining industry is not significant to the economy, and the country consumes more minerals and metals from imports than it produces. In 2012, mining and quarrying of industrial minerals, including dolomite, iodine, limestone, pyrophyllite, silica sand, and silica stone, was still being done but at a lower level of production than in previous years. Operating mines and employment in the mining industry had been in decline because of depleted ore reserves, high mining costs, and the availability of cheaper imports. Japan had, however, a world-class metallurgical industry for nonferrous metals. The mining and mineral-processing businesses were owned and operated by private companies (table 2).

Mineral Trade

Japan imported raw materials and mineral fuels for its industrial production and use. These imported mineral commodities were mostly bauxite, coal, coke, copper concentrate, iron ore, LNG, and petroleum. In 2012, Japan imported about 1.266 million metric tons (Mt) of copper, 464,000 metric tons (t) of zinc, and 91,000 t of lead, all in concentrates. The country imported 36,000 t of refined copper and exported 546,000 t. Japan imported 24,000 t of refined zinc and 29,000 t of refined lead and exported 136,000 t and 20,000 t, respectively. The country also imported 1,699,000 kilograms (kg) of silver and exported 3,297,000 kg. Japan imported a total of 210,000 kg of cadmium metal entirely from the Republic of Korea (100%) and exported a total of 952,000 kg of cadmium metal to China (68%), India (21%), and Sweden (10%) (Japan Mining Industry Association, 2013, p. 5, 10, 11–13, 28).

Commodity Review

Metals

Bauxite and Alumina and Aluminum.—In 2012, Japan produced 137,000 t of secondary aluminum, which was a decrease of 3.2% compared with the output of 2011, and 3,100 t of primary aluminum. The country exported 281,000 t of rolled aluminum products and 12,700 t of unwrought aluminum and imported 2.75 Mt of unwrought aluminum, 49,700 t of aluminum waste and scrap, and 107,000 t of rolled aluminum products. Domestic consumption of aluminum increased by 1.8% to 3.93 Mt, of which the transportation sector accounted for 42%; building and construction, 14%; fabricated metal, 11%; food packaging, 11%; and other uses, 22% (Japan Aluminum Association, 2013).

Sumitomo Chemical Co. Ltd. produced specialty alumina, such as high-purity alumina (more than 99.99%), which was used for lithium-ion battery materials. The company expanded the production capacity at its Ehime Works by 400 metric tons per year (t/yr) in 2010 and planned to increase it again to 3,200 t/yr by the second quarter of 2012. The increase was in response to the expanding market for hybrid cars and electric vehicles (Roberts, 2012).

Furukawa-Sky Aluminum Corp., which was a maker of rolled aluminum products, agreed to take over Sumitomo Light Metal Industries Ltd., which was a metal fabricator, to reduce costs, accelerate expansion, and strengthen its competitiveness in the international market. The deal was expected to be completed in October 2013 (Suga, 2012).

Antimony.—Nihon Seiko Co. Ltd. was a leading producer of antimony metal and antimony trioxide using imported material. A high-grade antimony deposit was found in the seabed off Amami-Oshama Island in Kagoshima Prefecture. The deposit could become a source of antimony if the supply of material on the world market remains tight (Clarke, 2012).

Cadmium.—At the beginning of 2012, Japan had stocks of 282 t of cadmium metal. With production of 1,856 t and imports of 210 t, the country's supplies increased to 2,348 t. With an apparent consumption of 1,099 t and exports of 952 t for a total demand of 2,051 t, Japan's stocks increased to 297 t at the end of 2012 (Japan Mining Industry Association, 2013, p. 28).

Copper.—The country produced about 1.3 Mt of anode and blister copper from primary sources and 304,000 t from scrap. It also produced about 1.52 Mt of refined copper primarily from imported ore (83.8%), scrap (10.4%), and other sources (5.8%). Consumption of refined copper was reported to be about 930,000 t. Refined copper was used in the manufacturing of wire (62%), brass (37%), and miscellaneous products (1%), such as copper alloys and copper alloy casting (table 3; Japan Mining Industry Association, 2013, p. 11).

Pan Pacific Copper Co. Ltd., which was part of JX Holdings Inc., suspended operations at its 200,000-t/yr copper smelter at Saganoseki in Oita Prefecture in January after a fire damaged an electrical substation. The company was reviewing if its two other copper smelters could make up for the lost production. The company had a smelter at Hitachi in Ibaraki Prefecture and another at Tamano in Okayama Prefecture and had a total combined output capacity of 710,000 t/yr. The company had planned to produce 272,000 t of refined copper from October 2011 to March 2012 (Inoue, 2012).

Gold and Silver.—In 2012, Japan produced about 104,000 kg of gold from imported ore, scrap, domestic ore, and other sources, which accounted for 64%, 13%, 8%, and 15% of the source material, respectively. The country also produced about 1,765,000 kg of silver from imported ore, scrap, domestic ore, and other sources, which accounted for 62%, 17%, 0.4%, and 20% of the source material, respectively. Silver was used for miscellaneous applications (37%), in photographic materials (30%), in extension materials (15%), in point connectors (10%), for silver solder (6%), and as a nitrate for other uses (3%) (table 3; Japan Mining Industry Association, 2013, p. 10).

Iron and Steel.—Nippon Steel Corp. planned to increase the internal capacity of the blast furnace No. 4 at its Yawata Works by 18%. The $410 million overhaul was expected to begin in 2013 and to be completed in 85 days. The overhaul would help solidify the company's competitive strength in the manufacture of iron and steel in the global market (SteelOrbis, 2012).

Japanese electric furnace carbon steel producers were expected to encounter cost increases in electricity and materials, such as graphite electrode, ferrous scrap, and crude oil, beginning in April 2012. Tokyo Electric Power Co. increased its price of electricity. Tokai Carbon decided to increase the sale price of graphite electrodes by 5% to 10%. The price of crude oil remained more than $100 per barrel. The steel producers were likely to launch a price hike to pass along the cost increases to their customers (Steelguru.com, 2012).

Lead.—Japan produced a total of 252,000 t of lead in 2012. Of this amount, 209,000 t was refined lead made from 91,000 t of primary ore (which included imported ore and domestic ore) and 118,000 t of scrap and material from other sources. The remaining 43,400 t was remelted lead. Consumption of refined lead was reported to be 207,000 t. Refined lead was used in batteries (87%), pipe and sheet (6%), chemicals (2%), solder (0.7%), and other miscellaneous applications (4%) (table 3; Japan Mining Industry Association, 2013, p. 12, 14).

Nickel.—Japan produced 41,900 t of refined nickel and 125,000 t of nickel in ferronickel and nickel oxide for a total of 167,000 t of nickel. The country also produced 2,360 t of nickel chemicals. At the end of 2012, stocks of refined nickel, which totaled 3,420 t, and those of ferronickel, which totaled 3,840 t, were stored at producers' facilities (Japan Mining Industry Association, 2013, p. 4, 7).

More than one-half of Japan's nickel supply came from Indonesia, which was scheduled to ban exports of unprocessed nickel in 2014. Sumitomo Metal Mining Co. planned to increase nickel imports from New Caledonia, the Philippines, and the Solomon Islands. The company owned 62.5% of the $1 billion Taganito nickel project in the Philippines, which was to start production in 2014 at a rate of 27,000 t/yr of nickel (90% of capacity) (Inoue and Obayashi, 2012).

Titanium.—Kobe Steel Ltd. planned to increase its production capacity for titanium alloys and pure titanium. At one of its two plants at Takasago in Hyogo Prefecture, the company planned to construct a new ring-rolling mill to double its ring-rolling capacity; at the other plant, the company planned to install heat-treatment and inspection equipment for large forged products to increase its processing capacity by threefold. Kobe Steel Tube Co. Ltd. had completed a titanium welded tube line to increase its production capacity by 25% to 30% at Shimonoseki in Yamaguchi Prefecture. Demand for titanium was expected to increase owing to strong worldwide demand for civilian aircraft and infrastructure, such as powerplants and desalination plants (Kobe Steel Ltd., 2012).

Zinc.—Japan produced a total of about 606,000 t of zinc. Of this amount, 571,000 t was refined zinc made from 459,000 t of primarily imported ore (80%) and 112,000 t of other sources (18%) and scrap (1.2%). The remaining 34,700 t was remelted zinc. Consumption of refined zinc was reported to be 355,000 t. Refined zinc was used mainly in galvanized sheet (49%), other types of galvanized products (15%), brass (14%), die-casting alloy (11%), chemicals (7%), and miscellaneous uses (1%) (table 3; Japan Mining Industry Association, 2013, p. 13, 15).

Zirconium.—Tosoh Corp. was expanding its production capacities of chemical manganese oxide, high-silica zeolite, and zirconia powder. The company planned to build a new chemical manganese oxide plant at Hyuga Prefecture to produce 5,000 t/yr of the material. When completed in March 2013, the company would have a total capacity of 64,000 t/yr. Tosoh also planned to expand its capacity to produce high-silica zeolite at Yokkaichi with completion scheduled for March 2013. The expansion would double the company's total capacity to produce high-silica zeolite. The company would expand its facilities to produce zirconia powder at Shunan in Yamaguchi Prefecture and at Yokkaichi in Mie Prefecture to increase the total capacity by 20% (Watts, 2012).

Industrial Minerals

Limestone.—Calcium carbonate is derived from limestone, and both its ground and precipitated forms are used by the paper industry for fillers and coatings as an alternative to kaolin. Japan was a leading producer of calcium carbonate in the world. Imerys SA of France shut down its 60,000-t/yr plant in Miyagi Prefecture as a result of the earthquake and tsunami that took place in March 2011, and it brought the plant back online in the beginning of 2012. Fimatec Ltd. also shut down its 60,000-t/yr plant in Fukushima Prefecture and its joint-venture 96,000-t/yr plant in Ishinomaki Prefecture, and it brought the Fukushima plant back online in May 2011 (Ollett, 2012).

Rare Earths.—An estimated resource of 6.8 Mt of rare-earth minerals was discovered 5,600 m deep within the seabed in a 2,600-square-kilometer area located 310 kilometers southwest of Minami-tori-shima Island. The deposit was estimated to be large enough to supply all the needs of Japan's high-tech manufacturing industry. Rare-earth minerals found in mud samples had an average concentration that ranged from 1,000 parts per million (ppm) to 1,700 ppm. The 30-m-thick mud beds could be drilled using oil extraction technology to access the rare-earth deposits. The technology needed to produce the rare-earth minerals on a commercial scale remained to be developed, however (Mining Engineering, 2012).

Japan had been trying to diversify its import sources of rare earths because of China's reduced exports of rare earths to the world market. During the 9 months from January to September 2012, China sold 9,967 t of rare earths to international buyers. Japan also tried to reduce consumption of rare earths by substituting other materials for them in manufacturing and by increasing its recycling of rare-earth materials. In accordance with an agreement signed with the Government of India early in 2012, Japan was expected to begin importing 4,000 t/yr of rare-earth minerals from India, which was about 15% of Japan's annual consumption (Syrett, 2012).

Toyota Tsusho Corp. and an Indian natural resource development company were expected to establish a joint venture for a project in the Indian State of Odisha. India controlled 17% of the world's total resources of mineral sands, but its mineral sand production accounted for only 6% of global output. The proposed plant was to use a byproduct mixture to produce rare earths after uranium and thorium had been extracted from monazite ore by Indian Rare Earths Ltd. Production was expected to begin in April 2012 and to supply between 3,000 t/yr and 4,000 t/yr of rare earths to Japan. Toyota Tsusho also had a project to develop a rare-earth deposit at Dong Pao in Vietnam. The processing plant was expected to be capable of producing 7,000 t/yr of rare earths in 2012 to meet one-fourth of Japan's demand. In Indonesia, the company planned to build a specialized plant using refining technology to recover rare earths from the slag left over from tin smelting on the Island of Bangka (Toyota Tsusho Corp., 2012).

Ensuring a stable supply of rare-earth elements was one of the Government's top priorities for promoting expansion and advancement of Japan's high-tech industry and driving technological innovation in the high-tech sector. Summit Atom Rare Earth Co. LLP, which was a joint venture between Sumitomo Corp. and National Atomic Co. of Kazakhstan, opened a plant at Stepnogorsk in Kazakhstan in November 2012 to recover rare-earth elements from uranium-ore tailings that Kazakhstan had mined in the past. The plant had set a total output target of 1,500 t/yr of rare-earth oxides (REOs) during the initial years and planned to export the REOs to Japan in 2013. Shin-Etsu Chemical Co. Ltd. would provide technological support. The plant also was expected to produce light rare earths (Sumitomo Corp., 2012).

Japan Metals & Chemicals Co., Ltd. and Honda Motor Co., Ltd. developed a process to extract rare-earth metals from various used parts (initially from nickel-metal hydride batteries) in Honda products by using a newly established technology. The new process, which was to be employed at an existing recycling plant, would be the world's first mass-recyling effort for rare earths. The extraction rate was about 80% of rare-earth metals contained in the used parts. The companies expected to expand rare-earth metals recycling in the future (Honda Motor Co., Ltd., 2012).

Mineral Fuels

Petroleum.—JX Nippon Oil & Energy Corp. resumed full operations of its 145,000-barrel-per-day (bbl/d) Sendai oil refinery in March 2012 (1 year after the earthquake and tsunami that struck in March 2011). In January, the refinery resumed partial operations at limited capacity. JX Nippon operated eight oil refineries in Japan with a total capacity of 1,625,500 bbl/d. Cosmo Oil Co.'s 220,000-bbl/d Chibba oil refinery was awaiting permission from the local government to restart the first of two crude distillation units (Maeda, 2012).

Reserves and Resources

Japan has large reserves of industrial minerals, including dolomite, iodine, limestone, pyrophyllite, silica sand, and silica stone. Limestone is the most abundant indigenous mineral resource. The country's reserves of nonferrous metals, such as lead, silver, and zinc, are small, with the exception of a medium gold reserve; gold deposits had been found and were being mined on a small scale in Kagoshima Prefecture on Kyushu Island owing to deep-seated occurrences. The country's coal reserves were reaching depletion. Japan's reserves of gas and oil are negligible (table 4).

Outlook

Owing to the expanding market for hybrid cars and electric vehicles, Japan is expected to increase production of specialty alumina used in lithium-ion batteries. Because of Indonesia's ban on the export of unprocessed nickel beginning in 2014, Japan is expected to source its nickel supplies from New Caledonia and the Philippines. Owing to strong worldwide demand for civilian aircraft and infrastructure, Japan is expected to increase its output of titanium products, such as alloys, metal, and tubes. Japan is expected to diversify its supply sources of rare earths by importing them from such countries as Australia, India, the United States, and Vietnam. With new technological advances, Japan is likely to be able to extract rare earths from the slag production during tin smelting in Indonesia and from uranium-ore tailings in Kazakhstan. In the energy sector, consumption of coal, LNG, and various types of oil may increase as backup powerplants that burn fossil fuels try to compensate for the loss of some nuclear-power-generating capacity. Imports of LNG are expected to increase in the long term in Japan.

References Cited

Clarke, Gerry, 2012, Antimony (trioxide) on the watch list: Industrial Minerals, no. 540, September, p. 53.
Honda Motor Co., Ltd., 2012, Honda to reuse rare earth metals contained in used parts: Honda Motor Co., Ltd. news release, April 17, 2 p.
Inoue, Yuko, 2012, Pan Pacific's copper smelter suspended after fire: Mineweb.com, January 10. (Accessed January 10, 2012, at http://www.mineweb.com/mineweb/veiw/mineweb/en/page504?oid=142775.)
Inoue, Yuko, and Obayashi, Yuka, 2012, Sumitomo eyes New Caledonia, Philippines for nickel: Thomson Reuters, March 26. (Accessed March 28, 2012, at http://www.reuters.com/article/2012/03/26/us-miningsummit-sumitomo-mining.)
Japan Aluminum Association, 2013, Aluminum statistics: Tokyo, Japan, Japan Aluminum Association, February 13, 6 p.
Japan Mining Industry Association, 2013, Monthly statistics: Tokyo, Japan, Japan Mining Industry Association, March 5, 29 p.
Kobe Steel Ltd., 2012, Kobe Steel expanding titanium capacity: Kobe Steel Ltd., Topics, March 8. (Accessed March 13, 2012, at http://www.kobelco.co.jp/english/topics/2012/1187276_11802.html.)
Maeda, Risa, 2012, Quake-hit JX refinery resumes full operations: Thomson Reuters, March 9. (Accessed March 10, 2012, at http://www.reuters.com/assets/print?aid=UST9E8DM03F20120309.)
Mining Engineering, 2012, Rare earths discovery in Japan: Mining Engineering, v. 64, no. 8, August, p. 11.
Ollett, John, 2012, Japan's Tohoku earthquake and tsunami tragedy—One year on: Industrial Minerals, no. 535, April, p. 22.
Roberts, Jessica, 2012, Aluminas—Pure and not so simple: Industrial Minerals, no. 537, June, p. 43.
Steelguru.com, 2012, Japanese EAF steelmakers suffer triple cost up in FY 2012: Steelguru.com, February 20. (Accessed February 21, 2012, at http://www.steelguru.com/international_news/Japanese_EAF_steelmakers_suffer_triple_cost_up_in_FY_2012/251129.html.)
SteelOrbis, 2012, Nippon Steel to modernize blast furnace No. 4: SteelOrbis, March 22. (Accessed March 28, 2012, at http://www.steelorbis.com/steel-news/latest-news/nippon-steel-to-modernize-blast-furnace-no-4--674594.htm.)
Suga, Masumi, 2012, Furukawa-Sky, Sumitomo Light Metal to merge as competition rises: Bloomberg.com, August 29. (Accessed August 30, 2012, at http://www.bloomberg.com/news/furukawa-sky-sumitomo-light-metal-to-merge-as-competition-rises.html.)
Sumitomo Corp., 2012, Starting rare earth production in Kazakhstan: Sumitomo Corp. press release, November 5. (Accessed November 13, 2012, at http://www.sumitomocorp.co.jp/english/news/2012/20121105_110009.html.)
Syrett, Laura, 2012, Japan to import Indian rare earths to buffer China island spat: Industrial Minerals, October 22. (Accessed October 22, 2012, at http://www.indmin.com/news.)
Toyota Tsusho Corp., 2012, Working on rare earths business: Toyota Tsusho Corp. Environment & Society. (Accessed March 2, 2012, at http://www.toyota-tsusho.com/english/csr/business/cases01.html.)
U.S. Central Intelligence Agency, 2013, Japan, in The world factbook: U.S. Central Intelligence Agency. (Accessed November 25, 2013, at https://www.cia.gov/library/publications/the-world-factbook/geos/ja.html.)
Watts, Mark, 2012, Tosoh expands in zirconia, manganese oxide and zeolites: Industrial Minerals, February 24. (Accessed February 27, 2012, at http://www.indmin.com/print.aspx?articleid=2985145.)

TABLE 1
JAPAN: PRODUCTION OF MINERAL COMMODITIES[1]

(Metric tons unless otherwise specified)

Commodity		2008	2009	2010	2011	2012
METALS						
Aluminum:						
Alumina[e]	thousand metric tons	320	310	300	280	250
Aluminum hydroxide[e]	do.	700	710	700	690	650
Metal:						
Primary:						
Regular grades[e]	do.	7 [2]	6	5	4	5
High-purity	do.	52	33	49	43	26
Secondary[3]	do.	149	111	126	142	137
Antimony:						
Oxide		6,954	4,884	6,846	7,000 [e]	6,900 [e]
Metal		325	239	304	435	143
Arsenic, trioxide[e]		40	40	40	45 [r]	45
Bismuth		480	423	454	460 [e]	470 [e]
Cadmium, refined		2,126	1,824	2,053	1,755	1,855
Chromium, metal[e]		600	650	700	800	750
Cobalt, metal		1,071	1,332	1,935	2,007	2,542
Copper, metal:						
Blister and anode:						
Primary		1,366,310	1,297,943	1,382,655	1,168,284	1,304,916
Secondary		259,060	243,859	260,245	269,748	303,900
Total		1,625,370	1,541,802	1,642,900	1,438,032	1,608,816
Refined:						
Primary		1,328,157	1,238,012	1,333,787	1,094,999 [r]	1,270,914
Secondary		211,681	201,831	214,901	233,289 [r]	245,440
Total		1,539,838	1,439,843	1,548,688	1,328,288 [r]	1,516,354
Gallium, metal:[e]						
Primary		7	7	6	6	6
Secondary		85	80	78	75	70
Germanium:[e]						
Oxide		50	50	45 [r]	50	50
Metal, polycrystal	kilograms	1,720	1,730	1,750	1,760	1,780
Gold:						
Mine output, Au content	do.	6,868	7,708	8,544	7,922 [r]	7,233
Metal:						
Primary	do.	81,399	89,281	98,398	95,549	74,735
Secondary[4]	do.	43,433	43,979	37,413	36,288	29,544
Total	do.	124,832	133,260	135,811	131,837	104,279
Indium, metal[e]	do.	65,000	67,000	69,000	70,000	71,000
Iron and steel, metal:						
Pig iron	thousand metric tons	86,171	66,943	82,283	81,028	81,405
Electric-furnace ferroalloys:						
Ferrochrome		13,888	7,698	16,208	17,217	19,392
Ferromanganese		431,181	361,375	453,265	456,798	436,171
Ferronickel		301,361	284,884	348,420	279,944	371,913
Silicomanganese		58,884	49,205	49,865	49,798	52,287
Ferromolybdenum		4,554	3,598	4,615	5,167	4,616
Ferrovanadium		3,477	2,560	4,190	3,980	4,403
Unspecified		14,478	12,957	16,374	20,913	19,364
Total		827,823	722,277	892,937	833,817	908,416
Steel, crude	thousand metric tons	118,739	87,534	109,599	107,601	107,232
Semimanufactures, hot-rolled:[e]						
Ordinary steels	do.	84,000	68,000	67,000	65,000	66,000
Special steels	do.	21,000	16,000	15,000	15,000	16,000

See footnotes at end of table.

TABLE 1—Continued
JAPAN: PRODUCTION OF MINERAL COMMODITIES[1]

(Metric tons unless otherwise specified)

Commodity		2008	2009	2010	2011	2012
METALS—Continued						
Lead, metal, refined:						
Primary		107,005	96,794	101,610	100,078	91,037
Secondary		117,900 r	95,402 r	114,218 r	114,986 r	117,957
Total		224,905 r	192,196 r	215,828 r	215,064 r	208,994
Magnesium, metal, secondary[e]		13,000	13,000	14,000	14,000	15,000
Manganese, oxide[e]		45,000	44,000	43,000	43,000	40,000
Molybdenum, metal		1,217	695	1,154	1,234	1,013
Nickel metal:						
Refined		34,861	29,351	40,228	41,290	41,944
Ni content of nickel oxide sinter		61,753 r	58,808 r	59,011 r	50,437 r	52,000 e
Ni content of ferronickel		59,259	54,491	64,349	62,773 r	73,248
Ni content of chemical		2,333	1,669	2,497	2,383	2,362
Total[e]		158,000 r	144,000	166,000	157,000	170,000
Platinum-group metals:						
Palladium, metal	kilograms	7,526	6,675	6,107	7,534	8,052
Platinum, metal	do.	1,442	1,417	1,331	1,765	1,735
Selenium, metal		754	709	754	750 e	755 e
Silicon, multicrystalline		7,471	8,633	8,700 e	12,133	10,964
Silver:						
Mine output, Ag content	kilograms	2,043	1,500 e	1,200 e	4,486	3,577
Metal:						
Primary	do.	2,042,604	1,865,936	1,898,208	1,724,218	1,764,533
Secondary[4]	do.	253,374	326,487	313,931	325,373	348,620
Total	do.	2,295,978	2,192,423	2,212,139	2,049,591	2,113,153
Tantalum, metal[e]		95	95	95	100 r	98
Tellurium, metal		47 r	49 r	47 r	40 r	45 e
Tin, metal, smelter		956	757	841	947	1,133
Titanium:						
Dioxide		225,228	161,928	207,561	214,417	185,320
Metal[e]		45,000	35,000	38,000	40,000	38,000
Tungsten, metal		3,446	1,400	3,361	3,299	3,025
Vanadium, metal[e, 5]		560	560	560	560	580
Zinc:						
Oxide[e]		77,000	75,000	72,000	66,325 r, 2	58,896 [2]
Metal:						
Primary		502,910	435,905	470,057	444,446	459,322
Secondary		112,623 r	104,699 r	103,951 r	100,228 r	111,990
Total		615,533 r	540,604 r	574,008 r	544,674 r	571,312
Zirconium, oxide[e]		11,000	12,000	13,000	13,000	14,000
INDUSTRIAL MINERALS						
Bromine[e]		20,000	20,000	25,000 r	25,000 r	30,000
Cement, hydraulic	thousand metric tons	62,810	54,800	51,526	51,291	54,737
Clays:[e]						
Bentonite		435,000	432,000	430,000	425,000	420,000
Fire clay, crude		450,000	440,000	440,000	430,000	430,000
Kaolin		11,000	12,000	12,000	13,000	13,000
Diatomite[e]		115,000	110,000	110,000	100,000	100,000
Feldspar and related materials[e]		120,000 r	115,000 r	110,000 r	104,109 r, 2	100,000
Gypsum[e]	thousand metric tons	5,800	5,750	5,700	5,600	5,500

See footnotes at end of table.

TABLE 1—Continued
JAPAN: PRODUCTION OF MINERAL COMMODITIES[1]

(Metric tons unless otherwise specified)

Commodity		2008	2009	2010	2011	2012
INDUSTRIAL MINERALS—Continued						
Iodine		9,500 e	8,232	9,216	9,277	9,315
Lime, quicklime	thousand metric tons	9,528	6,746	8,547	8,005	7,581
Nitrogen, N content of ammonia	do.	1,244	1,021	1,178	1,211	1,055
Perlite e		230,000	220,000	210,000	200,000	200,000
Rare-earth oxides[6]		8,435	5,121	10,699	10,700 e	10,800 e
Salt, all types	thousand metric tons	1,200 e	1,200 e	1,122	978	925
Silica:						
Sand	do.	3,664	2,856	3,078	3,003	2,877
Stone, quartzite	do.	10,682	9,189	9,159	9,543	9,306
Sodium compounds, n.e.s.:e, 7						
Soda ash		400,000	400,000	410,000 r	430,000 r	450,000
Sulfate, anhydrous		140,000	142,000	142,000	145,000	147,000
Stone, crushed:						
Dolomite	thousand metric tons	3,370	3,122	3,438	3,492	3,361
Limestone	do.	156,813	132,350	133,974	134,176	140,038
Sulfur:						
Byproduct of metallurgy e	do.	1,300	1,350	1,400	1,450	1,500
Byproduct of petroleum	do.	2,034	1,864	1,892	1,755	1,747
Talc and related materials:e						
Talc		26,000	25,000	24,000	24,000	25,000
Pyrophyllite		350,000	340,000	340,000	350,000	340,000
Vermiculite e		6,000	6,000	6,000	6,200 r	6,200
MINERAL FUELS AND RELATED MATERIALS						
Carbon black	thousand metric tons	821	575	729	681	638
Coal, bituminous e, 8	do.	1,300	1,100	1,000	900	700
Coke, including breeze, metallurgical	do.	38,568	32,587	37,447	35,379	34,743
Gas, natural:						
Gross[9]	million cubic meters	3,735	3,539	3,396	3,298	3,276
Marketed e	do.	3,900	3,700	3,600	3,500	3,500
Petroleum:						
Crude	thousand 42-gallon barrels	6,200	5,795	5,491	5,235	4,995
Refinery products:e						
Gasoline:						
Aviation	do.	50	50	50	50	60
Other	do.	360,000	362,000	360,000	355,000	350,000
Asphalt and bitumen	do.	33,000	32,000	31,000	30,000	30,000
Distillate fuel oil	do.	250,000	245,000	242,000	240,000	240,000
Jet fuel	do.	82,000	78,000	76,000	75,000	80,000
Kerosene	do.	170,000	165,000	166,000	165,000	165,000
Liquefied petroleum gas	do.	58,000	56,000	55,000	55,000	60,000
Lubricants	do.	17,000	16,000	14,000	15,000	15,000
Naphtha	do.	139,000	136,000	135,000	130,000	135,000
Paraffin, wax	do.	800	800	750	750	750
Petroleum coke	do.	4,500	4,400	4,300	4,200	4,200
Refinery fuel and losses[10]	do.	150,000	150,000	150,000	150,000	160,000
Residual fuel oil	do.	330,000	300,000	320,000	350,000	355,000
Unfinished oils	do.	50,000	50,000	50,000	50,000	55,000
Total	do.	1,640,000	1,600,000	1,600,000	1,620,000	1,650,000

eEstimated; estimated data are rounded to no more than three significant digits; may not add to totals shown. r Revised. do. Ditto.

[1]Table includes data available through November 4, 2013.
[2]Reported figure.
[3]Includes alloyed and unalloyed aluminum ingot.
[4]Includes metal recovered from scrap and waste.

TABLE 1—Continued
JAPAN: PRODUCTION OF MINERAL COMMODITIES[1]

[5]Represents metal content of vanadium pentoxide recovered from petroleum residues, ashes, and spent catalysts.

[6]Includes oxide of cerium, europium, gadolinium, lanthanum, neodymium, praseodymium, samarium, terbium, and yttrium.

[7]Not elsewhere specified.

[8]All major coal mines had closed by January 2002, but eight smaller mines were still in operation in 2012.

[9]Includes output from gas wells and coal mines.

[10]May include some additional unfinished oils.

TABLE 2
JAPAN: STRUCTURE OF THE MINERAL INDUSTRY IN 2012

(Thousand metric tons unless otherwise specified)

Commodity		Major operating companies and major equity owners	Location of main facilities	Annual capacity
Cement		Aso Cement Co., Ltd.	Tagawa and Kanda, Fukuoka Prefecture	2,400
Do.		Daiichi Cement Co., Ltd.	Kawasaki, Kanagawa Prefecture	1,169
Do.		Denki Kagaku K.K.	Omi, Niigata Prefecture	2,762
Do.		Hachinohe Cement Co., Ltd.	Hachinohe, Aomori Prefecture	1,533
Do.		Hitachi Cement Co., Ltd.	Hitachi, Ibaraki Prefecture	941
Do.		Mitsubishi Materials Corp.	Higashidori, Shimokita-gun, Apmori Prefecture; Higashiyama, Higashiiwai-gun, Iwate Prefecture; Yokoze, Saitama Prefecture; Kurosaki, Kyushu, and Higashitani, Fukuoka Prefecture	13,467
Do.		Mitsui Mining Co. Ltd.	Togawa, Fukuoka Prefecture	2,075
Do.		Myojo Cement Co., Ltd.	Itoigawa, Niigata Prefecture	2,482
Do.		Nippon Steel Chemical Co., Ltd.	Tobata, Kitakyushu, Fukuoka Prefecture	855
Do.		Nittetsu Cement Co., Ltd.	Muroran, Hokkaido Prefecture	1,589
Do.		Ryukyu Cement Co. Ltd.	Yabu, Nago, Okinawa Prefecture	722
Do.		Sumitomo Osaka Cement Co. Ltd.	Tamura, Fukushima Prefecture; Aso, Tochigi Prefecture; Motosu, Gifu Prefecture; Sakata, Shiga Prefecture; Ako, Hyogo Prefecture; and Susaki, Kochi Prefecture	14,402
Do.		Taiheiyo Cement Corp.	Ofunato, Iwate Prefecture; Kumagaya and Saitama, Saitama Prefecture; Fujiwara, Mie Prefecture; Tsukumi, Oita Prefecture; and Kamiiso, Hokkaido Prefecture	28,800
Do.		Tokuyama Cement Co. Ltd.	Nanyo, Yamaguchi Prefecture	5,936
Do.		Tosoh Corp.	Shin Nanyo, Yamaguchi Prefecture	2,869
Do.		Tsuruga Cement Co. Ltd.	Tsuruga, Fukui Prefecture	1,710
Do.		Ube Industries Ltd.	Ube and Isa, Yamaguchi Prefecture, and Kanda, Fukuoka Prefecture	10,736
Coal		Kushiro Coal Mine Co. Ltd.[1]	Kushiro, Hokkaido Prefecture	750
Cobalt, refined	metric tons	Sumitomo Metal Mining Co. Ltd. (SMM)	Niihama, Ehime Prefecture	1,000
Copper, refined	do.	Mitsubishi Materials Corp.	Naoshima, Kagawa Prefecture	225,600
Do.	do.	Onahama Smelting and Refining Co. Ltd. (Mitsubishi Materials Corp., 49.29%; Dowa Mining Co. Ltd., 31.15%; Furukawa Co. Ltd., 8.31%; Furukawa Electric Co. Ltd. and Mitsubishi Cable Industries Ltd., 4.17% each; others, 2.91%)	Onahama, Fukushima Prefecture	300,000
Do.	do.	Pan Pacific Copper Co., Ltd. (JX Nippon Mining & Metals Co., Ltd., 66%, and Mitsui Mining and Smelting Co., Ltd., 34%)	Saganoseki, Oita Prefecture; Hitachi, Ibaraki Prefecture; and Tamano, Okayama Prefecture[2]	710,000
Do.	do.	Sumitomo Metal Mining Co. Ltd. (SMM)	Besshi/Toyo (Saijyo), Ehime Prefecture	410,000
Do.	do.	Kosaka Smelting and Refining Co. Ltd. (wholly owned subsidiary of Dowa Mining Co. Ltd.)	Kosaka, Akita Prefecture	72,000
Gold:				
In concentrate	kilograms	Sumitomo Metal Mining Co. Ltd. (SMM)	Hishikari, Kagoshima Prefecture	9,000
Refined	do.	Kosaka Smelting and Refining Co. Ltd. (wholly owned subsidiary of Dowa Mining Co. Ltd.)	Kosaka, Akita Prefecture	24,000
Do.	do.	Mitsui Mining and Smelting Co., Ltd.	Takehara, Hiroshima Prefecture	22,000
Do.	do.	Mitsubishi Materials Corp.	Naoshima, Kagawa Prefecture	60,000
Do.	do.	JX Nippon Mining & Metals Co., Ltd.	Hitachi, Ibaraki Prefecture	30,000
Do.	do.	Sumitomo Metal Mining Co. Ltd. (SMM)	Niihama, Ehime Prefecture	36,000

See footnotes at end of table.

TABLE 2—Continued
JAPAN: STRUCTURE OF THE MINERAL INDUSTRY IN 2012

(Thousand metric tons unless otherwise specified)

Commodity		Major operating companies and major equity owners	Location of main facilities	Annual capacity
Iodine, crude	metric tons	Ise Chemical Industries Co. Ltd. (Asahi Glass Co. Ltd., 52.4%, and Mitsubishi Corp., 11.2%)	Oami-Shirasato, and Ichinomya, Chiba Prefecture; and Sadowara, Miyazaki Prefecture	3,600
Do.	do.	Godo Shigen Sangyo Co. Ltd. (Kanto Natural Gas Development Co. Ltd., 11%, and Mitsui & Co. Ltd., 10%)	Chosei, Chiba Prefecture	2,400
Do.	do.	Kanto Natural Gas Development Co. Ltd. (Mitsui Chemicals, Inc., 21.9%, and Godo Shigen Sangyo Co. Ltd., 14.3%)	Mobara, Chiba Prefecture	1,200
Do.	do.	Nihon Tennen Gas Co. Ltd. (Kanto Natural Gas Development Co. Ltd., 50%, and Tomen Corp., 41%)	Shirako and Yokoshiba, Chiba Prefecture	1,200
Do.	do.	Toho Earthtech, Inc. (Itochi Corp., 34.1%; Mitsubishi Gas Chemical Co. Ltd., 32.2%; Nippon Light Metal Co. Ltd., 31.1%)	Kurosaki, Niigata Prefecture	720
Do.	do.	Nippon Chemicals Co. Ltd. (Nippon Shokubai Co. Ltd., 17%; Takeda Chemical Industries Ltd., 16.4%; Chugai Boyeki Co. Ltd., 13.6%)	Isumi, Chiba Prefecture	720
Lead, refined	do.	Kamioka Mining and Smelting Co. Ltd.	Kamioka, Gifu Prefecture[3]	33,600
Do.	do.	Mitsui Mining and Smelting Co., Ltd.	Takehara, Hiroshima Prefecture	43,800
Do.	do.	Toho Zinc Co. Ltd.	Chigirishima, Hiroshima Prefecture	120,000
Do.	do.	Sumitomo Metal Mining Co. Ltd. (SMM)	Harima, Hyogo Prefecture	30,000
Do.	do.	Kosaka Smelting and Refining Co. Ltd.	Kosaka, Akita Prefecture	25,200
Do.	do.	Hosokura Smelting and Refining Mining Co. Ltd. (wholly owned subsidiary of Mitsubishi Materials Corp.)	Hosokura, Miyagi Prefecture[3]	22,200
Limestone		Mitsubishi Materials Corp.	Higashitani, Fukuoka Prefecture	10,000
Do.		Nittetsu Mining Co., Ltd.	Torigatayama, Kochi Prefecture; Oita, Oita Prefecture; and Shiriya, Aomori Prefecture	23,000
Do.		Sumikin Mining Co., Ltd.	Hachinohe Sekkai, Aomori Prefecture	5,500
Do.		Sumitomo-Osaka Cement Co. Ltd.	Ibuku, Shiga Prefecture, and Karazawa, Tochigi Prefecture	4,000
Do.		Shuho Mining Co., Ltd.	Sumitomo Cement Shuho, Yamaguchi Prefecture	8,200
Do.		Taiheiyo Cement Co. Ltd.	Ofunato, Iwate Prefecture; Ganji and Tsukumi, Oita Prefecture; Garo, Hokkaido Prefecture; Kawara, Fukuoka Prefecture, Tosayama, Kochi Prefecture; Taiheiyo Buko, Saitama Prefecture; and Shigeyasu, Yamaguchi Prefecture	46,000
Do.		Todaka Mining Co. Ltd.	Todaka-Tsukumi, Otia Prefecture	12,000
Do.		Ube Kosan Co. Ltd.	Ube Isa, Yamaguchi Prefecture	9,000
Manganese, electrolytic dioxide		Mitsui Mining and Smelting Co., Ltd.	Takehara, Hiroshima Prefecture	24
Do.		Tosoh Corp.	Hyuga, Miyazaki Prefecture	34
Nickel:				
In ferronickel	metric tons	Hyuga Smelting Co. Ltd. [wholly owned subsidiary of Sumitomo Metal Mining Co. Ltd. (SMM)]	do.	22,000
Do.	do.	Yakin Oheyama Co. Ltd.	Oheyama, Kyoto Prefecture	12,720
Do.	do.	Pacific Metals Co. Ltd.	Hachinohe, Aomori Prefecture	40,800
In oxide	do.	Tokyo Nickel Co. Ltd.	Matsuzaka, Mie Prefecture	60,000
Refined	do.	Sumitomo Metal Mining Co. Ltd. (SMM)	Niihama, Ehime Prefecture	36,000

See footnotes at end of table.

TABLE 2—Continued
JAPAN: STRUCTURE OF THE MINERAL INDUSTRY IN 2012

(Thousand metric tons unless otherwise specified)

Commodity		Major operating companies and major equity owners	Location of main facilities	Annual capacity
Pyrophyllite		Goto Kozan Co. Ltd.	Goto, Nagasaki Prefecture	204
Do.		Ohira Kozan Co. Ltd.	Ohira, Okayama Prefecture	132
Do.		Sankin Kogyo Co. Ltd.	Otsue, Hiroshima Prefecture	72
Do.		Shinagawa Shirenga Co. Ltd.	Mitsuishi, Okayama Prefecture	180
Do.		Shokozan Kogyosho Co. Ltd.	Yano-Shokozan, Hiroshima Prefecture	180
Do.		Showa Kogyo Co. Ltd.	Showa-Shokozan, Hiroshima Prefecture	60
Steel, crude		JFE Steel Corp. (wholly owned subsidiary of JFE Holdings Inc.)	Chiba, Chiba Prefecture; Kawasaki (Keihin), Kanagawa Prefecture; Nishinomiya, Hyogo Prefecture; Handa, Aichi Prefecture; Fukuyama, Hiroshima Prefecture; and Kurashiki, Okayama Prefecture	33,835
Do.		Kobe Steel Ltd.	Kakogawa and Kobe, Hyogo Prefecture	8,943
Do.		Nippon Steel & Sumitomo Metal Corp.	Oita, Oita Prefecture; Kawata, Fukuoka Prefecture; Kimitsu, Chiba Prefecture; and Nagoya, Aichi Prefecture	33,199
Do.		do.	Kashima, Ibaraki Prefecture; Kokura, Fukuoka Prefecture; and Wakayama, Wakayama Prefecture	12,820
Do.		Nisshin Steel Co. Ltd.	Kuri, Hiroshima Prefecture; Osaka City; Shunan, Yamaguchi Prefecture; and Toyo, Ehime Prefecture	4,000
Tantalum	metric tons	Japan New Metals Co. Ltd.	Akita, Akita Prefecture	95
Do.	do.	Mitsui Mining and Smelting Co. Ltd.	Miyama, Fukuoka Prefecture	NA
Titanium:				
In sponge metal		Sumitomo Titanium Corp. (Sumitomo Metal Industries, Ltd., 75.2%, and Kobe Steel Ltd., 24.8%)	Amagasaki, Hyogo Prefecture	24
Do.		Toho Titanium Co. Ltd. (JX Nippon Mining & Metals Co., Ltd., 47%; Mitsui & Co. Ltd., 20%; others, 33%)	Chigasaki, Kanagawa Prefecture	15
In dioxide	metric tons	Fuji Titanium Industry Co. Ltd. (Ishihara Sangyo Kaishia Ltd., 24.8%, and others, 75.2%)	Kobe, Hyogo Prefecture	17,400
Do.	do.	Ishihara Sangyo Kaisha Ltd.	Yokkaichi, Mie Prefecture	154,800
Do.	do.	Sakai Chemical Industries Co. Ltd.	Onahama, Fukushima Prefecture	60,000
Do.	do.	Tayca Corp.	Saidaiji, Okayama Prefecture	60,000
Do.	do.	Titan Kogyo Kabushiki Kaisha	Ube, Yamaguchi Prefecture	16,800
Zinc, refined	do.	Akita Smelting Co. Ltd. [Dowa Mining Co. Ltd., 57%; JX Nippon Mining & Metals Co., Ltd., 24%; Sumitomo Metal Mining Co. Ltd. (SMM), 14%; Mitsubushi Materials Corp., 5%]	Iijima, Akita Prefecture	200,400
Do.	do.	Hachinohe Smelting Co. Ltd. (Mitsui Mining and Smelting Co. Ltd., 57.7%; JX Nippon Mining & Metals Co., Ltd., 27.8%; Toho Zinc Co. Ltd. and Nisso Smelting Co. Ltd., 14.5%)	Hachinohe, Aomori Prefecture	117,600
Do.	do.	Hikoshima Smelting Co. Ltd.	Hikoshima, Yamaguchi Prefecture	84,000
Do.	do.	Kamioka Mining and Smelting Co. Ltd.	Kamioka, Gifu Prefecture	72,000
Do.	do.	Toho Zinc Co. Ltd.	Annaka, Gunma Prefecture	139,200
Do.	do.	Sumitomo Metal Mining Co. Ltd. (SMM)	Harima, Hyogo Prefecture	90,000

Do., do. Ditto. NA Not available.

[1]Coal mining operations continued, but output has been in decline.

[2]Saganoseki Smelter and Refinery and Hitachi Refinery (Nikko Smelting & Refining Co. Ltd.) [450,000 metric tons per year (t/yr)] and Tamano Smelter and Refinery (Hibi Kyoto Smelting Co. Ltd.) (260,000 t/yr).

[3]Secondary lead smelter and refinery.

TABLE 3
JAPAN: SUPPLY AND DEMAND FOR SELECTED NONFERROUS METALS

(Metric tons unless otherwise specified)

	Refined copper			Refined lead		
	2010	2011	2012	2010	2011	2012
Stocks at the beginning of the year	104,409	110,279	124,578	17,303	16,521	15,168
Production	1,548,688	1,328,288 ʳ	1,516,354	215,828	215,064 ʳ	208,994
Imports	45,912	126,569	35,876	11,126	22,049	28,869
Total supply	1,699,009	1,565,136 ʳ	1,676,808	244,257	253,634 ʳ	253,031
Exports	528,384	437,247	545,908	40,461	33,367	19,831
Reported consumption	1,104,823 ʳ	1,058,646 ʳ	930,241	192,710 ʳ	191,801 ʳ	206,995
Total demand	1,633,207 ʳ	1,495,893 ʳ	1,476,149	233,171 ʳ	225,168 ʳ	226,826
Stocks at the end	110,279	124,578 ʳ	145,128	16,521	15,168	19,510
Apparent consumption[1]	1,060,346	1,003,311 ʳ	985,772	187,275	205,099 ʳ	213,690
	Refined zinc			Silver (kilograms)		
	2010	2011	2012	2010	2011	2012
Stocks at the beginning of the year	73,600	65,555	92,265	872,325	1,062,722	999,120
Production	574,008	544,674 ʳ	571,312	1,898,208	1,724,218	1,764,533
Remelting	NA	NA	NA	313,931	325,373	348,620
Imports	31,855	77,881	23,960	2,087,621	1,929,204	1,698,639
Total supply	679,463	688,110 ʳ	687,537	5,172,085	5,041,517	4,810,912
Exports	97,745	95,278	135,560	2,733,284	2,837,596 ʳ	3,297,120
Reported consumption	389,036 ʳ	351,848 ʳ	354,789	1,878,619 ʳ	1,737,564 ʳ	1,048,971
Total demand	486,781 ʳ	447,126 ʳ	490,349	4,611,903 ʳ	4,575,160 ʳ	4,346,091
Stocks at the end	65,555	92,265 ʳ	72,389	1,062,722	999,120 ʳ	1,037,918
Apparent consumption[1]	516,163	500,567 ʳ	480,588	1,376,079	1,204,801 ʳ	475,874

ʳRevised. NA Not applicable.

[1]Apparent consumption is total supply less exports and stocks at yearend.

Source: Japan Mining Industry Association, 2013.

TABLE 4
JAPAN: RESERVES OF MAJOR MINERAL COMMODITIES IN 2012

(Thousand metric tons unless otherwise specified)

Commodity		Exploitable reserves
Coal[1]		773,000
Dolomite		913,000
Gold ore, Au content	kilograms	159,000
Iodine		5,000 ᵉ
Limestone		40,400,000
Pyrophyllite		59,700
Silica sand		73,600
Silica stone, white		462,000

ᵉEstimated.

[1]Recoverable reserves, including brown coal.

Source: Natural Resources and Fuel Department, Agency of Natural Resources and Energy.

THE MINERAL INDUSTRY OF NORTH KOREA

By Lin Shi

The economy of North Korea (the Democratic People's Republic of Korea) continued to grow in 2012. According to the Bank of Korea, North Korea's real annual gross domestic product (GDP) increased by 1.3% in 2012 compared with that of the previous year. This growth was attributed mainly to growth in the agriculture, forestry, and fishing sector, for which output increased by 3.9%; growth in the manufacturing sector, for which production increased by 1.6%; and a 1.6% increase in power generation in the energy sector (Bank of Korea, 2013).

Government Policies and Programs

In 2012, North Korea renewed its special economic zones agreement with China, negotiated a new payment structure with Russia to settle its $11 billion Soviet-era debt, and proposed new agricultural and industrial policies to boost domestic production. Despite international economic sanctions, the Government sought to attract foreign investment to improve the overall standard of living in North Korea (U.S. Central Intelligence Agency, 2013).

Minerals in the National Economy

The North Korean Government controlled the Nation's economy, including its mining activity and financial market. The Government directed a large share of the GDP to the military sector, which affected the country's overall economic performance. Although the mineral industry's production capacity was limited by the county's restricted financial and technical resources, the mineral industry supported the country's military sector expenses, as well as met the country's industrial requirements for raw materials. Detailed or updated mining and mineral production data have not been available in the country's official reports in the previous 5 years, and the production estimates in table 1 are based on the historic production data and (or) extracted from public media reports.

Production

Mining output was estimated to have increased by 0.8% in 2012 compared with that of 2011 and included a 1.2% increase in coal production but a 0.5% decrease in metallic mineral production. The output of the manufacturing sector increased by 1.6% in 2012 compared with that of 2011 and included a 4.7% increase in light industrial production (mainly because of increases in the production of food and tobacco), and a 0.2% increase in heavy and chemical industrial production combined (mainly because of increases in the production of chemical products and transportation equipment) (Bank of Korea, 2013).

Structure of the Mineral Industry

North Korea's mineral industry included a coal mining sector, a ferrous and nonferrous metals mining and processing sector, and an industrial minerals mining and processing sector. Most of the large-scale mining and mineral processing enterprises in North Korea were owned and operated by the central Government. Provincial and local governments owned and operated various small- and medium-scale mining and mineral processing facilities. Companies from China, the Republic of Korea, and other countries participated in joint ventures with North Korea for the development and operation of the cement, coal, copper, graphite, iron ore, lead and zinc, magnesite, molybdenum, and precious metals production facilities in North Korea (table 2).

Mineral Trade

The value of North Korean exports in 2012 increased by 3.3% to $2.88 billion from $2.79 billion in 2011. Of this amount, the export value of minerals decreased by 0.2%. The value of the country's total imports increased by 10.2% to $3.93 billion from $3.53 billion (revised) in 2011. Of this amount, the import value of transportation equipment increased by 39.6%. Total North Korean exports to the Republic of Korea decreased in value by 6.2% from that of 2011. Bilateral trade between the Republic of Korea and North Korea increased by 15.0% to $1.97 billion in 2012. The Kaesong Industrial Complex accounted for 99.5% of this trade, and electronic goods were the primary trading items. North Korea exported more than 2 metric tons of gold to China in 2012 to earn $100 million (Bank of Korea, 2013; Business Insider, 2013).

Commodity Review

Metals

Copper.—Wanxiang Resources Co. Ltd. of China held a 51% stake in the Hyesan copper mine and had been investing in the mine since 2007. Wanxiang and the Ministry of Mining Industries of North Korea set up the Hyesan-China Joint Venture Mineral Co. to operate the Hyesan copper mine, which is located in Yanggang Province near Changbai City in China's Jilin Province. The mine had a designed output capacity of 50,000 to 70,000 metric tons per year of copper concentrate, and it resumed production in the fourth quarter of 2011. Wanxiang was unable to obtain significant profit from copper ore shipped to China in 2012.

A major difficulty for investing in North Korea is that contracts or agreements have limited binding power on North Korean partners, which are often unpredictable in their policies. No copper production from the Hyesan copper mine was reported for 2012. Although North Korea has rich mineral deposits, the country lacks the technology to mine and refine the copper ore, as well as a transportation network to ship the ore (Yonhap News, 2012; PEdaily.cn, 2013; WantChinaTimes.com, 2013).

Industrial Minerals

Rare Earths.—According to Asia Times Online, North Korea has deposits of about 20 Mt of rare-earth metals (REMs) in the eastern and western parts of country but does not have the technology to explore the REM resource nor to manufacture rare-earth-based products. North Korea would likely need international assistance through joint projects to explore the REM resources and to produce and (or) possibly export REMs. The Republic of Korea had expressed interest in working with North Korea to explore and mine the REM deposits and to use the REMs to manufacture industrial products. An REM reprocessing plant was located in Hamhung but was not fully operational (Petrov, 2012a, b).

Outlook

Because of the mineral industry's significance to the country's economy, the Government of North Korea is likely to continue working to attract international investment in the country's mineral sector. The increasing international demand for minerals, especially the demand from China and Russia, is expected to stimulate North Korea's production of mineral commodities, such as coal, iron ore, magnesite, molybdenum, nickel, sand, and zinc.

In March 2010, when a North Korean submarine sunk the Republic of Korea's Navy ship *Cheonan*, the Republic of Korea stopped all trade with North Korea except at the Kaesong Industrial Complex, and in November 2010, there was a serious cross-border clash between the two countries. North Korea's economic cooperation with the Republic of Korea is expected to continue recovering slowly in the future after the 2010 interruption.

The country's mining and associated business infrastructure, such as international banking, Internet connections, mobile phone coverage, and power supply are expected to be established, improved, and regulated gradually, beginning in the economic free zones in the border areas, to improve communications and promote economic cooperation with China, the Republic of Korea, Russia, and even Western European countries. North Korea's political control will constrain the country's economic system and further economic development or reform, however (U.S. Central Intelligence Agency, 2013).

References Cited

Bank of Korea, 2013, Gross domestic product estimates for North Korea for 2012: Bank of Korea, July 12. (Accessed September, 25, 2013, at http://www.bok.or.kr/contents/total/eng/boardView.action?menuNaviId=1959&boardBean.brdid=12202&boardBean.menuid=1959.)

Business Insider, 2013, China buys North Korea gold reserves as South Korea increased gold reserves by 30% (16 tons); Paraguay buys first time: Business Insider, January 22. (Accessed September 25, 2013, at http://www.businessinsider.com/china-buys-north-korea-gold-reserves-as-south-korea-increased-gold-reserves-by-30-16-tonnes-paraguay-buys-first-time-2012-9.)

PEdaily.cn, 2013, Wan Xiang mining trapped in North Korea—A bottomless pit 560 million investment nowhere to retreat: Beijing, China, Zero2IP Group, March 7. (Accessed September 25, 2013, at http://pe.pedaily.cn/201303/20130307784777_2.shtml.)

Petrov, Leonid, 2012a, Rare earth metals—North Korea's new trump card: East Asia Forum, August. (Accessed September 25, 2013, at http://www.eastasiaforum.org/2012/08/21/rare-earth-metals-north-koreas-new-trump-card/.)

Petrov, Leonid, 2012b, Rare earths bankroll North Korea's future: Asia Times Online, August 8. (Accessed August 31, 2012, at http://www.atimes.com/atimes/Korea/NH08Dg01.html.)

WantChinaTimes.com, 2013, WangXiang North Korean copper hole goes deeper, and deeper: WantChinaTimes.com, March 17. (Accessed September 25, 2013, at http://www.wantchinatimes.com/news-subclass-cnt.aspx?id=20130317000009&cid=1102.)

U.S. Central Intelligence Agency, 2013, Korea, North, *in* The world factbook: U.S. Central Intelligence Agency. (Accessed September 25, 2013, at https://www.cia.gov/library/publications/the-world-factbook/geos/kn.html.)

Yonhap News, 2012, N. Korea, China agree to jointly develop three mines in North: Yonhap News, August 9. (Accessed August 31, 2012, at http://english.yonhapnews.co.kr/northkorea/2012/08/09/98/0401000000AEN20120809004600315F.HTML.)

TABLE 1
NORTH KOREA: ESTIMATED PRODUCTION OF MINERAL COMMODITIES[1, 2]

(Metric tons unless otherwise specified)

Commodity[3]		2008	2009	2010	2011	2012
METALS						
Cadmium metal, smelter[4]		200	200	200	200	200
Copper:						
Mine output, Cu content[4]		12,000	12,000	12,000	12,000	12,000
Metal, primary and secondary:[4]						
Smelter		15,000	15,000	15,000	15,000	15,000
Refinery		15,000	15,000	15,000	15,000	15,000
Gold, mine output, Au content[4]	kilograms	2,000	2,000	2,000	2,000	2,000
Iron and steel:						
Iron ore and concentrate, marketable:						
Gross weight	thousand metric tons	5,316[5]	5,300	5,300	5,300	5,300
Fe content	do.	1,488[5]	1,500	1,500	1,500	1,500
Metal:						
Pig iron[4]	do.	900	900	900	900	900
Ferroalloys, unspecified[4]	do.	10	10	10	10	10
Steel, crude	do.	1,279[5]	1,300	1,300	1,300	1,300
Lead:[4]						
Mine output, Pb content		13,000	13,000	13,000	13,000	13,000
Metal, primary and secondary:						
Smelter		13,000	13,000	13,000	13,000	13,000
Refinery		9,000	9,000	9,000	9,000	9,000
Silver, mine output, Ag content[4]		20	20	20	20	20
Tungsten, mine output, W content		270	100	100	100	100
Zinc:[4]						
Mine output, Zn content		70,000	70,000	70,000	70,000	70,000
Metal, primary and secondary		75,000	75,000	75,000	75,000	75,000
INDUSTRIAL MINERALS						
Cement, hydraulic	thousand metric tons	6,415[5]	6,400	6,400	6,400	6,400
Fluorspar[4]		12,500	12,500	12,500	12,500	12,500
Graphite[4]		30,000	30,000	30,000	30,000	30,000
Magnesite, crude[4]		150,000	150,000	150,000	150,000	150,000
Nitrogen, N content of ammonia[4]	thousand metric tons	100	100	100	100	100
Phosphate rock, P_2O_5 equivalent[4]		300,000	300,000	300,000	300,000	300,000
Salt, all types[4]		500,000	500,000	500,000	500,000	500,000
Sulfur[4]	thousand metric tons	42	42	42	42	42
Talc, soapstone, pyrophyllite		50,000	50,000	50,000	50,000	50,000
MINERAL FUELS AND RELATED MATERIALS						
Coal, anthracite	thousand metric tons	25,060[5]	36,000	41,000	41,000	41,492[5]
Coke[4]	do.	2,000	2,000	2,000	2,000	2,000

do. Ditto.

[1]Estimated data are rounded to no more than three significant digits.

[2]Table includes data available through September 25, 2013.

[3]In addition to the commodities listed, crude construction materials, such as sand and gravel and other varieties of stone, and refined petroleum products and rare-earth-based products presumably are produced, but available information is inadequate to make reliable estimates of output.

[4]Because of the lack of official reported data, most of the mineral commodity production numbers have been estimated for the past 5 years.

[5]Reported figure by Bank of Korea.

TABLE 2
NORTH KOREA: STRUCTURE OF THE MINERAL INDUSTRY IN 2012

(Thousand metric tons unless otherwise specified)

Commodity	Major operating companies and major equity owners	Location of main facilities	Annual capacity[e]
Cement	Sunchon Cement Complex	Sunchon, Pyongannam Province	3,000
Do.	Samgwong Cement Complex	Samgwong, Kangwon Province	2,500
Do.	Gomusan Cement Factory	Cheongjin, Hamgyongbuk Province	2,000
Do.	Cheonnaeri Cement Factory	Cheonae, Hamgyongnam Province	1,000
Coal	Anju Coal Mining Complex and Sunchon Coal Mining Complex	Anju, Kaechon, Pukchang, Sunchon, and Tokechon, South Pyongan (Pyongannam) Province; and North Pyongan (Pyonganbuk) Province	9,500
Do.	Saebyol Coal Mining Complex and Northern Coal Mine Enterprise	Saebyo, North Hamgyong (Hamgyongbuk) Province	6,000
Copper, mine output, Cu content	Hyesan Youth Copper Mine (51% owned by Luanhe Industrial Group and another unnamed Chinese company)	Hyesan, Yanggang Province	13
Gold, mine output, kilograms Au content	Gumsan (Kumsan) Joint Venture Co.	Sierra near Changjin northwest of Hamgyongbuk Province	530
Do. do.	Daebong Mine	Yanggang Province	150
Graphite	Yeongchon Graphite Mine (Joint venture of Korea Resources Corp. and the Government of North Korea)	Yeongchon, Yonan County, South Hwanghae Province	3
Iron ore, concentrate, gross weight	Ministry of Metal and Machinery, Department of Mines, Musan Iron Ore Mine Complex	Near the town of Musan, Hamgyongbuk Province	10,000
Do.	Unryul Mine	Unryul, Hwanghaenam Province	1,000
Lead:			
In concentrate	Korea Zinc Industrial Group	Komdok, near Tancheon, Hamgyongnam Province	20
Refined	do.	Munpyong, Kangwon Province	32
Magnesite, concentrate, gross weight	Korea Magnesia Clinker Industry Group (KMCIG)	Daehung and Yongyang, Hamgyongnam Province; Paek Bai near Kim Chaeck, Hamgyongbuk Province	2,500
Magnesia clinker	Korea Magnesia Clinker Industry Group (KMCIG) and Quintermina AG	Danchon and Daehung, Hamgyongnam Province; Song Jin, Hamgyongbuk Province	1,200
Steel, crude			
Do.	Kim Chaek Iron and Steel Complex (Ministry of Metal and Machinery)	Chongjin, Hamgyongbuk Province	2,400
Do.	Hwanghae (Hwanghai) Iron Works	Songjin, North Hamgyong Province	1,500
Do.	Kangson Works	Kangson, Hwanhaebuk Province	960
Do.	Chollima Steel Works	Cholliam District, Nampo City, Pyungnam Province	760
Zinc:			
In concentrate	Korea Zinc Industrial Group	Komdok near Tancheon and Sankok near Kowon, Hamgyongnam Province; Nakyong, Hwanhaenam Province	80
Refined	do.	Munpyong, Kangwon Province; Tancheon, Hamgyongnam Province	100

[e]Estimated; estimated data are rounded to no more than three significant digits. Do., do. Ditto.

Sources: Korea Resources Corp. and North Korea Resource Institute.

THE MINERAL INDUSTRY OF THE REPUBLIC OF KOREA

By Lin Shi

The Republic of Korea's real gross domestic product (GDP) growth rate was 2.0% in 2012 compared with 3.7% (revised) in 2011. The country's exports continued to increase in the first half of the year, then slowed because of a decrease in external demand; domestic consumption and sales revenues also decreased. The inflation rate was 2.2% in 2012 compared with a rate of 4% in 2011. The mining and quarrying sector grew by 1.3% in 2012 compared with negative growth of 4.8% (revised) in 2011. The growth of the manufacturing sector was 2.2% in 2012 compared with 7.3% in 2011. The production of manufactured nonmetallic mineral products decreased by 8.4%, and the production of metals increased by 4.6% (Bank of Korea, 2013, p. 20–29).

Minerals in the National Economy

The Republic of Korea's mining and quarrying sector was not a significant contributor to the country's economic development in terms of monetary value; it accounted for only 0.2% of the GDP in 2012. Many of the country's mineral commodity requirements were met through imports (Bank of Korea, 2013, p. 20–29).

Government Policies and Programs

In 2012, the Bank of Korea, the Financial Services Commission, the Financial Supervisory Service, and the Ministry of Strategy and Finance worked together to enhance their policy cooperation and information sharing to help advance the country's financial stability. The Korea–U.S. Free Trade Agreement (KORUS FTA) entered into force. Also, the Republic of Korea hosted the 2012 Nuclear Security Summit (U.S. Department of State, 2012).

Production

In 2012, production of sand increased by about 80% compared with that of 2011; talc, by about 39%; cadmium, by about 30%; diatomite, by about 17%; and quartzite and silver, by about 16% each. Production of some mineral commodities decreased, including that of kaolin, which decreased by about 36%; mica, by about 18%; and salt, by 17% (table 1).

Structure of the Mineral Industry

The Ministry of Commerce, Industry, and Energy supervised the country's coal mining, natural gas, petrochemical, and petroleum refining companies, all of which were state owned. The rest of the mining, quarrying, and mineral processing companies were privately owned and operated (table 2).

Mineral Trade

According to the Bank of Korea, the country's exports of goods and services increased by 4.2% in 2012, led by exports of (in order of value) semiconductors, electrical equipment and devices, and automobiles. The country's imports of goods and services increased by 2.5% in 2012. There were significant flows of manufactured goods, services, and technology between the Republic of Korea and the United States. The Republic of Korea exported goods and services valued at $58.9 billion to the United States in 2012, which was an increase of about 4% from the $56.7 billion exported to the United States from the Republic of Korea in 2011. These exports included $10.6 billion in passenger cars, $3.7 billion in petroleum products (other than fuel oil), $1.6 billion in computer accessories, $1.5 billion in drilling and oilfield equipment and platforms, and $1.4 billion in semifinished iron and steel products. The Republic of Korea imported from the United States goods and services valued at $42.3 billion, which was a decrease of about 3% from the $43.4 billion imported from the United States in 2011. These imports included $4.1 billion in semiconductors, $1.2 billion in steelmaking materials, $1.1 billion in petroleum products (other than fuel oil), $609 million in nonferrous metals, $577 million in precious metals, $426 million in alumina and aluminum, and $311 million in copper (Bank of Korea, 2013, p. 20–29; U.S. Census Bureau, 2014a, b).

Commodity Review

Metals

Copper.—The Republic of Korea did not produce copper ore or copper concentrate in 2012, although the refineries produced about 0.6 million metric tons (Mt) of primary and secondary copper metal. The country relied on imports to meet the raw material requirements for its copper smelters, which are located in Changhang and Onsan. LS-Nikko Copper Inc., which was a Korean and Japanese joint venture based in Seoul, held a 50.1% share of the company's stock and was engaged in the import of copper ore and in metal recycling, smelting, and refining. LS-Nikko Copper's copper refinery had a 60% market share in electrolytic cathode copper. The company also owned the Danyang factory at the Maepo Agricultural and Industrial Complex in Choongchungbuk-do through its affiliate GRM Inc. The Danyang factory had the capacity to produce 73,300 metric tons (t) of recycled metals, including 17,600 t of copper (LS-Nikko Copper Inc., 2011, 2014).

Gold.—The Republic of Korea produced 47,992 kilograms of refined gold in 2012. The country held about 700 Mt of gold ore. The Korea Exchange Inc. established a trading system and planned to begin gold trading in 2014 to allow investors easy access to the market and to provide a new financial opportunity for the Government to generate revenue. About 100 t of gold was traded domestically each year, and small local smelters could recycle about $4 billion worth of gold scrap and jewelry into bars each year (Hur, 2013; Korea Institute of Geoscience and Mineral Resources, 2013, p. 7).

Iron and Steel.—In 2012, the Republic of Korea relied heavily on imports to meet its iron ore requirements, and the domestic iron ore production was about 0.6 Mt. The country produced about 69 Mt of crude steel, which was an increase of 1% from that of 2011. Domestic steel consumption was about 52 Mt, which was a decrease of about 5% from that of 2011. According to the Korea Iron and Steel Association, the country exported about 30 Mt of steel in 2012, which was a decrease of about 4% from the previous year. The Republic of Korea's steel exports had decreased to about 20.5 Mt in 2009 from 20.8 Mt in 2008 owing to the international financial crisis; steel exports increased from 2010 to 2012, however, and totaled 24.9 Mt, 29.1 Mt, and 30 Mt, respectively (BusinessKorea Co. Ltd., 2013; Korea Institute of Geoscience and Mineral Resources, 2013, p. 7).

Lead, Nickel, Tin, and Zinc.—Tin and nickel were used for making alloys to produce electronic goods and kitchenware, Korea Zinc was the Republic of Korea's leading lead and zinc producer, and zinc was used for making steel for automobiles and construction materials. Korea Zinc also produced copper, gold, and silver. Korea Zinc had subsidiaries and affiliates in Australia, Canada, and Thailand. Zinc production accounted for about 40% of Korea Zinc's sales revenue, and silver and lead each accounted for about 20% of the company's revenue (Pohang Iron and Steel Co. Ltd., 2012; Korea Zinc Co. Ltd., 2014).

Tungsten.—The Sangdong tungsten-molybdenum project is located on the east coast, 187 kilometers from Seoul, and the mine and plant together were expected to become the world's leading tungsten production site and to account for 7% to 10% of global tungsten output. Woulfe Mining Corp. of Canada signed a memorandum of understanding with the Shinhan Bank in July to fund its Sangdong tungsten/molybdenum project. The Shinhan Bank, which was the Republic of Korea's largest bank, provided about $104 million in loans for the development of the project at less than 6% annual interest, which had a significant positive effect on the project. Woulfe reported that the Sangdong project's net present value was about $400 million in 2012 and that the internal rate of return was 46% (Lazenby, 2012; Woulfe Mining Corp., 2012).

Industrial Minerals

Rare Earths.—The Republic of Korea relied on imports to meet its requirement for rare-earth elements. In 2012, the country announced the discovery of a major deposit of rare-earth elements that was estimated to be adequate to supply the country's domestic demand for rare earths for about 50 years. The Korea Institute of Geoscience and Mineral Resources analyzed the soil samples collected from 11 locations around the country and reported that there were resources totaling 23 Mt containing 147,000 t of rare-earth elements in deposits spanning areas of Chungju, North Chungcheong Province, and Hongcheon, Gangwon Province. The rare-earth minerals are hosted in iron ore. According to the China Daily, China exported 72 t of rare-earth material to the Republic of Korea in September. The price of rare earths decreased by 8.6% in October, and the Republic of Korea increased its rare-earth imports from China by about 19% beginning in October (KBS World, 2011; China Daily, 2012).

Mineral Fuels and Related Materials

Natural Gas and Petroleum.—In 2012, the Russian state-owned natural gas company OAO Gazprom was discussing the railway junction, gas pipeline, and power line projects that were to run from Russia to the Republic of Korea by way of North Korea with the Governments of the Republic of Korea and North Korea. Gazprom was under a contract to supply at least 1.5 Mt of liquefied natural gas (LNG) to the Republic of Korea each year until 2025, and the volume was expected to increase greatly to meet the country's future energy needs. Gazprom sent gas to Japan directly through the Sakhalin-2 pipeline, but it had to send LNG to the Republic of Korea by ship. The Republic of Korea and Russia thought it was important for all three countries to have a gas pipeline that goes through North Korea from Russia to the Republic of Korea. The countries were negotiating a deal that would enable North Korea to earn $100 million per year from the trans-Korean pipeline when it opens in 2017 (Blank, 2012; UPI.com, 2013).

Uranium.—Stonehenge Metals Ltd. of Australia was engaged in exploration for uranium in the Republic of Korea and was focusing on its 100%-owned Daejon project in the areas of Chubu, Kolnami, and Yokwang. The company discovered a vanadium resource in the project area and expected to discover a vanadium deposit. Woulfe owned the Ogshon uranium project; the project is located within the Okcheon metamorphic belt, which extends through the middle of the country. Woulfe had been granted five registered mining rights (RMRs) and four mining applications (MAs) in the Daejeon-Geumsan area. Korea Resources Corp. conducted the U_3O_8 resource drilling program for Woulfe in Daejeon 29, Daejeon 39, and Daejeon 48. Woulfe had also been granted one RMR and two MAs in the Miwon area, and the company's sampling and testing programs were in process (Australia's Paydirt, 2012, p. 71; Woulfe Mining Corp., 2012).

Outlook

According to the Korea Iron and Steel Association, the Republic of Korea's steel production, domestic consumption, and exports are expected to increase by about 4%, 1%, and 4%, respectively, in 2014. The Republic of Korea's GDP is dependent mainly on exports. The Bank of Korea projects that the country's economy will grow in 2013 but at only a slow pace, and that the value of the country's exports will also increase only slowly because of geopolitical risks in certain regions of the world and because of global economic conditions. Domestic demand, however, is projected to remain relatively steady as real purchasing power increases owing to improved working conditions and stable oil prices (Bank of Korea, 2013, p. i–vi, 14–17).

References Cited

Australia's Paydirt, 2012, Stonehenge looks for Korean domestic fit: Australia's Paydirt, v. 1, no. 193, p. 71.

Bank of Korea, 2013, 2012 annual report: Seoul, Republic of Korea, Bank of Korea, March, 156 p.

Blank, Stephen, 2012, Russia's Lorea projects gather dust: Asia Times Online, October 26. (Accessed November 30, 2012, at http://www.atimes.com/atimes/Central_Asia.html.)

BusinessKorea Co., Ltd., 2013, Korea's steel exports dropped this year: BusinessKorea Co., Ltd., December 26. (Accessed January 15, 2014, at http://www.businesskorea.co.kr/article/2746/steel-industry-korea%E2%80%99s-steel-exports-dropped-year.)

China Daily, 2012, S. Korea rare earth import from China grows 19%: Xin Hua News Agency, October 18. (Accessed January 17, 2014, at http://www.chinadaily.com.cn/business/2012-10/18/content_15828842.htm.)

Hur, Jae, 2013, Korea exchange targets gold trade as Park hunts taxes: Bloomberg, December 3. (Accessed January 14, 2014, at http://www.bloomberg.com/news/2013-12-03/korea-exchange-targets-gold-trading-as-park-hunts-tax-revenue.html.)

KBS World, 2011, Rare earth mineral deposits found in S. Korea: Korea Communications Commission, June 29. (Accessed June 30, 2011, at http://world.kbs.co.kr/english/news/news_Ec_detail.htm?No=82621&id=Ec.)

Korea Institute of Geoscience and Mineral Resources, 2013, 2012 mineral demand and supply information: Daejeon, Republic of Korea, Korea Institute of Geoscience and Mineral Resources (KIGAM), 310 p.

Korea Zinc Co. Ltd., 2014, Company profile: Korea Zinc Co. Ltd. (Accessed January 15, 2014, at http://www.koreazinc.co.kr/english/company/page/summary.aspx.)

Lazenby, Henry, 2012, Woulfe Mining obtains additional funding for Korean project: Mining Weekly, July 12. (Accessed January 17, 2014, at http://www.miningweekly.com/print-version/woulfe-mining-obtains-additional-funding-for-korean-project-2012-07-12.)

LS-Nikko Copper Inc., 2011, LS-Nikko Copper finished construction of a recycling factory: LS-Nikko Copper Inc., June 13. (Accessed January 14, 2014, at http://www.lsnikko.com/english/html/pressroom/news_view.aspx?idx=100007&page=2&Category=S.)

LS-Nikko Copper Inc., 2014, Company overview: LS-Nikko Copper Inc. (Accessed January 14, 2014, At http://www.bnamericas.com/company-profile/en/ls-nikko-copper-inc-ls-nikko-copper.)

Pohang Iron and Steel Co. Ltd., 2012, Daewoo International builds on the success of its inroads into Africa; vows to expand: Pohang Iron and Steel Co. Ltd. press release, February 9. (Accessed January 16, 2014, at http://www.posco.com/homepage/docs/eng2/jsp/prcenter/news/s91c1010035p.jsp?idx=1983.)

UPI.com, 2013, Putin mulls gas pipeline to South Korea through North Korea: United Press International, November 13. (Accessed January 17, 2014, at http://www.upi.com/Business_News/Energy-Resources/2013/11/13/Putin-mulls-gas-pipeline-to-South-Korea-through-North-Korea/UPI-91981384345624/.)

U.S. Census Bureau, 2014a, U.S. exports from S. Korea by 5-digit end-use code 2003–2012: U.S. Census Bureau. (Accessed January 13, 2014, at http://www.census.gov/foreign-trade/statistics/product/enduse/exports/c5800.html.)

U.S. Census Bureau, 2014b, U.S. imports from S. Korea by 5-digit end-use code 2003–2012: U.S. Census Bureau. (Accessed January 13, 2014, at http://www.census.gov/foreign-trade/statistics/product/enduse/imports/c5800.html.)

U.S. Department of State, 2012, South Korea: U.S. Department of State Fact Sheet, December 17. (Accessed January 9, 2014, at http://www.state.gov/r/pa/ei/bgn/2800.htm.)

Woulfe Mining Corp., 2012, On the move in South Korea: Woulfe Mining Corp., November 29. (Accessed January 17, 2014, at http://www.woulfemining.com/s/Home.asp.)

TABLE 1
REPUBLIC OF KOREA: PRODUCTION OF MINERAL COMMODITIES[1]

(Metric tons unless otherwise specified)

Commodity		2008	2009	2010	2011	2012
METALS						
Bismuth, metal		210	300	498	480	437
Cadmium, smelter		3,090	2,500	4,166	3,005	3,904
Copper:						
Mine output, Cu content		4	14	9	NA[r]	NA[r]
Metal:						
Refined, primary and secondary		537,925	531,701	564,600	595,447	591,000
Gold:						
Mine output, Au content	kilograms	175	274	235	209	336
Metal, refined	do.	37,989	51,186	54,540	49,550	47,992
Iron and steel:						
Iron ore and concentrate:						
Gross weight	thousand metric tons	366	455	513	542	593
Fe content	do.	205	255[r]	287[r]	303[r]	332
Metal:						
Pig iron	do.	31,043	27,405[r]	31,228	42,213	NA[r]
Ferroalloys:						
Ferromanganese		251,125	216,400[r]	286,259	355,047	364,800[r]
Ferrosilicomanganese		76,184	151,100[r]	120,779	195,650	184,700[r]
Total		327,309	367,500	407,038	550,697	549,500
Steel, crude	thousand metric tons	53,493	48,752[r]	58,914	68,519	69,073
Lead:						
Mine output, Pb content		449	2,064	1,168	2,577	3,879
Metal, smelter		244,137	216,918	197,900	256,851	280,000
Nickel:						
Ferronickel		2,506[2]	21,609[2]	20,512[r]	19,011[1]	20,858[r]
Metal		28,653	NA	NA	NA	NA
Silver:						
Mine output, Ag content	kilograms	NA[r]	NA[r]	2,025	2,649	2,925
Metal	do.	1,461,886	1,740,078	1,735,535	2,197,409	2,547,315
Zinc:						
Mine output, Zn content		3,672	4,441[r]	710	1,486	2,868
Metal, primary		406,542[r]	751,179	717,100	828,735	875,000
INDUSTRIAL MINERALS						
Cement, hydraulic	thousand metric tons	51,653	50,127	47,420	48,300	NA[r]
Clays, kaolin	do.	955	659	764	799	515
Diatomaceous earth		2,540	2,440	2,200	5,150	6,000
Feldspar		344,257	622,700	496,511	384,221	360,413
Graphite, all types		73	48	34	NA[r]	NA[r]
Mica, all grades		49,474	27,078	36,486	31,260[r]	25,594
Salt		384,304	382,270	222,509	372,230	308,847
Stone, sand and gravel:						
Limestone	thousand metric tons	87,282[r]	81,612[r]	83,628[r]	86,945[r]	86,912
Quartzite	do.	3,325	3,536	3,603	3,603	4,184
Sand, including glass sand	do.	1,757	455	535	394	709
Talc and related materials:						
Pyrophyllite		892,625	617,411	673,936	510,708	483,133
Talc		6,439[r]	5,997[r]	5,729	15,608	21,625
Zeolites		217,691	235,226	242,190	231,420	245,285

See footnotes at end of table.

TABLE 1—Continued
REPUBLIC OF KOREA: PRODUCTION OF MINERAL COMMODITIES[1]

(Metric tons unless otherwise specified)

Commodity		2008	2009	2010	2011	2012
MINERAL FUELS AND RELATED MATERIALS						
Carbon black[e]		484,000	500,000	500,000	500,000	500,000
Coal, anthracite[e]	thousand metric tons	2,773 [2]	2,519 [2]	2,084 [r, 2]	2,084 [r, 2]	2,000 [e]
Fuel briquets, anthracite briquets[e]	do.	2,320	2,000	2,000	2,000	2,000
Petroleum, refinery products[e, 3]	thousand 42-gallon barrels	747,827 [2]	750,000	750,000	750,000	750,000

[e]Estimated; estimated data are rounded to no more than three significant digits; may not add to totals shown. [r]Revised. do. Ditto.
NA Not available.
[1]Table includes data available through January 16, 2014.
[2]Reported figure.
[3]Includes bunker oil C-type, diesel oil, gasoline, kerosene, liquefied petroleum gas, and naphtha.

Sources: Ministry of Commerce, Industry and Energy, Korea Institute of Geoscience and Mineral Resources; Mineral Commodity Summaries 2012; Korea Mineral Information; U.S. Geological Survey Minerals Questionnaires 2008–12. World Bureau of Metal Statistics, December 2011; The Bank of Korea Monthly Statistical Bulletin, table 41, Exports by principal commodity, and table 42, Imports by principal commodity, May 2011, p. 132–135.

TABLE 2
REPUBLIC OF KOREA: STRUCTURE OF THE MINERAL INDUSTRY IN 2012

(Thousand metric tons unless otherwise specified)

Commodity		Major operating companies and major equity owners	Location of main facilities	Annual capacity
Bismuth, metal	metric tons	Korea Zinc Co. Ltd.	Onsan refinery	500
Cadmium	do.	do.	do.	2,100
Do.	do.	Young Poong Corp.	Sukpo refinery	2,100
Cement		Ssangyong Cement Industrial Co. Ltd.	Plants at Tonghae, Kwang Yang, Munkyung, Pukpyong, and Yeongwol	15,040
Do.		Sung Shin Cement Manufacturing Co. Ltd.	Tanyang plant	13,700
Do.		Tong Yang Major Corp.	Plants at Pukpyong and Samchok	11,580
Do.		Lafarge Halla Cement Corp.	Plants at Kwang Yang and Okkye	9,500
Do.		Hyundai Cement Co. Ltd.	Plants at Tanyang and Yongwol	8,600
Do.		Hanil Cement Manufacturing Co.	Plants at Chungbuk and Tanyang	7,200
Do.		Asia Cement Manufacturing Co. Ltd.	Plants at Daegu and Jaechon	4,600
Coal		Korea Coal Corp.	Mines at Changsung, Dogae, and Hwasoon	2,000
Copper, metal, primary		Korea Zinc Co. Ltd.	Onsan	20
Do.		LS-Nikko Copper Inc.	Changhang	60
Do.		do.	Onsan	510
Gas, natural		Korea National Oil Corp.	Ulleung Basin	NA
Gold:				
In concentrate	kilograms	Hangum Co. Ltd.	Muguk Mine, Haenam, Jeonnam (South Cholla) Province	1,600
Refined	do.	Korea Zinc Co. Ltd.	Onsan	50,000
Do.	do.	LS-Nikko Copper Inc.	do.	60,000
Graphite		Kaerion Graphite Ltd.	Kangwon	NA
Do.		Wolmyong Mining Co.	do.	NA
Indium, metal	kilograms	Korea Zinc Co. Ltd.	do.	55,000
Iron ore		NA	Mines at Sinyemi, Gangwon Province	600
Lead, metal, primary		Korea Zinc Co. Ltd.	Kangwon	200

See footnotes at end of table.

TABLE 2—Continued
REPUBLIC OF KOREA: STRUCTURE OF THE MINERAL INDUSTRY IN 2012

(Thousand metric tons unless otherwise specified)

Commodity		Major operating companies and major equity owners	Location of main facilities	Annual capacity
Magnesium[1]		Pohang Iron and Steel Co. Ltd.	Magnesium refinery plant, Gangneung City, Gangwon Province	10
Do.		do.	Magnesium metal sheet plant, Suncheon City, Jeonnam (South Jeolla) Province	3
Molybdenum[1]	metric tons	Korea Resources Corp. (KORES)	Mine at Uljin; Smelter at Yeosu, Jeonnam (South Jeolla) Province	6,000
Do.	do.	NMC Resource Corp.	Moland Mine, at Daejang-ri, Geumseongmyeon, Jecheon-si, Chungcheongbuk-do District	2,000
Nickel:				
Ferronickel		Pohang Iron and Steel Co. Ltd.	Gwangyang ferronickel plant	30
Metal		Korea Nickel Corp.	Onsan nickel refinery	48
Petroleum, refinery products	thousand 42-gallon barrels per day	SK Corp.	Ulsan	817
Do.	do.	LG-Caltex Corp.	Yocheon (Yosu)	650
Do.	do.	Hyundai Oil Refinery Co.	Daesan and Inchon	589
Do.	do.	S-Oil Corp.	Onsan	520
Pyrophyllite		NA	Wan-Do, Sungsan, Hwansan, Okmesan, Dae-Do, and Chin-Do Mines in Haenam	446
Do.		NA	Nilyang, Yangsan, Kimhae, Pusan, and Kyong-Nam Mines in Dong-Nae	446
Silver:				
In concentrate	kilograms	Hangum Co. Ltd.	Haenam, Jeonnam (South Cholla) Province	3,700
Refined	metric tons	Korea Zinc Co. Ltd.	Onsan	1,000
Do.	do.	LS-Nikko Copper Inc.	do.	370
Steel, crude		Pohang Iron and Steel Co. Ltd.	Kwangyang (Gwangyang) Works	15,000
Do.		do.	Pohang Works	13,000
Do.		Hyundai Steel Co. Ltd.	Inchon Plant	4,800
Do.		do.	Pohang Plant	3,200
Do.		Dongkuk Steel Mill Co. Ltd.	Inchon Works	1,450
Do.		do.	Pohang Works	3,600
Do.		Korea Iron and Steel Co. Ltd.	Masan and Changwon Works	1,200
Talc		IL Shin Industrial Co. Ltd.	Choong Ju, Chungbuk Province	160
Do.		Korea Zinc Co. Ltd.	Onsan	430
Do.		Young Poong Corp.	Sukpo	280
Zinc		Korea Zinc Co. Ltd.	Onsan refinery	445
Do.		Young Poong Corp.	Sukpo refinery	303

Do., do. Ditto. NA Not available.

[1]Production data are not available.

The Mineral Industry of Laos

By Yolanda Fong-Sam

In 2012, Laos produced a variety of mineral commodities. These included barite, gold, iron ore, lead, and silicon metal, for which production increased significantly compared with that of 2011. On the other hand, significant decreases in production were reported for anthracite (19.8%) and gypsum (15.7%) (table 1).

On October 26, 2012, the General Council of the World Trade Organization (WTO) approved Laos to join the WTO, 15 years after the country first applied for membership. On January 3, 2013, the Government of Laos informed the WTO that it had ratified its membership, and, on February 2, Laos officially became the 158th member of the organization. During the 15 years of negotiations, Laos implemented significant domestic reforms and adjustments for its WTO accession; additionally, the Government passed more than 90 laws and regulations in the areas of customs valuation, import licensing, intellectual property rights, investments, technical barriers to trade, and trading rights. Laos' accession to the WTO opened the door for the country to join the international trading community, backed up with a legal platform; gave it to access other WTO members' markets; and enabled it to become part of a transparent international trading system. By accepting WTO membership, Laos became subject to market access commitments for its goods and services, subsidy limits in agriculture, and a tariff ceiling on goods. Laos also committed to engage in bilateral and multilateral negotiations on partnership agreements with nine other WTO members, including the United States, Australia, Canada, China, Chinese Taipei (defined by the WTO as the separate customs Territory of Taiwan, Kinmen, Matsu, and Penghu), the European Union, Japan, the Republic of Korea, and Ukraine (World Trade Organization, 2012a, b; 2013a, b).

In October, the Laos Ministry of Energy and Mining announced that it had agreed with the Government of China on the financing of a railway designed to ship raw materials from Laos' capital city of Vientiane to Yunnan Province in China. China agreed to provide a $7 billion loan to finance the project, which would consist of approximately 420 kilometers (km) of rail line and was expected to be commissioned by 2017. By 2020, China expected to secure 5 million metric tons per year (Mt/yr) of mineral commodities by way of the rail line, which was expected to include such raw materials as bauxite, copper, gold, iron ore, lead, potash, and zinc, as well as agricultural and timber products (Gronholt-Pedersen, 2012; Railway Gazette, 2012).

In November, the Ministry of Energy and Mines announced its plans to continue with the construction of the Xayaburi Dam on the Mekong River; the river is shared with the neighboring countries of Cambodia, Thailand, and Vietnam. In the past, Laos had encountered opposition from some neighboring countries regarding the environmental impact of such a project and the threat to the livelihood of the population living down the river. With the Xayaburi Dam and other proposed mainstream dams on the Mekong River, Laos intended to triple its hydroelectric power production in order to sell it to its neighbors. The international consortium for the project, Xayaburi Power Co., had started infrastructure construction in the area. The total cost for the Xayaburi Dam was estimated to be $3.8 billion. No projected completion date for the dam had yet been released (Otto, 2012).

Minerals in the National Economy

In 2012, the industrial sector as a whole, which included the construction, electricity, manufacturing, and mining and quarrying sectors, contributed 28.3% to Laos' gross domestic product (GDP) (at constant 2002 prices) compared with 27.5% in 2011. Specifically, the mining and quarrying sector contributed about 7.0% to Laos' GDP in both 2011 and 2012. The construction sector contributed 6.7% to the GDP compared with 6.1% in 2011. Foreign direct investment in the mineral industry amounted to $662.5 million in 2012 (Bank of the Lao PDR, 2012, p. 8, 10, 48, 65).

Government Policies and Programs

The mineral sector in Laos is governed by the Mining Law of 2008, which had been in the implementation stage since the law's adoption. Mining projects, however, had been under a moratorium imposed by the Government since 2009, and no new mining licenses had been issued since the moratorium went into effect. The moratorium was implemented in 2009 as details emerged on licensed projects that had been inactive for an extended period of time and findings that mining operations were causing environmental damage and social concerns. The Government updated its projected timeframe to start granting new licenses to between 2015 and 2016 (it was previously estimated for 2014), once an effective monitoring system and an appropriate land management system are implemented in the country. In addition, the Government was working towards reforming its legal system to a more standardized and plan-based one, which would affect mainly Laos' trading system and development in the private sector. The Government was also reviewing the laws regarding natural resource management. The Ministry of Natural Resources and the Environment was in charge of the compilation of a geologic database to more efficiently manage its oversight of exploration and mining projects. Although a moratorium remained in place, the Government did allow for land assessments and surveys (Lao Voices, 2011; Mining Journal, 2012a).

The Mining Law of 2008 establishes the requirements for obtaining a mining license to develop a mineral project. The Department of Mines issues the mining licenses. The Department of Geology and the Department of Mines jointly oversee the implementation of the Mining Law, which includes inspecting and monitoring mineral development activities in the country; they also assist in the negotiation of mining contracts and in mineral exploration and mining licensing activities, promote investment in the mining sector, maintain other geologic databases, and provide mineral exploration support and data analyses (Department of Geology and Mines of Laos PDR, 2010a, b).

Structure of the Mineral Industry

As of yearend 2011 (the latest year for which data were available), a total of 152 mining companies were operating in Laos, of which 70 were locally owned and 82 were foreign enterprises. During the past few years, the Government had authorized mining companies to operate on 256 mining concessions across the country (Lao Voices, 2011). Laos has a variety of undeveloped mineral resources, and the Government recognized mining as a critical sector of the economy and continued to support the development of the sector by promoting domestic and foreign investment. In 2012, the main producers of copper, gold, and silver in Laos were Lane Xang Minerals Ltd. (MMG LXML), which was a subsidiary of MMG Ltd. of Hong Kong (90% interest) and the Government (10% interest), and PanAust Ltd. of Australia. The country's major mineral industry facilities and their capacities are listed in table 2.

Mineral Trade

In 2012, total trade in Laos was reported to be $4.7 billion, which represented an increase of 3.1% compared with that of 2011. The country's total exports were valued at about $2.3 billion, which was an increase of 3.6% compared with that of 2011. The value of mineral commodity exports amounted to $947 million (41.7% of total exports) and that of electricity exports amounted to $502 million (22.1% of total exports). Mineral commodity exports included copper (valued at $683 million, or 30.1% of total exports) and gold (valued at $151 million, or 6.6% of total exports). The total value of imports in 2012 increased by 2.6% to $2,467 million, of which electricity imports amounted to $63 million (an increase of 55% compared with that of 2011) and gold and silver imports combined amounted to $8.55 million (Bank of the Lao PDR, 2012, p. 34, 61).

Commodity Review

Metals

Bauxite and Alumina.—The Laos Bolaven Plateau bauxite project, which is located in the southern part of the country, was being developed by Sino Australian Resources (Laos) Co., Ltd. (SARCO). SARCO was a joint venture between China Nonferrous Metals Industry's Foreign Engineering and Construction Co., Ltd. of China (NFC) (51% interest) and ORD River Resources Ltd. of Australia (49% interest). SARCO had two tenements in the property for a total of 487 square kilometers (km^2). They were the LSI tenement, which covers a 66-km^2 area, and the Yuqida tenement, which covers a 421-km^2 area. In April, after an extensive drilling campaign during 2010 and 2011, ORD announced that the resource estimate at the Bolaven Plateau project was upgraded in accordance with the Australasian code for reporting of exploration results, mineral resources, and ore reserves [the Joint Ore Reserves Committee (JORC) Code]. Total resources for the project were 226 million metric tons (Mt) at grades of 26.7% alumina and 2.62% silica content; the previously reported estimate was 130 Mt at average grades of 31.7% alumina and 3.2% silica. The total measured resource was estimated to be 69 Mt at grades of 28.1% alumina and 2.09% silica. In September, SARCO started two studies on the Bolaven Plateau bauxite project to support the company's application for a mining license, which SARCO was expecting to be approved by mid-2013. After the acquisition of the license, SARCO planned to conduct a feasibility study to develop a 600,000-metric-ton-per-year (t/yr)-capacity alumina refinery (ORD River Resources Ltd., 2012a, b).

Copper, Gold, and Silver.—In 2012, the Sepon copper-gold project produced a total of 86,295 metric tons (t) of copper cathodes, which exceeded the targeted production of 84,000 t for the year, and 2,186 kilograms (kg) of gold. This production represented an increase of about 9.4% for copper and a decrease of about 5.7% for gold, respectively, compared with production in 2011. The increase in copper production was mainly owing to productivity improvements implemented during the fourth quarter of 2012. MMG LXML expected to produce between 83,000 and 88,000 t of cathode during 2013. On the other hand, the decrease in gold production was caused mainly by a decrease in ore availability and lower gold grades. MMG LXML, which was the Sepon project's operator, expected to produce between 1,400 and 1,700 kg of gold in 2013, although in late 2011 it was determined that the oxide gold ore supply in Sepon would be depleted by 2013 (MMG Ltd., 2013).

PanAust Ltd., through its wholly owned subsidiary Pan Mekong Exploration Pty. Ltd., owned a 90% interest in the Lao-registered company Phu Bia Mining Ltd. (PBM) (which managed the Phu Kham copper-gold operation) and the Government owned the remaining 10%. The Government and PBM had a mineral exploration and production agreement that regulated exploration and mining within the company's contracted 2,636-km^2 area. In 2012, the Phu Kham copper-gold operation, which is located approximately 140 km north of the capital city of Vientiane, had produced 63,285 t of copper concentrate, which was an increase of 5.7% compared with that of 2011 (table 1). The mine also produced 1,851 kg of gold, which was an increase of about 11% compared with the 1,667 kg produced in 2011; and 14,617 kg of silver, which was a decrease of 12.7% compared with the 16,738 kg of silver produced in 2011. The Phu Kham upgrade project, which was commissioned in the third quarter of 2012, increased its designed processing capacity to 16 Mt/yr of ore from its original capacity of 12 Mt/yr. The company expected the Phu Kham operation to produce up to 65,000 t of copper concentrate in 2013 (PanAust Ltd., 2012, p. 9, 11).

PanAust's mineral resource interests in Laos also included the Ban Houayxai gold-silver project and the Phonsavan copper-gold project. The Ban Houayxai project, which was completed at a cost of $208 million, was commissioned in April 2012. Commercial production started on June 1 and the operation reached full capacity by the fourth quarter of 2012. The Ban Houayxai project, which is located approximately 25 km west of the Phu Kham copper-gold operation, was an open pit operation that produced 2,378 kg of gold and 4,564 kg of silver in 2012. The company was targeting production of approximately 3,100 kg of gold in 2013 (PanAust Ltd., 2012, p. 11, 13).

The Phonsavan copper-gold project consisted of two copper deposits named KTL and Tharkhek that are located within 5 km of each other. During 2012, the project was under a prefeasibility study, which PanAust anticipated would be completed by September 2013. The study focused on the development of an open pit mining operation with an annual processing capacity of 7 Mt and output of approximately 25,000 t/yr of copper and 622 kilograms per year of gold. The project was estimated to have a mine life of 10 years (PanAust Ltd., 2012, p.13).

Zinc.—During 2012, Padaeng Industry Public Co. Ltd. of Thailand, through its wholly owned subsidiary Padaeng Industry (Laos) Co. Ltd., announced that it had stopped exploratory work at its 400-km² zinc exploration project in Vientiane Province. The decision to stop the work was based on uncertainty regarding the company's exploration agreements with the Government (Padaeng Industry Public Co. Ltd., 2012, p. 25).

Industrial Minerals

Cement.—Cement production in Laos was projected to reach 1.5 Mt/yr by 2014. The main drivers for the increase were thought to be the infrastructure developments proposed by the Government, which included major hydroelectric plants, the development of a railway system to connect Laos with China, and the expansion of the Wattay International Airport located in Vientiane. The construction of the 420-km railway was estimated to require about 3.5 Mt of cement (Bell, 2012, p. 108).

Lao Cement Co. Ltd. planned to build its third cement plant in the town of Vang Vieng, Vientiane Province; the plant would have the capacity to produce from 0.7 to 1.0 Mt of cement and was expected to be commissioned by 2013. The new plant was being built at a projected cost of $77 million. During 2012, Luang Prabang Cement was building its second cement factory in Nam Bark District, Luang Prabang Province. The proposed 270,000-t/yr plant had a cost of $22 million; no estimated commission date for the plant had yet been released (Bell, 2012, p. 108).

Potash.—In 2012, Sinohydro Mining Co. (Lao) Ltd. was developing a potash mine with a production capacity of 120,000 t/yr. The company planned to start operations in 2013, and the project's mine life was estimated to be 30 years (Mining Journal, 2012b).

Outlook

Laos' accession to the WTO on February 2, 2013, is expected to enhance the country's economic development, and the increased interaction with other WTO members could help attract investment interest from other members of the international trading community. As Laos continues to develop its infrastructure and expands its economic cooperation with other Southeast Asian countries, including by creating business ties with neighboring countries, its demand for construction materials, fertilizers, metals, and fuel minerals is expected to increase. Starting in 2013, the cement industry of Laos is expected to increase its production in order to meet the demand for the construction of the infrastructure projects proposed by the Government, which include hydroelectric plants, a railway system, and the expansion of the Wattay International Airport.

Similarly, the production of copper, gold, and silver is expected to increase starting in 2013 as major projects and expansions were commissioned in 2012, such as the Sepon copper project (which implemented productivity improvements at its facilities to increase the production of copper cathode) and PanAust's Ban Houayxai gold-silver project. Increases in the production of potash are also expected within the next few years as mines and plants that have been in the construction phase in recent years are completed.

References Cited

Bank of the Lao PDR, 2012, Annual economic report 2012: Vientiane, Laos, Bank of the Lao PDR, 73 p. (Accessed October 17, 2013, at http://www.bol.gov.la/together_use/new Annual Report 2012.pdf.)

Bell, Peter, 2012, Emerging market challenges: International Cement Review, July, p. 104–110.

Department of Geology and Mines of Laos PDR, 2010a, Department's functions: Department of Geology and Mines of Laos PDR. (Accessed January 27, 2010, at http://www.dgm.gov.la/.)

Department of Geology and Mines of Laos PDR, 2010b, Policy and legislation: Department of Geology and Mines of Laos PDR. (Accessed January 27, 2010, at http://www.dgm.gov.la/.)

Gronholt-Pedersen, Jacob, 2012, Laos, China revive deal for rail link: The Wall Street Journal, October 25, p. A10.

Lao Voices, 2011, Lao mining industry booms as global economy recovers: Lao Voices, May 3. (Accessed January 10, 2013, at http://laovoices.com/lao-mining-industry-booms-as-global-economy-recovers/.)

Mining Journal, 2012a, Indochina & Thailand—Committed to growth: Mining Journal. (Accessed December 3, 2013, at http://www.mining-journal.com/reports/indochina-and-thailand-committed-to-growth.)

Mining Journal, 2012b, Potash and phosphates—Cooling off period: Mining Journal. (Accessed December 3, 2013, http://www.mining-journal.com/reports/potash-and-phosphates-cooling-off-period.)

MMG Ltd., 2013, Fourth quarter production report for the three months ended 31 December 2012: Melbourne, Victoria, Australia, MMG Ltd., January 31, 17 p. (Accessed November 22, 2013, at http://www.mmg.com/en/Investors-and-Media/Reports-and-Presentations/~/media/Files/Exchange Announcements/Investors and Media/News/2013/01/31/e_2013-01-31_Quarterly Production Report pdf.ashx.)

ORD River Resources Ltd., 2012a, Annual report year ended 30 June 2012: Sydney, New South Wales, Australia, ORD River Resources Ltd., June 30, 78 p. (Accessed November 13, 2013, http://www.ord.com.au/wp-content/uploads/reports/Annual Report 2012.pdf.)

ORD River Resources Ltd., 2012b, Upgrade of JORC resources to 226mt including 69mt in measured category: Sydney, New South Wales, Australia, ORD River Resources Ltd., April 2, 16 p.

Otto, Ben, 2012, Laos dam kicks off controversial Mekong plans: The Wall Street Journal, November 8, p. A18.

Padaeng Industry Public Co. Ltd., 2012, Annual report 2012: Bangkok, Thailand, Padaeng Industry Public Co. Ltd., 106 p. (Accessed November 22, 2013, at http://www.padaeng.com/files/en/report/2013_09/pdf/PDI_Annual_Report2012_Eng.pdf.)

PanAust Ltd., 2012, Annual review 2012: PanAust Ltd., 36 p. (Accessed November 13, 2013, at http://www.panaust.com.au/sites/default/files/reports/PANAUST_AnnualReview2012.pdf.)

Railway Gazette, 2012, Chinese loan agreements revive trans-Laos project: Railway Gazette, October 30. (Accessed November 2, 2012, at http://www.railwaygazette.com/news/single-view/view/chinese-loan-agreements-revive-trans-laos-project.html.)

World Trade Organization, 2012a, General Council accepts Laos' membership, only ratification left: Geneva, Switzerland, World Trade Organization, October 26. (Accessed December 2, 2013, at http://www.wto.org/english/news_e/news12_e/acc_lao_26oct12_e.htm.)

World Trade Organization, 2012b, Protocol on the accession of the Lao People's Democratic Republic to the Marrakesh agreement establishing the World Trade Organization done on 26 October 2012—Notification of acceptance: Geneva, Switzerland, World Trade Organization, November 2, Reference no. WLI–101, 1 p.

World Trade Organization, 2013a, Laos ratifies membership package, will join WTO on 2 February: Geneva, Switzerland, World Trade Organization, January 3. (Accessed December 2, 2013, at http://www.wto.org/english/news_e/news13_e/acc_lao_08jan13_e.htm.)

World Trade Organization, 2013b, Protocol on the accession of the Lao People's Democratic Republic to the Marrakesh agreement establishing the World Trade Organization done on 26 October 2012—Notification of acceptance and entry into force: Geneva, Switzerland, World Trade Organization, January 9, reference WLI–101, 1 p.

TABLE 1
LAOS: PRODUCTION OF MINERAL COMMODITIES[1]

(Metric tons unless otherwise specified)

Commodity[2]		2008	2009	2010	2011	2012
Antimony		--	887	530	1,456	1,042
Barite[e]		1,000 [r]	12,460 [r, 3]	17,500 [3]	2,500 [r]	21,900
Cement[e]		400,000	400,000	400,000	400,000	400,000
Clay, common		NA [r]	1,783,567 [r, 3]	1,901,530 [r]	609,840 [r]	512,587
Coal:						
Anthracite		186,468	167,447	211,721	166,609	133,583
Lignite		379,273 [r]	466,082 [r]	501,622 [r]	511,700 [r]	578,068
Copper:						
Concentrate		24,929	54,019	67,806	59,897	63,285
Metal, refined		64,075	67,561	64,241	78,859	86,295
Gold	kilograms	4,333	5,033	5,061	3,984	6,415
Gypsum		337,304 [r]	761,331 [r]	553,396 [r]	686,150 [r]	578,543
Iron ore:						
Gross weight		18,000	42,000	50,900	42,700	48,400
Fe content (62%)		11,219	26,095	31,565	26,471	30,000 [e]
Lead, mine output		2,950 [r]	2,000 [r]	2,270 [r]	2,921 [r]	4,510
Limestone		911,658 [r]	1,488,070 [r]	3,106,724 [r]	997,591 [r]	1,014,000
Potash		NA	NA	NA	NA	42,798
Salt, rock		NA	6,536 [r]	13,421 [r]	23,395 [r]	11,980
Sandstone		NA	752,781	3,695,838	339,331	1,214,668
Silicon, metal		3,026	7,350	7,768	3,001	15,281
Silver	kilograms	6,700	14,726	15,788	16,738	19,181
Tin, mine output, Sn content		551 [r]	598 [r]	925 [r]	674 [r]	762
Zinc, mine output, Zn content		8,597 [r]	4,000 [r]	5,000	5,320 [r]	5,250

[e]Estimated; estimated data are rounded to no more than three significant digits. [r]Revised. NA Not available. -- Zero.

[1]Table includes data available through February 20, 2014.

[2]In addition to the commodities listed, sapphire and crude construction materials, such as sand and gravel and varieties of stone, were produced irregularly.

[3]Reported figure.

Sources: U.S. Geological Survey Minerals Questionnaire for Laos (2012); Pan Australian Resources Ltd., Annual Report 2009–12; Minerals and Metals Group Lane Xang Minerals Ltd. (MMG LXML) Quarterly Reports 2009–12.

TABLE 2
LAOS: STRUCTURE OF THE MINERAL INDUSTRY IN 2012

(Metric tons unless otherwise specified)

Commodity		Major operating companies and major equity owners	Location of main facilities	Annual capacity[e]
Barite		Barite Mining Co., Inthavong Mining Co., Lao Development Construction Co., Phethongkham Co., Oravan Barite Co., and Singphooufar Co.	Muongfuong and Sanakham, Vientiane Province	30,000 [1]
Cement		Lao Cement Co. Ltd., a joint venture between China Yunnan Corp. for International Techno-Economic Cooperation and Lao State Agricultural Industry Development Enterprise Imp-Exp & General Service	Vang Vieng Cement Plant No. I, Vientiane Province	100,000
Do.		do.	Vang Vieng Cement Plant No. II, Vientiane Province	240,000
Do.		Lao Cement Industry Co. Ltd	Thakhek Cement Plant, Khammouane Province	850,000
Do.		Luang Prabang Cement Co. Ltd.	Luang Prabang Cement Plant, Luang Prabang Province, 340 kilometers from Vientiane, Vientiane Province	100,000
Do.		Wanrong Cement I	Vangvieng, Vientiane Province	78,000
Do.		Wanrong Cement II (Yunnan Industrial Economic Co., 60%, and Agricultural and Forestry Development and Service Co. of Laos, 40%)	do.	200,000
Do.		Wanrong Cement III	Savannakhet Province	200,000
Do.		Zhongyayici Co.	Saravan Cement Plant, Saravan Province	450,000
Coal:				
Anthracite		Agriculture Industry Development Enterprises	do.	60,000
Lignite		Viengphoukha Coal Mine Co. Ltd.	Viengphoukha, Luangnamtha Province	300,000
Copper:				
Mined ore output, Cu content		Lane Xang Minerals Ltd. (MMG LXML) (MMG Ltd.), 90%, and Government, 10%	Sepon, Vilabouly District, Savannakhet Province	80,000
Refined		do.	do.	88,000
Do.		Phu Bia Mining Ltd. (wholly owned subsidiary of PanAust Ltd.), 90%, and Government, 10%	Phu Kham copper-gold operation located in Xaisomboun special zone, 120 kilometers north of Vientiane, Vientiane Province	65,000
Gemstone (sapphire)	carats	Bokeo Mining Co. Ltd.	Bokeo Province	300,000
Do.	do.	Buhae Industrial Corp.	Houaxay District, Bokeo Province	500,000
Do.	do.	Lao International Trade and Service	do.	400,000
Gold, mine output, Au content	kilograms	Lane Xang Minerals Ltd. (MMG LXML) (MMG Ltd.), 90%, and Government, 10%	Sepon, Vilabouly District, Savannakhet Province	7,500
Do.	do.	Phu Bia Mining Ltd. (wholly owned subsidiary of PanAust Ltd.), 90%, and Government, 10%	Phu Kham copper-gold operation located in Xaisomboun special zone, 120 kilometers north of Vientiane, Vientiane Province	2,000
Do.	do.	PanAust Ltd.	Ban Houayxai gold-silver project, located approximately 25 km west of the Phu Kham copper-gold operation	3,100
Gypsum		Lao State Gypsum Mining Co. Ltd.	Champhon District, Savannakhet Province	200,000
Do.		Mining Development Economy Cooperation (OEDCD)	Tha Kect District, Khammouane Province	150,000
Do.		Savan Gypsum Mining Co. Ltd.	Champhon District, Savannakhet Province	70,000
Do.		LAVICO Co. Ltd. (a Laos-Vietnam joint venture)	Xebangfay District, Khammouane Province	100,000
Limestone		Laos Cement Co. Ltd. (a Laos-China joint venture)	Vangvieng, Vientiane Province	250,000
Do.		Agriculture Industry Development Enterprises	do.	150,000
Do.		V.S.K. Co. Ltd.	Tha Kect District, Khammouane Province	150,000
Do.		Phanangnon Co. Ltd.	do.	100,000

See footnotes at end of table.

TABLE 2—Continued
LAOS: STRUCTURE OF THE MINERAL INDUSTRY IN 2012

(Metric tons unless otherwise specified)

Commodity		Major operating companies and major equity owners	Location of main facilities	Annual capacity[e]
Potash		SinoAgri Mineral Resources Exploration Ltd. (a joint venture between Beijing Jiang Zhi Yuan Investment Ltd. and China National Agricultural Means of Production Group Corp.)	Plant in Sakhon Nakon Basin in Khammouane Province	100,000
Do.		Laotian Potash Mining Industry Ltd. Co.	Potassium chloride plant, Vientiane Province	50,000
Silver	kilograms	Phu Bia Mining Ltd. (wholly owned subsidiary of PanAust Ltd.), 90%, and Government, 10%	Phu Kham copper-gold operation located in Xaisomboun special zone, 120 kilometers north of Vientiane, Vientiane Province	14,000
Do.	do.	PanAust Ltd.	Ban Houayxai gold-silver project located approximately 25 km west of the Phu Kham copper-gold operation	20,000
Tin, mine output, Sn content		Lao-North Korea Tin Mines	Hinboune District, Khammouane Province	120
Do.		S V Mining Co. Ltd.	do.	300
Zinc, mine output, Zn content		Padaeng Industry Public (Laos) Co. Ltd. [Majority interest owned by Padaeng Industry (Public) Co. Ltd., and minority interest owned by the Government]	Kaiso, Vangvieng, Vientiane Province	5,000

[e]Estimated. Do., do. Ditto.

[1]Estimated combined capacity of the six local barite mining companies.

THE MINERAL INDUSTRY OF MALAYSIA

By Pui-Kwan Tse

Malaysia's economy was dependent on exports of manufactured goods and on the service sector. The slow recovery in the world economy affected the Malaysian economy, which grew at a moderate pace in 2012. The country's gross domestic product (GDP) increased by 5.6% compared with an increase of 5.1% in 2011. Malaysia's economic growth was driven mainly by domestic demand (business and household spending). Private investment increased by 7.7% and accounted for the major share of the GDP growth. The manufacturing sector grew by only 4.8% as demand remained weak for manufactured products in most of the industrialized countries in the Western Hemisphere. The growth rate of the construction sector increased by 16.5% as a result of the Government startup of several infrastructure projects in 2012. The output value of the mining and quarrying sector increased by 1.4% compared with a decrease by 5.7% in 2011, reflecting an increase in production of crude oil and condensate (Bank Negara Malaysia, 2013, p. 15–25).

Minerals in the National Economy

Malaysia has identified mineral resources of barite, bauxite, clays, coal, copper, gold, ilmenite, iron ore, limestone, monazite, natural gas, petroleum, silica, silver, struverite (tantalum), tin, and zircon. During the 20th century, mineral production played an important role in Malaysia's national economy; after many years of exploitation, however, such minerals as barite, bauxite, copper, ilmenite, iron ore, and tin were either depleted or the capacities to produce them had decreased significantly. In terms of its contribution to the country's economy, the mining and quarrying sector accounted for 8.8% of the GDP in 2012 (Bank Negara Malaysia, 2013, p. 1; Department of Statistics, 2013a, p. 26).

Government Policies and Programs

In Malaysia, mineral sector activity is governed by the Mineral Development Act 1994 and the State Mineral Enactment. The Mineral Development Act 1994 defines the power of the Federal Government to regulate mineral exploration, mining, and related activities, including the authority to conduct inspections. The State Mineral Enactment gives the States the power to issue mineral prospecting and exploration licenses and mining leases. Apart from paying a corporate tax to the Federal Government, mine and quarry operators are required to pay value-based royalties to the State in which their operation is located. Royalty rates depend on the mineral commodity and on the assessment of each of the individual States.

The Government amended the Safeguard Act 2006 (Act 657) in 2012. The Safeguard Act (amendment) 2012 allows the Government to take action to prevent serious injury to domestic industry from lower priced imported materials; the Act was to take effect on September 1, 2013. These measures do not allow the Government to target imports from a particular country but import quotas can be placed upon supplying countries (Southeast Asia Iron and Steel Institute, 2013).

Production

Malaysia produced bauxite, coal, feldspar, gold, ilmenite, iron ore, mica, natural gas, petroleum, struverite (tantalum), tin, and zircon. Malaysia had been one of the major tin-producing countries in the world; owing to depleted reserves and lower ore grades, however, tin concentrate production had decreased in recent years. The country depended on imported tin concentrates and crude tin mainly from Australia and Indonesia to meet its demand for feedstocks for its smelter and refinery. In 2012, production of such commodities as feldspar, iron ore, manganese, rutile, and mined tin increased by more than 10% whereas production of kaolin and zirconium decreased by more than 10% (Department of Statistics, 2013c, p. 71–84).

Structure of the Mineral Industry

Malaysia's mineral industry consisted of a small mining sector for coal and ferrous and nonferrous metals. Metallic and nonmetallic mineral processing facilities were operated by private companies incorporated in Malaysia. Oil and gas exploration, production, and processing activities and facilities were owned and operated by Petroliam Nasional Berhad (Petronas), which was a state-owned company, and by joint ventures of Petronas and foreign companies. Foreign investors were permitted to have a 100%-equity stake in companies operating in Malaysia or to form joint ventures with local companies (table 2).

Mineral Trade

Malaysia's major export products were automotive parts, chemicals, electronics, and machinery. The volume of mineral commodity exports had declined in recent years. In 2012, total trade increased to $436.3 billion; of that amount, exports increased by 0.7% to $234.0 billion and imports increased by 5.9% to $202.3 billion. Electrical and electronic products continued to be Malaysia's leading export category and accounted for 36.5% of the total exports. The export share of liquefied natural gas (LNG) and petroleum products was 7.9% and 4.6%, respectively. Malaysia exported 23.8 million metric tons (Mt) of LNG, which was a decrease of 4.3% from that of 2011. LNG was exported to (in descending order of export value) Japan, the Republic of Korea, and China and accounted for 91% of the country's total LNG exports in 2012. Malaysia exported 11.9 Mt of crude oil, which had a total value of $10.6 billion and was a decrease of 5.1% from the value in 2011. Crude oil was exported to (in descending order of export value) Australia, India, Thailand, Japan, China, New Zealand, and the Republic of Korea, which together accounted for 93% of the country's total crude oil exports in 2012. Malaysia's major import category was machinery and transport equipment, which accounted for 39.1% of the country's total imports. China continued to be Malaysia's leading trading partner in 2012 followed by Singapore and Japan (Department of Statistics, 2013b, p. 1–30).

Commodity Review

Metals

Aluminum.—Malaysia did not have an aluminum refinery, and most of its bauxite output was exported to other Asian countries. Press Metal Sarawak Sdn Bhd (a subsidiary of Press Metal Berhad) completed the construction of a 120,000-metric-ton-per-year (t/yr) aluminum smelter in Mukah in the State of Sarawak; the smelter was fully operational by the second half of 2012. Press Metal had chosen the Aluminum Corp. of China Ltd. (Chalco) as its technical partner for the first phase of the aluminum smelter project. Chalco's Guiyang Aluminum and Magnesium Research Institute installed the prebaked cells. The company signed a memorandum of understanding with Sarawak Energy Berhad to provide 510 megawatts (MW) of electricity-generating capacity for the smelter in 2010. After the smelter was fully operational, the company started the construction of its second potline. The second potline would be equipped with 400-kiloampere prebaked anode cells and was scheduled to be completed in 2013; it would increase the smelter's total output capacity to 300,000 t/yr. Sumitomo Corp. of Japan acquired a 20% share in the second-phase aluminum project and had sales rights to some of the output (Press Metal Berhad, 2013a, p. 9; 2013b).

Copper.—Without any refined copper production, Malaysia relied on imported copper to meet its demand. In 2011 (the latest year for which data were available), Malaysia imported 226,017 metric tons (t) of refined copper and copper alloys and 15,972 t of copper scrap and exported 10,264 t of refined copper and copper alloys and 29,969 t of copper scrap. In 2006, Malaco Mining Sdn Bhd explored for copper and gold in the State of Pahang. After 3 years of exploration, Malaco discovered a copper deposit at Sri Jaya, and the Mengapur Mine began production in 2009; however, no production was recorded in 2010 and 2011. Monument Mining Ltd. of Canada, through its Malaysian subsidiary Monument Mengapur Sdn Bhd, acquired a 70% interest in the Mengapur polymetallic mine in February 2012. The Mengapur Mine is located 130 kilometers (km) from Monument's wholly owned Selinsing gold mine. In October 2012, Monument announced that the company had resolved an issue concerning iron ore material that was covering the skarn at the designated area of the Mengapur project with Cermat Aman Sdn Bhd, Phoenix Lake Sd. Bhd, and ZCM Aman Sdn Bhd. Monument did not own the iron in the "free-digging materials" in the oxide zone on the mining lease area when Monument acquired the interest in Malaco Mining Sdn Bhd.

The parties agreed to form a technical committee to define the technical boundary and methods to separate the iron ore material from other metals. Monument planned to invest a significant amount of capital to develop the open pit mine and processing facilities based on base- and precious-metals production. The company planned to put the production plant into operation in 2013 (Minerals and Geoscience Department, 2012, p. 19–22; Monument Mining Ltd., 2012b, d).

Gold.—Approximately 17 gold mines were operating in Malaysia; all were located in the States of Kelantan, Pahang, and (or) Terengganu. More than 90% of mined gold was from the State of Pahang, mainly the Penjom gold mine at Penjom, the Selinsing gold mine in Bukit Selinsing Koyan, and Raub Australian Gold Mining Sdn Bhd's gold mine in Raub. The Penjom gold mine was a leading gold producer in the country. Owing to depleting resources within the mining lease area, gold production from the Penjom gold mine decreased to 1,748 kilograms (kg) (56,203 troy ounces) in 2012. The Selinsing gold mine ranked second for gold production and produced 1,487 kg (47,811 troy ounces) (Monument Mining Ltd., 2012a, p. 4; 2013; PT J Resources Asia Pasifik Tbk, 2013, p. 12).

Monument completed the treatment plant expansion at its Selinsing gold mine in 2012. The overall milling and treatment capacity increased to 1 million metric tons per year (Mt/yr) of ore from 400,000 t/yr. The total cost of the expansion was $8.6 million and included installation of an additional crusher and enlargement the tailings storage facility. The increased capacity would enable the company to increase the amount of ore to be processed at the mill when the ore grades become lower in the future. Also, Monument had identified gold mineralization on the north and south zones of the Buffalo Reef and the Selinsing mining areas (Monument Mining Ltd., 2012c).

Iron and Steel.—Malaysian's iron ore production was from small-scale mines located in the States of Johor, Pahang, Perak, and Terengganu. The low-grade iron ores were consumed by the pipe-coating industry that supplied cement plants and the oil and gas sector. The high-grade iron ore was exported to China. In 2011, Malaysia exported 5.7 Mt of high-grade iron ore to China and imported 2.8 Mt of high-grade iron ore from, in descending order of amount received, Brazil and Bahrain. The State Government of Terengganu approved iron ore mining concessions for Eastern Steel Sdn Bhd and Perwaja Holdings Bhd in Bukit Besi. In Terengganu, geologists estimated that more than 50 Mt of iron ore resources was located in the Bukit Besi area and that the iron content was about 70%. Perwaja planned to invest about $130 million to build a pelletizing plant in Kemaman, where its direct-reduced iron plant was located. Perwaja intended to mine 2 Mt/yr of iron ore and to produce 1.2 Mt/yr of pellets during its first year of operation; it planned to increase the pellet production to 2.4 Mt/yr in 2013. The company estimated that production costs would be about $50 per metric ton of pellet, which was lower than the $90 to $120 per metric ton cost of imported pellet (Southeast Asia Iron and Steel Institute, 2012).

Manganese.—Malaysia's manganese resources were located in Johor, Kelantan, Pahang, and Terengganu, and the manganese content was usually less than 50%. The amount of manganese output from Malaysia depended on the price of manganese in the world markets. Since 2005, with an increase in manganese prices in the world, Malaysia's manganese output had gradually increased. Without much domestic demand for manganese, the country exported nearly all its output to China. Pertama Ferroalloys Sdn Bhd [formerly known as ANL Manganese (Malaysia) Sdn Bhd] [a joint venture between Asia Mineral Ltd. (ANL), 51%, and customers from Japan Steel Group, Korea Steel Group, and a local Malaysian company, 49%] planned to build a manganese ferroalloys plant in the Samalaju Industry Park in Bintulu in the State of Sarawak. The ferroalloys plant was designed to produce 350,000 t/yr of ferromanganese and ferrosilicon alloys beginning in 2014. ANL signed an agreement with Sarawak Energy for the provision of 270 MW of electricity-generation capacity for 20 years. Raw materials would be sourced from Brazil, South Africa, and local mines (Minerals and Geoscience Department, 2012, p. 37).

South Africa's Assmang Ltd. and African Rainbow Minerals Ltd., China Steel Corp. of Taiwan, and Sumitomo Corp. of Japan jointly completed a feasibility study on the construction of a 163,000-t/yr ferromanganese plant in Sarawak. China Steel would invest $62.5 million to secure a 19% share of the joint-venture project and to obtain between 30,000 and 32,000 t/yr of ferromanganese alloys. Construction of the plant was scheduled to start in 2014, and production was expected to begin in 2016 (Metal-Pages Ltd., 2013).

Tin.—Malaysia's tin mines produced about 3,000 t/yr during the past several years. Resources were depleted and ore grades were lower after more than 100 years of active mining operations. The country imported tin concentrates from other countries in Asia and Africa to meet its demand. Solder production was the leading tin-consuming sector in Malaysia, followed by tinplate and pewter. Tin consumption in Malaysia decreased to less than 3,000 t/yr during the past 3 years. The decrease in tin consumption was mainly the result of a decrease in demand from the solder and pewter sectors; consumption by other consumers remained at the same level during that period. Malaysia Smelting Corp. Bhd. (MSC) was Malaysia's sole integrated tin producer; it produced 37,792 t of refined tin at its Butterworth smelter in 2012, which was about 6% less than it produced in 2011. The decrease in tin production was a result of MSC's inability to source raw material from the African region because tin exports from the African region were required to meet standards set by the ITRI Tin Supply Chain Initiative (iTSCi) to ensure that tin exports from the African region are accountable and transparent. In 2012, Malaysia imported 29,719 t of tin concentrates compared with 33,031 t in 2011. Malaysia's refined tin exports decreased to 37,191 t in 2011 from 42,302 t in 2011 and went mainly to China, Japan, the Republic of Korea, Singapore, and Taiwan (Department of Statistics, 2013b, p. 27; Malaysian Tin Bulletin, 2013; Malaysia Smelting Corp. Bhd., 2013, p. 20, 166).

Industrial Minerals

Cement.—Malaysia's cement sector was dominated by three companies: Cement Industries of Malaysia Bhd, Lafarge Malaysia Cement Bhd, and YTL Cement Bhd; together, these companies accounted for about 78% of the country's total cement output capacity. Cement demand in Malaysia had fluctuated between 16 and 17 Mt/yr during the past 5 years. West Malaysia had one of the most developed infrastructures in the country, but east Malaysia remained relatively undeveloped. Under the 10th Malaysia Plan and Economic Transformation program, the Government planned to build the east coast highway from Jabur to Kuala Terenggaru and to improve rural infrastructure. Together with ongoing construction of commercial properties, the demand for cement was expected to increase during the next several years. KHD Humboldt Wedag International AG was awarded a contract by YTL Group to build an integrated cement plant, which would be located near Kuantan. The new plant would have a design capacity to produce 5,000 metric tons per day. The $130 million cement plant would have the latest environmentally friendly equipment (KHD Humboldt Wedag International AG, 2012).

Rare Earths.—Globally, the production and resources of rare earths were dominated by China. Lynas Corp. Ltd. of Australia mined the rare-earth deposit at Mount Weld in Western Australia and shipped rare-earth concentrates to Malaysia for further processing. Lynas secured approval from the Malaysian Government to build an advanced materials plant in the Gebeng III Industrial Area, which is located near the Port of Kuantan in the State of Pahang. The construction of the plant was scheduled to be completed in late 2011, but the completion date was postponed to 2012. The plant would have an initial output capacity of 11,000 t/yr of rare-earth-oxide-equivalent products. Local residents objected to the construction of the rare-earth plant in their area because they worried about the safety of storing low-level radioactive waste that could cause lasting environmental damage. They feared that Lynas's plant would be a repeat of the Mitsubishi Chemical rare-earth plant in the area, which was shut down in 1992. The Malaysia Parliamentary Committee approved the issuance of a temporary operating license to Lynas. A local environmental group applied to the High Court of Malaysia for an injunction to block Lynas's license, but the court decided to allow the rare-earth separation plant to begin operating in late 2012. The company faced technical problems on its cracking and leaching units at its Malaysian rare-earth plant. As a result, the volume of output was much less than its designed capacity. The company planned to complete the debottlenecking of these technical problems by the end of 2013. Lynas started the construction of the phase 2 expansion project to increase the output capacity to 22,000 t/yr; the expansion project was planned to be completed in late 2013 (Lynas Corporation Ltd., 2013).

Mineral Fuels

Coal.—Malaysia's coal resources are located in the States of Perak, Perlis, Sabah, Sarawak, and Selangor. Coal was produced from the areas of Bintulu, Merit-Pila, Silantek, and Tutoh in the State of Sarawak. The country has coal resources of about 1.9 billion metric tons (Gt), of which 281 Mt was measured, 378 Mt was indicated, and 1.3 Gt was inferred. About 1.5 Gt of the country's coal resource is located in Sarawak, and more than 300 Mt is located in Sabah. Owing to the lack of infrastructure, most of the coal in the interior areas of the country had not been exploited. Coal resources located in Sabah were in the Maliau Basin Conservation area, which the Government had designated as a protected area. Mining and exploration for coal were conducted only in Sarawak.

Power-generating plants consumed about 70% of the total supply of coal (domestic production and imports), and the remaining supply was consumed by the cement and iron and steel sectors. Despite Malaysia's position as a natural gas exporter, Tenaga Nasional Berhad planned to decrease the use of natural gas at its powerplants to 49% from 72% and to shift to the use of coal because of a shortage in the supply of natural gas in the domestic market. Overall, coal demand for powerplants was likely to increase; as a result, coal consumption was expected to increase to 24 Mt in 2013.

Coal consumption was expected to increase to 36 Mt by 2020 because the demand for electricity was expected to increase, and in expectation of this increase in demand, the Government planned to build another coal-fired powerplant. The supply of domestic coal would likely not be sufficient to meet the expected increase in demand for coal, and the country was expected to increase coal imports to fill the gap. In 2011 (the latest year for which data were available), Malaysia imported 22.0 Mt of coal, which was about 2 Mt more than in 2010. Coal imports from Indonesia accounted for 72.2% of the total imports followed by Australia, 14.1%; South Africa, 11.8%; and others, 1.9% in 2011 (Minerals and Geoscience Department, 2012, p. 102–106).

Natural Gas and Petroleum.—Malaysia remained a net exporter of natural gas and crude oil. The increase in natural gas production was caused by the growth of external demand for LNG from China and Japan. The Malaysian Government offered incentives for companies to explore deeper and less-profitable fields in a bid to increase reserves as energy demand increases. Eight new fields were brought onstream in 2012, which included the Berantai field in Peninsular Malaysia and the Gumusut-Kakap and Kanowit deepwater fields in Sarawak (Petroliam Nasional Berhad, 2013, p. 43–45).

Outlook

Malaysia's economy is projected to grow at a slower rate during the next 3 years than in the previous several years because of the projected slow recovery of the global economy. Private and public spending, however, will likely continue to support economic growth. The Government is aware of the country's need to reduce its dependence on external markets and to produce a more-diversified range of goods for export. To improve the investment climate and build a more-competitive economy, the Government plans to privatize state-owned companies, sell Government land, and reassess Government subsidies. The Government plans to further relax some rules regarding foreign investment in Malaysian companies and properties, initial public offerings, and the financial sector. The construction sector is expected to expand as a result of increased investment by the Government in infrastructure under the Tenth Malaysia Plan, and the demand for construction steel products will also likely increase. Several natural gas and oil projects are set to come onstream to replace maturing fields during the next several years.

References Cited

Bank Negara Malaysia, 2013, Annual report 2012: Kuala Lumpur, Malaysia, Bank Negara Malaysia, 140 p.

Department of Statistics [Malaysia], 2013a, Malaysian economy in brief: February 2013: Kuala Lumpur, Malaysia, Department of Statistics, 45 p.

Department of Statistics [Malaysia], 2013b, Malaysia external trade statistics December 2012: Kuala Lumpur, Malaysia, Department of Statistics, 58 p.

Department of Statistics [Malaysia], 2013c, Monthly manufacturing statistics: Kuala Lumpur, Malaysia, Department of Statistics, March 2013, 98 p.

KHD Humboldt Wedag International AG, 2012, KHD awarded €100 million project in Malaysia: Cologne, Germany, KHD Humboldt Wedag International AG press release, March 30, 2 p.

Lynas Corporation Ltd., 2013, Quarterly report for the period ending 30 September 2013: Sydney, New South Wales, Australia, Lynas Corporation Ltd., 9 p.

Malaysian Tin Bulletin, 2013, Malaysian refined tin production import of tin-in-concentrates and export of tin metal: Malaysian Tin Bulletin, April, p. 9.

Malaysia Smelting Corp. Bhd., 2013, Annual report 2012: Kuala Lumpur, Malaysia, Malaysia Smelting Corp. Bhd., 170 p.

Metal-Pages Ltd., 2013, Feasibility study completed on manganese smelting facility for Malaysia: Metal-Pages Ltd., June 20. (Accessed June 21, 2013, at http://Metal-pages.com/news/.../feasibility-study-completed-on-manganese-smelting-facility-for-malaysia.)

Minerals and Geoscience Department [Malaysia], 2012, Malaysian minerals yearbook 2011: Kuala Lumpur, Malaysia, Minerals and Geoscience Department, 114 p.

Monument Mining Ltd., 2012a, Annual report 2012: Vancouver, British Columbia, Canada, Monument Mining Ltd., 95 p.

Monument Mining Ltd., 2012b, Monument announces 2013 plans for Mengapur project: Vancouver, British Columbia, Canada, Monument Mining Ltd. press release, December 18, 3 p.

Monument Mining Ltd., 2012c, Monument completed Selinsing gold plant expansion: Vancouver, British Columbia, Canada, Monument Mining Ltd. press release, August 16, 1 p.

Monument Mining Ltd., 2012d, Monument signs harmonization agreement on Mengapur project: Vancouver, British Columbia, Canada, Monument Mining Ltd. press release, October 9, 3 p.

Monument Mining Ltd., 2013, Monument reports second quarter fiscal 2013 results: Vancouver, British Columbia, Canada, Monument Mining Ltd. press release, March 4, 3 p.

Petroliam Nasional Berhad, 2013, Petronas annual report 2012: Kuala Lumpur, Malaysia, Petroliam Nasional Berhad, 251 p.

Press Metal Berhad, 2013a, Annual report 2012: Selangor Daral Ehsan, Malaysia, Press Metal Berhad, April 13, 158 p.

Press Metal Berhad, 2013b, Memorandum of understanding between Press Metal Berhad and Sumitomo Corporation: Selangor Daral Ehsan, Malaysia, Press Metal Berhad announcement, April 13, 2 p.

PT J Resources Asia Pasifik Tbk, 2013, Annual report 2012: Jakarta, Indonesia, PT J Resources Asia Pasifik Tbk, 158 p.

Southeast Asia Iron and Steel Institute, 2012, Terengganu awards Bukit Besi iron ore concession to Perwaja: Selangor Danul Ehsan, Malaysia, SEAISI Newsletter, December, p. 8.

Southeast Asia Iron and Steel Institute, 2013, Govt can now restrict import on certain products under new Act: Selangor Danul Ehsan, Malaysia, SEAISI Newsletter, September, p. 6.

TABLE 1
MALAYSIA: PRODUCTION OF MINERAL COMMODITIES[1]

(Metric tons unless otherwise specified)

Commodity[2]		2008	2009	2010	2011	2012
METALS						
Aluminum:						
Bauxite, gross weight		295,176	263,432	124,274	188,141	121,873
Aluminum		--	--	--	--	100
Copper, mine output, Cu content		--	240	--	--	--
Gold, mine output, Au content[3]	kilograms	2,489	2,794	3,765	4,219 r	4,625
Iron and steel:						
Iron ore, gross weight		981,932	1,470,186	3,465,895	8,077,879 r	10,277,849
Pig iron, direct-reduced iron, and hot-briquetted iron	thousand metric tons	1,957	2,388	2,390	2,876	2,329
Steel, crude	do.	6,423	5,354	5,693	5,941	5,612
Magnesium metal e		--	--	--	200	5,000
Manganese, gross weight		536,675	468,963	899,703	597,917 r	1,099,585
Niobium (columbium)-tantalum metals, struverite, gross weight		216	176	84	110	262
Silver, mine output, Ag content[3]	kilograms	349	367	436	459	1,678
Tin:						
Mine output, Sn content		2,605	2,412	2,668	3,340 r	3,726
Metal, smelter		31,691	36,407	38,737	40,267	37,792
Titanium:						
Ilmenite concentrate, gross weight		36,779	15,983	19,036	28,782	22,275
Rutile		1,834	1,502	7,567	10,810	20,008
Zirconium, zircon concentrate, gross weight		984	1,145	1,267	1,685	442
INDUSTRIAL MINERALS						
Barite		4,372	22,390	1,000	--	--
Cement, hydraulic	thousand metric tons	19,629	19,457	19,762	21,198 r	21,726
Clays and earth materials	do.	25,065	22,966	27,543	28,384 r	28,163
Feldspar		457,377	410,053	455,497	379,628	482,906
Kaolin		506,462	487,632	530,331	442,500	393,068
Mica		5,593	4,323	4,515	4,245 r	3,967
Rare earths, monazite and xenotime, gross weight		233	25	732	779	179
Sand and gravel	thousand metric tons	24,472	17,382	30,678	37,339 r	40,000 e
Silica sand		1,466,904	630,394	932,159	1,340,013 r	931,880
Stone:						
Aggregate	thousand metric tons	75,883	86,497	101,809	118,510 r	100,000 e
Dolomite		57,900	49,000	50,900	50,000 e	50,000 e
Limestone	thousand metric tons	35,228	35,808	32,398	34,300 r	35,000 e
MINERAL FUELS AND RELATED MATERIALS						
Coal		1,166,525	2,138,390	2,397,340	2,915,788	2,951,124
Gas, natural:						
Gross	million cubic meters	68,000	65,000	72,000	73,000	74,000
Net[4]	do.	61,004	58,560	61,136	61,400	62,000
Liquefied natural gas	thousand metric tons	23,422	22,452	24,363	25,822 r	23,986
Petroleum:						
Crude and condensate	thousand 42-gallon barrels	251,811	240,479	232,100	207,696 r	212,979
Refinery products e	do.	210,000	200,000	210,000	215,000	215,000

e Estimated; estimated data are rounded to no more than three significant digits; may not add to totals shown. r Revised. do. Ditto. -- Zero.

[1] Table includes data available through October 5, 2013.

[2] In addition to the commodities listed, a variety of materials, which include ammonia, fertilizers, lead (secondary), and salt, are produced but not reported, and information is inadequate to make reliable estimates of output.

[3] Includes byproduct from tin mines in Peninsular Malaysia and gold mines in Peninsular Malaysia and the State of Sarawak.

[4] Includes production from Peninsular Malaysia and the States of Sabah and Sarawak.

Sources: Ministry of Primary Industry, Minerals and Geoscience Department (Kuala Lumpur), Malaysian Minerals Yearbook 2011; U.S. Geological Survey Minerals Questionnaire, 2013; and Southeast Asia Iron and Steel Institute, Steel Statistical Yearbook, 2011.

TABLE 2
MALAYSIA: STRUCTURE OF THE MINERAL INDUSTRY IN 2012

(Thousand metric tons unless otherwise specified)

Commodity			Major operating companies and major equity owners	Location of main facilities	Annual capacity
Aluminum, metal			Press Metal Sarawak Sdn Bhd (Press Metal Berhad)	Mukah, Sarawak	120.
Bauxite			Johore Mining and Stevedoring Co. Sdn. Bhd.	Teluk Rumania and Sg. Rengit, Johor	400.
Cement[1]			Cement Industries of Malaysia Bhd. (United Engineers Malaysia Bhd., 53.97%, and others, 46.03%)	Kangar, Perlis	2,000 cement; 1,650 clinker.
Do.			do.	Bahau, Negeri Sembilan	1,580 cement; 1,300 clinker.
Do.			CMS Cement Sdn Bhd (subsidiary of Cahya Mata Sarawak Bhd)	Bintulu, Sarawak	750 cement.
Do.			do.	Kuching, Sarawak	1,000 cement.
Do.			Holcim (Malaysia) Sdn Bhd (Holcim Ltd.)	Pasir Gudang, Johor	1,300 cement.
Do.			Lafarge Malaysia Cement Bhd. (subsidiary of Lafarge S.A.)	Rawang, Selangor	6,810 cement; 4,900 clinker.
Do.			do.	Kanthan, Perak, Langkawi, Kedah	5,370 cement; 3,300 clinker.
Do.			do.	Pasir Gudang, Johor	770 cement.
Do.			YTL Cement Berhad (subsidiary of YTL Group)	Bukit Sagu, Pahang	1,300 cement; 1,200 clinker.
Do.			do.	Padang Rengas, Perak	3,400 cement; 3,000 clinker.
Do.			do.	Pasir Gudang and Westport, Johor	1,000 cement.
Do.			Tasek Corp. Bhd (publicly owned company)	Ipoh, Perak	2,300 cement; 2,300 clinker.
Copper, mine			Monument Mengapur Sdn Bhd (subsidiary of Monument Mining Ltd.)	Sri Jaya, Pahang	4.
Gas:					
Natural		million cubic meters per day	ExxonMobil Exploration and Production Malaysia, Inc.	Offshore Terengganu	45.
Do.		do.	Sabah Shell Petroleum Co. Ltd.	Offshore Sabah	3.
Do.		do.	Sarawak Shell Bhd.	Offshore Sarawak	80.
Liquefied			Malaysia LNG Sdn. Bhd. [Petroliam Nasional Berhad, (Petronas) 65%; Shell Gas N.V., 15%; Mitsubishi Corp., 15%; Sarawak State government, 5%]	Tanjung Kidurong, Bintulu, Sarawak	8,100.
Do.			Malaysia LNG Dua Sdn. Bhd. [Petroliam Nasional Berhad, (Petronas) 60%; Shell Gas N.V., 15%; Mitsubishi Corp., 15%; Sarawak State government, 10%]	do.	7,800.
Do.			Malaysia LNG Tiga Sdn. Bhd. [Petroliam Nasional Berhad, (Petronas) 60%; Shell Gas N.V., 15%; Nippon Oil LNG (Netherlands) BV, 10%; Sarawak State government, 10%; Diamond Gas Netherlands BV, 5%]	do.	6,800.
Gold, refined		kilograms	PT J Resources Asia Pasifik Tbk (J&Partners, L.P., 100%)	Penjom, Pahang	4,000.
Do.		do.	Raub Australian Gold Mining Sdn. Bhd (Peninsular Gold Ltd., 100%)	Raub, Pahang	500.
Do.		do.	Monument Mining Ltd. of Canada	Bukit Selinsing Koyan, Pahang	1,500.
Iron and steel:					
Direct-reduced iron			Lion DRI Sdn Bhd (Lion Group)	Banting, Selangor	1,540.
Do.			Perwaja Steel Sdn. Bhd. (Kinsteel Bhd, 51%, and Maju Holdings Sdn. Bhd., 49%)	Kemaman, Terengganu	1,800.
Hot-briquetted iron			Amsteel Mills Sdn Bhd (Lion Group)	Labuan Island, offshore Sabah	880.

See footnotes at end of table.

TABLE 2—Continued
MALAYSIA: STRUCTURE OF THE MINERAL INDUSTRY IN 2012

(Thousand metric tons unless otherwise specified)

Commodity		Major operating companies and major equity owners	Location of main facilities	Annual capacity
Iron and steel—Continued:				
Crude steel		Amsteel Mills Sdn Bhd (Lion Group)	Banting, Selangor	1,250.
Do.		do.	Klang, Selangor	750.
Do.		Ann Joo Steel Bhd (Ann Joo Group)	Prai, Penang	900.
Do.		Antara Steel Sdn. Bhd. (Lion Grop)	Pasir Gudang, Johr	600.
Do.		Kinsteel Sdn Bhd	Kuantan, Pahang	500.
Do.		Megasteel Sdn Bhd (Lion Group)	Banting, Selangor	700.
Do.		Malaysia Steel Works Bhd	Bukit Raja, Selangor	450.
Do.		Perwaja Steel Sdn. Bhd. (Kinsteel Bhd, 51%, and Maju Holdings Sdn. Bhd., 49%)	Kermaman, Terengganu	1,500.
Do.		Southern Steel Bhd. [Camerlin (a member of Hong Leong Group Malaysia), 40.75%; Natsteel Ltd., 27.03; others, 32.22%]	Prai, Penang	1,300.
Magnesium, metal		CVM Minerals Ltd.	Kamunting Raya, Perak	15,000.
Nitrogen, ammonia		Asean Bintulu Fertilizer Sdn. Bhd. (Petroliam Nasional Berhad, (Petronas) 63.5%; P.T. Pupuk Sriwidjaja Indonesia, 13%; Thai Ministry of Finance, 13%; Philippines National Development Co., 9.5%; Singapore Temasek Holdings Pte. Ltd., 1%)	Bintulu, Sarawak	395.
Do.		Petronas Fertilizer Kedah Sdn. Bhd. [wholly owned subsidiary of Petroliam Nasional Berhad (Petronas)]	Gurun, Kedah	378.
Do.		Petronas Ammonia Sdn. Bhd. (wholly owned subsidiary of Petroliam Nasional Berhad)	Kerth, Terengganu	370.
Petroleum, crude	thousand 42-gallon barrels per day	ExxonMobil Exploration and Production Malaysia, Inc.	Offshore Terengganu	390.
Do.	do.	Sabah Shell Petroleum Co. Ltd.	Offshore Sabah	100.
Do.	do.	Sarawak Shell Bhd.	Offshore Sarawak	184.
Do.	do.	Petronas Carigali Sdn. Bhd.	Offshore Terengganu	22.
Do.	do.	Murphy Sarawak Oil Co. Ltd.	Offshore Sarawak	15.
Rare earths (REO equivalent)		Lynas Corp. Ltd. of Australia	Kuantan, Pahang	11.
Tin:				
Concentrate		Delima Industries Sdn. Bhd.	Dengkil, Selangor	1.1.
Do.		Maiju Sama Sdn. Bhd.	Puchong, Selangor	1.6.
Do.		New Lahat Mines Sdn. Bhd.	Lahat, Perak	0.3.
Do.		Omsam Telecommunication Sdn. Bhd.	Bakap and Batu Gajah, Perak	0.5.
Do.		Rahman Hydraulic Tin Bhd.	Klian Intan, Perak	3.
Do.		S.E.K. (M) Sdn. Bhd.	Kampar, Perak	0.4.
Do.		Tasek Abadi Sdn Bhd.	Senudong and Kampar, Perak	0.5.
Refined		Malaysia Smelting Corp. Bhd. (The Straits Trading Co. Ltd., 37.44%; Malaysia Mining Corp., 37.44%; others, 25.12%)	Butterworth, Penang	35.
Titanium dioxide		Huntsman Trioxide Sdn. Bhd. (a subsidiary of Huntsman Trioxide)	Kemaman, Terengganu	56.

Do., do. Ditto.

[1] All companies operated integrated plants.

The Mineral Industry of Mongolia

By Susan Wacaster

Mongolia has large proven reserves of coal, copper, and fluorspar. Mineral deposits with copper, gold, molybdenum, tin, and tungsten are common in Mongolia. The central region of Mongolia is characterized by basins and complex folded geologic structures where mineralized zones are frequently encountered. As of 2008, it was estimated that about 30% of all the near-surface territory of Mongolia hosted granitoids, most of which have greisens and vein mineralization associated with calc-alkaline magmas (Gotovsuren and others, 2012).

Turquoise Hill Resources Ltd. of Canada's Oyu Tolgoi Mine, which is part of one of the world's largest known porphyry deposit systems in terms of reserves and resources, has total estimated copper reserves of 13.1 million metric tons (Mt) and about 1 million kilograms (kg) of gold. Estimated resources (including indicated and inferred resources) accounted for about another 14 Mt of copper and more than 400,000 kg of gold. The primary hypogene mineralization within the porphyry is chalcopyrite, and it contains lesser amounts of bornite and magnetite; the secondary supergene mineralization is dominantly chalcocite (Ivanhoe Mines Ltd., 2012b, p. 13).

The Oyu Tolgoi Mine was in the preproduction stage in 2012, and Turquoise Hill Resources began ramping up for commercial operation of the mine in the first half of 2013. The value derived from Oyu Tolgoi alone was expected to account for one-third of Mongolia's gross domestic product (GDP) by 2020. Mongolia's year-on-year real GDP growth rate was estimated to have decreased in 2012 relative to that of 2011, but direct investment in mining, primarily at the Oyu Tolgoi project, resulted in Mongolia having had the fastest growing GDP in the world in 2011 at 17.5% compared with just 6.4% in 2010 and -1.3% in 2009 (Commonwealth of Australia, 2011, p. 93–94).

Minerals in the National Economy

According to the National Statistical Office of Mongolia, preliminary estimates of the 2012 GDP (at current prices) indicated a year-on-year increase of 25.8% compared with that of 2011, or a 12.3% increase at constant prices (with a base year 2005). The economy grew at a slightly slower rate than had been predicted by economic analysts as a result of decreased demand from China, which was Mongolia's primary trade partner. Mongolia's fiscal deficit for 2012 had increased to 8.4% of its GDP in 2012 (World Bank, The, 2013, p. 3; Xinhuanet.com, 2013).

Mongolia's mining and quarrying sector was the primary contributor to the country's economy. In 2012, the output from the mining and quarrying sector accounted for 18.6% of the GDP compared with 21% in 2011. Economic growth in Mongolia in 2012 was also supported by a 21.3% increase in the value of output from the agricultural sector, an 11.4% increase in the output value of the transportation sector, and a 9.2% increase in the output value of the sector composed of retail and wholesale trade and repair of motor vehicles and motorcycles compared with the output values in 2011 (National Statistical Office of Mongolia, 2013, p. 20–21, 61).

Economic growth in Mongolia, however, outpaced the production capacity of the economy. Major infrastructure projects were proposed for the following sectors: construction and urban development; fuel and energy; industrial manufacturing; mining, metallurgical, and chemical industries; and roads and transportation. Financing for the infrastructure projects were to be sourced, in part, from Mongolia's international debt instrument, the Chinggis bonds, which were initially offered at the end of 2012. The initial offering raised $1.5 billion for the Government. Had a larger supply of bonds been available, the Government could potentially have raised as much as $15 billion, as the demand for the bonds was reportedly 10 times greater than the available supply in the first offering (World Bank, The, 2013, p. 3).

Government Policies and Programs

Mineral resources in Mongolia are the property of the state. The Minerals Law of Mongolia regulates the prospecting and exploration for and mining of minerals within the country's territory. Numerous other laws, guidelines, and procedures govern prospecting, exploration, and mining of minerals, including the Constitution of Mongolia, the Environmental Protection Law, the Land Law, the National Security Law, the Subsoil Law, and the Water and Forest Law, among others (Mineral Resources Authority of Mongolia, 2011).

Mineral deposits are grouped into one of three classifications in Mongolia—strategic, common, and conventional. A strategic deposit is one with the potential to affect the national security and economic and social development of the country at the national and regional levels; a deposit also is considered strategic if it accounts for, or has the potential to account for, greater than 5% of the total GDP in a given year. Examples of strategic deposits in Mongolia include coal and copper deposits. Common minerals are those minerals whose concentrations are abundant in sediments and rocks and that might be used as construction material; an example is iron ore. Conventional minerals are those minerals that are not of strategic importance and are not classifiable as common minerals (Ernst & Young Global Ltd., 2012, p. 9).

The 15 deposits that the Government of Mongolia has classified as being of strategic importance include deposits of coal, copper, gold, phosphorite, silver, and zinc. If the Government funds the exploration work that determines proven reserves at a strategic deposit, the Government may claim up to 50% ownership in a joint venture formed to exploit the mineral deposit. If proven reserves in a strategic deposit have been determined through private funding sources, however, the Government may own only up to 34% of the shares of the investment made by the license holder (Ernst & Young Global Ltd., 2012, p. 9).

In 2011, Mongolia implemented the Fiscal Stability Law, which establishes fiscal policies designed to shield the economy from mining sector-related booms and busts. In December 2011, the country also passed the Integrated Budget Law, which defines the country's entire budget process and contains measures to support fiscal sustainability, and, in particular, the successful implementation of the Fiscal Stability Law (World Bank, The, 2013, p. 25).

On May 17, 2012, a law titled The Regulation of Foreign Investment in Business Entities Operating in Sectors of Strategic Importance (BESI) was introduced by the Mongolian Parliament. The law applies to businesses of strategic importance (primarily companies operating in the banking and finance, media and telecommunications, and mining sectors). The law requires Government approval for (a) transactions that involve the acquisition of, or the right to acquire, 33% or more of the shares of a BESI; (b) transactions that involve a foreign investor that has the right (as a result of a proposed acquisition in a BESI, irrespective of the percentage of equity interest) to appoint the executive management or a majority of the board, to veto decisions of the executive management or board of directors, or to determine or implement management decisions and (or) operations; and (c) transactions (as a result of a proposed acquisition involving a foreign investor, irrespective of the percentage of equity interest) that may potentially give rise to a monopoly over mineral products on international or domestic commodity markets, that may directly or indirectly influence the market price of mineral products exported from Mongolia, or that may result in a potential reduction in the shareholding interest of a foreign investor (Ernst & Young Global Ltd., 2012, p. 20).

Production

In 2012, cement production decreased by 17.8% compared with that of 2011, and coal production decreased by 7.7%. The gross weight of iron ore production and the iron content contained in iron ore increased by 33.2% and 33.4%, respectively. Production of crushed stone increased by 148%; lime, by 51.1%; petroleum, 42.6%; zinc content of mine output, 13.8%; crude steel, 13.5%; and salt, 12.7% (table 1; National Statistical Office of Mongolia, 2013, p. 131–132).

Structure of the Mineral Industry

Table 2 is a list of major mineral industry facilities. Most of the producing mining companies in Mongolia have mixed ownership between private international companies and the Mongolian Government or are primarily state owned, but there are also some that are wholly owned by foreign investors.

In 2012, a reported 61,661 people were employed by the country's industrial division compared with 57,506 in 2011. The industrial division included the mining and quarrying sector (which includes petroleum extraction), the manufacturing sector (which includes the production of unspecified nonmetallic mineral products, base metals, and fabricated metal products except machinery), and the power and water supply sector. The mining and quarrying sector employed 19,217 people in 2012 compared with 17,209 in 2011. Of this amount, 8,268 were engaged in the mining of metal ores compared with 8,320 in 2011; 8,138 were engaged in the mining of coal and lignite and the extraction of peat compared with 5,887 in 2011; 1,963 were engaged in other unspecified mining and quarrying compared with 1,839 in 2011; and 848 were engaged in the extraction of crude petroleum compared with 1,163 in 2011 (National Statistical Office of Mongolia, 2013, p. 30, 107).

In 2012, a total of 430 active and 261 inactive mining and (or) quarrying establishments were registered with the Government compared with 383 and 130 establishments, respectively, in 2011. Fifty-seven of the establishments were reported not to have started activities compared with 41 in 2011; 177 had temporarily stopped activities compared with 41 in 2011, and 10 had permanently stopped activities compared with 3 in 2011; the status of the remaining 17 inactive establishments was reported as unknown. Of the 430 active establishments, about 60% were reported to have 1 to 9 employees, 13% had 10 to 19 employees, 10% had 20 to 49 employees, and about 16% had 50 or more employees (National Statistical Office of Mongolia, 2013, p. 107).

Mineral Trade

In 2012, Mongolia had trade relations with 146 nations. China received virtually the total volume of exported coal, copper concentrate, iron ore, and zinc ores and concentrate. China was also the leading recipient of Mongolian exports in terms of value, accounting for about $4.06 billion compared with $4.4 billion in 2011, or 92.5% of the total value in 2012. China was followed by Canada and Russia, which accounted for $117 million and $79.6 million, respectively, or, when combined, about 4.5% of the total. The country had a trade deficit in 2012 of about $2.3 billion compared with $1.8 billion in 2011 (National Statistical Office of Mongolia, 2013, p. 93–95).

The value of industrial output in 2012, increased owing primarily to increases of between 0.1% and 42.6% in the value of exports of mine and quarry products, including concentrates and (or) ores of copper, fluorspar, gold, iron, petroleum, and zinc. The value of exported (unspecified) mineral products decreased by 9.06% to $3.9 billion from $4.3 billion in 2011 (National Statistical Office of Mongolia, 2013, p. 23–25; 93–95).

A total of 20.9 Mt of coal valued at $1.9 billion was exported in 2012 compared with 21.3 Mt valued at $2.3 billion in 2011. Of the coal exported in 2012, 98.9% was bituminous coal, which accounted for 98.2% of the total value. The volume of Mongolian copper exports remained practically unchanged in 2012, but their value decreased by 13.4% to $839 million as 574,500 metric tons (t) of copper concentrate was exported compared with 575,900 t valued at $969 million in 2011.

Other notable mineral exports in 2012 included 2.1 Mt of refined copper and copper alloys valued at $16.6 million compared with 2.4 Mt valued at $21 million in 2011; 428,900 t of fluorspar valued at $102.5 million compared with 407,100 t valued at $95.5 million in 2011; 2,800 kg of gold valued at $122 million compared with 2,600 kg valued at $109.8 million in 2011; and 4,300 t of molybdenum ore and concentrate valued at $38.2 million compared with 4,200 t valued at $46.7 million in 2011 (National Statistical Office of Mongolia, 2013, p. 76–97).

In 2012, China received 80.4% of Mongolia's molybdenum exports and the Republic of Korea received 17.4%, whereas in 2011 China received 40.6% and the Republic of Korea received 55.4%. Mexico received the remainder in both years. In 2012, Russia received 60.8% of Mongolia's fluorspar exports compared with 69.3% in 2011; China received 36.7% compared with 28%; Ukraine received 1.2% compared with 2.0%; and the Republic of Korea received 0.3% compared with 0.2% (National Statistical Office of Mongolia, 2013, p. 93–95).

Commodity Review

Metals

Copper and Gold.—The Oyu Tolgoi copper and gold mine is located in the Southern Gobi region and, in 2012, continued to be developed by Turquoise Hill Resources as the operation reached preproduction stage. The mine is based on a series of porphyry deposits containing copper, gold, silver, and lesser amounts of molybdenum that stretch for 26 kilometers (km) from the Hugo North deposit in the north through the adjacent Hugo South deposit, down to the Southern Oyu deposit and the Heruga deposits. Oyu Tolgoi was the country's flagship strategic deposit, and it was also ranked among the top three copper and gold mines in the world in terms of total reserves. The concentrator at the mine was commissioned in 2012. Development at the mine, however, had been mired in delays in recent years as the project moved towards commercial production from an original startup date of early 2009 (Ivanhoe Mines Ltd., 2012b, p. 20).

The Oyu Tolgoi mining district was initially explored by a joint Mongolian and Russian team that conducted a regional geochemical survey when the central deposit was first identified as a molybdenum anomaly. The team found evidence of alteration and copper mineralization in 1983. BHP Billiton Group (BHP) (then BHP Minerals) conducted exploration on the central and southern deposits between 1997 and 1999 but discontinued exploration owing to cutbacks in the company's exploration budget (Turquoise Hill Resources Ltd., 2013).

In May 2000, Ivanhoe Mines Ltd. (Ivanhoe) of Canada agreed to acquire the project from BHP by investing $6 million in exploration in 7 years and by paying BHP $5 million in cash. BHP retained the right to back in or to retain a 2% net smelter return (NSR). By February 2002, having closed a private placement of $15 million, Ivanhoe had doubled its landholdings to a total of 33,600 square kilometers (km^2) through the acquisition of mineral licenses. Ivanhoe earned 100% interest in the Oyu Tolgoi project after completing a $3 million phase 1 exploration program and paying BHP $5 million. By May 2002, the inferred resource estimate was 821 Mt grading 0.52 gram per metric ton (g/t) gold and 0.38% copper (Ivanhoe Mines Ltd., 2001; 2002a–c; 2003c, p. 9).

In June 2002, BHP's back-in rights to Turquoise Hill expired, leaving it with only the 2% NSR option. BHP's back-in rights to Turquoise Hill would have been exercisable, however, if 250 Mt of supergene copper mineralization grading at least 1% copper, or 300 Mt of hypogene mineralization grading at least 1% copper, had been identified by June 7. Ivanhoe completed more than 50,000 m of drilling at Turquoise Hill in 2002, but it was not until November that drilling revealed mineralization in excess of 1% copper. By April 2002, the indicated and inferred resources for all four deposits at Oyu Tolgoi were estimated to be 1.55 billion metric tons (Gt) grading 1.02% copper and 0.34 g/t gold. In December, Ivanhoe reported that the Mineral Resources Authority of Mongolia (MRAM) had issued four mining licenses to the company for Oyu Tolgoi, which covered a total of 238 km^2. In late September 2005, Ivanhoe estimated that it would cost $1.2 billion to develop the mine and that the mine could produce more than 450,000 metric tons per year (t/yr) of copper and greater than 10,000 kilograms per year (kg/yr) of gold during a 40-year mine life; production was expected by early 2009 (Ivanhoe Mines Ltd., 2003a, b; 2005, p. 1; Turquoise Hill Resources Ltd., 2013).

In January 2006, Ivanhoe upgraded the southern Oyu open pit resources to proven and probable reserves totaling 930 Mt grading 0.5% copper and 0.36 g/t gold. Development drilling was ongoing to upgrade the Hugo Dummett inferred resource to measured and indicated. In April, about 100 citizens of Mongolia staged a sit-in protest regarding Ivanhoe's mining contract, demanding that the Government of Mongolia retain 51% ownership of the Oyu Tolgoi Mine instead of Ivanhoe owning the mine outright (as was the case at the time). In July 2006, the Mongolian Parliament approved the revised Minerals Law, which gave the Parliament the authority to acquire interests in mineral deposits classified as strategic (Ivanhoe Mines Ltd., 2006a, c; Pravda.ru, 2006).

In early September 2006, the Government established a working group to negotiate with Ivanhoe on a formal 30-year investment agreement that would confirm the tax, fiscal, and legal framework for the development of Oyu Tolgoi, but in mid-September, it was reported that the Government would seek to become a partner in Oyu Tolgoi in order to acquire up to a 34% share as a result of the recent legal changes. On October 18, 2006, Rio Tinto plc (Rio Tinto) agreed to acquire a 19.9% interest in Ivanhoe and Oyu Tolgoi for $691 million. The companies would jointly develop the project. Under the agreement, Rio Tinto could increase its holding in Ivanhoe to 33.35% for a total of $1.5 billion, and could eventually acquire up to 40% of the company (Ivanhoe Mines Ltd. 2006b, d).

As of 2009, the Mongolian Parliament had authorized the Government of Mongolia to conclude the Oyu Tolgoi investment agreement with Ivanhoe and Rio Tinto. During the intervening years, the companies had advanced the planning for the project's development, including capital estimates, procurement, design of the underground shaft, and other construction activities, and increased the project's mineral resource estimate, but projected startup of the mine was delayed. Under the investment agreement, the Government would become a partner in the development of the project and acquire a 34% interest in the Ivanhoe subsidiary OT LLC, which held the Oyu Tolgoi mining licenses. The Government's interest was to be held through the state-owned sovereign wealth resources company Erdenes MGL LLC, and Ivanhoe would own a 66% indirect interest in OT LLC (Ivanhoe Mines Ltd., 2012b, p. 2).

Ivanhoe and the Government agreed that because of the extent of the mineral discoveries and the potential for additional discoveries, the approved investment agreement should conform with the provision of Mongolia's Minerals Law, which, at that time, specified that certain deposits of strategic importance qualify for 30 years of stabilized tax rates and regulatory provisions, with an option of extending the term of the investment agreement for an additional 20 years. Major taxes and rates that were to be stabilized for the life of the agreement included the corporate income tax, customs duty, excise tax, exploration and mining licenses, and immovable property and (or) real estate tax, royalties, and value-added tax.

In May 2010, Ivanhoe released an updated integrated development plan for Oyu Tolgoi and forecast annual production of 544,000 t/yr of copper and greater than 20,000 kg/yr of gold for the first 10 years of operation. The mine life was then estimated to be 27 years based only on reserves or 59 years based on reserves and an upgrade of the inferred resources. In June, Rio Tinto confirmed that it would increase its ownership in Ivanhoe to 29.6% by acquiring 46 million shares at a cost of $393 million, and full-scale construction was commenced at Oyu Tolgoi. Rio Tinto increased its stake in Ivanhoe to 34.9% in September 2010 (Ivanhoe Mines, Ltd., 2010a; 2010b, p. 1–5; Politics.co.uk, 2010).

In December 2010, as part of a new financial agreement between Ivanhoe and Rio Tinto, Rio Tinto would manage the Oyu Tolgoi project with the aim of bringing it online in late 2012. The companies continued to work on completing a project financing package for Oyu Tolgoi of up to $3.6 billion. By mid-December, Rio Tinto reported a total capital cost projection of $5.9 billion for phase 1 of the project. The revised agreement, which was signed by Ivanhoe and Rio Tinto in December, established that Rio Tinto would manage the core operations and that Ivanhoe would manage the exploration activities in the noncore area (defined as the area beyond a 200-m buffer from resources defined within the investment agreement as core operations) (Rio Tinto plc., 2010a, b; 2011a, p. 10; Ivanhoe Mines Ltd., 2012b, p. 2, 4).

In March 2011, Rio Tinto agreed to purchase 15 million shares in Ivanhoe for $232.4 million. The financing provided by Rio Tinto was being used for ongoing development at Oyu Tolgoi and to secure long-lead-time equipment. The same month, Ivanhoe reported that Oyu Tolgoi was expected to produce an average of greater than 93,000 kg/yr of silver during its first 10 years of commercial production. Rio Tinto increased its stake in Ivanhoe to 48.5% in August. Ivanhoe reported that Oyu Tolgoi's phase 1 mine was on track to begin initial production from the open pit in late 2012 and to ramp up to commercial production in the first half of 2013 (Business-Mongolia.com, 2010; Ivanhoe Mines, Ltd., 2011; Rio Tinto plc, 2011b, c).

In late September 2011, media reports quoted a Minister of the MRAM as indicating that Ivanhoe and Rio Tinto would receive a letter from the Government asking the companies to consider entering into discussions regarding a change to the investment agreement that would increase the Government's share in Oyu Tolgoi to 50% by purchasing an additional 16% at fair market value at some future point, after Ivanhoe and Rio Tinto recouped their capital investments in the project. Ivanhoe stated that the existing investment agreement for the Oyu Tolgoi project was a legally binding contract and that the company would not agree to the Government's plans to increase its interest in the project. In October 2011, Ivanhoe and Rio Tinto advised the Mongolian Government that they would not renegotiate the terms of the Oyu Tolgoi investment agreement. The companies had also written to the members of Mongolia's National Security Council requesting assistance to ensure the Government's full and immediate support for the agreement. Later reports in January 2012 indicated that the Government might suspend development of the Oyu Tolgoi deposit (as well as the Tavan Tolgoi coal project) until the summer of 2012 to avoid political and economic problems associated with the projects, as the country was scheduled to have Parliamentary elections in June (Kosich, 2007; French and Ferreia-Marques, 2012; Ivanhoe Mines Ltd., 2012a).

In March 2012, Ivanhoe completed an integrated development and operations plan for commercial production, which was to begin in the 4th quarter of 2012 with the ore sourced mainly from the Southern Oyu open pit. Underground infrastructure and mine development would continue for the Hugo North underground block cave deposit. Stockpiling would allow the higher grade ore from Hugo North to displace the open pit ore gradually as the underground production ramped up to 85,000 metric tons per day (t/d). The production rate following the phase 2 expansion would be 58 Mt/yr, and production was projected to total 11.3 Mt of copper and nearly 386,000 kg of gold during 27 years of mine life (Ivanhoe Mines Ltd., 2012b, p. 398).

As of September 2012, initial production from the open pit operation was expected to begin in late 2012. Commercial production was anticipated to start in the first half of 2013, with full production, including underground mining, set to begin in 2018. The company was completing a feasibility study for the underground portion of the project. Completion of the underground project would boost mill capacity to 160,000 t/d from 100,000 t/d.

In October 2012, Ivanhoe rejected another request from the Government of Mongolia to renegotiate the Oyu Tolgoi investment agreement. In November 2012, Turquoise Hill Resources announced that it had signed a power purchase agreement with the Inner Mongolia Power Corp. to supply power to the Oyu Tolgoi Mine. In December 2012, the Oyu Tolgoi concentrator was commissioned. The company expected to process the first ore through the concentrator by yearend, followed by concentrate production 1 month later and commercial production in another 3 to 5 months (Turquoise Hill Resources, 2012a, b).

Mineral Fuels and Related Materials

Coal.—In 2012, coal production accounted for about 30% of the mineral sector's contribution to the GDP; 75% of the extracted coal was exported. The predominant type of coal in Mongolia is lignite, which is found in the eastern and middle portions of the country. Bituminous coal is found in southern and western Mongolia, and subbituminous coal is found in the central and northern regions. Mongolian coal ranges in age from late Carboniferous to Early Cretaceous. Late Carboniferous coal is found in the west whereas Middle Jurassic coal is found throughout the country, and late Permian to Early Cretaceous coal is found in the central, eastern, and northern portions of Mongolia (Erdenetsogt and others, 2009; National Statistical Office of Mongolia, 2013, p. 61).

Mongolia has at least 200 coal deposits and occurrences and perhaps as many as 300. The combined estimated resources at 26 of the known deposits account for about 12.9 billion metric tons (Gt) of coal. Another of the country's coal deposits is the unexploited Tavan Tolgoi coal deposit. This deposit alone contains a reported 6.4 Mt of coking coal resources and is classified as a strategic mineral deposit (Coal Mongolia, 2012, p. 21–23).

Reports released in 2011, indicated that an initial public offering (IPO) for Tavan Tolgoi would take place by yearend 2011 or in early 2012. The state-owned mining company Erdenes Tavan Tolgoi Co. had sought to secure investors for infrastructure development for the proposed project, including the coal handling and processing plants, transport facilities, and water supply. China's Shenhua Group Corp. had won a 40% stake to develop the project before the Government of Mongolia suspended the sale and decided to retain ownership of 50% of the project. As a result, the West Tsankhi area of the Tavan Tolgoi deposit had been at the center of protracted negotiations between the Government and companies that included Peabody Energy Corp. of the United States, OAO Russian Railways of Russia, and Shenhua Group. Reviews of the agreement that would have given Shenhua a 40% share, Peabody a 24% share, and Russian Railways a 36% share led to controversy within the country, with politicians calling for the Government to develop the coalfield itself, as well as among companies from other countries, including Japan and the Republic of Korea, that were interested in developing the project. As of yearend 2012, the Government indicated that a decision on the granting of rights to develop Tavan Tolgoi's western block would not be made before 2013 (Thomson Reuters, 2010; Asia Miner, The, 2012).

Outlook

In its April 2013 Mongolia Economic Update, the World Bank reported that it had revised downward its estimate of the growth of Mongolia's GDP for 2013 to 13% based on negative export growth, decreased foreign direct investment inflows to Mongolia, and a sluggish first-quarter economic recovery in China. Uncertainty about the success of, or any potential delays or disruptions in, the rampup of production at the Oyu Tolgoi Mine could have a large adverse impact on Mongolia's economic outlook (World Bank, The, 2013, p. 3).

Public investment projects that were planned to be funded using proceeds from the Chinggis bonds were another source of uncertainty with respect to the country's economic growth. These large-scale projects would require significant time for planning and feasibility assessment and the likelihood that significant upfront investment using the bond proceeds seems small. It was thought that any large extra public investment expenditure using the proceeds from the Chinggis bonds could add to inflationary pressure and stifle private investment, thereby limiting growth (World Bank, The, 2013, p. 3).

In February 2013, talks were expected to take place between Rio Tinto and the Mongolian Government to resolve concerns that spending at Oyu Tolgoi was ballooning and that the country was not benefiting sufficiently from the development of the project. The cost of developing the Oyu Tolgoi Mine had increased to $6.6 billion, and the Government sought to exercise increased control.

References Cited

Asia Miner, The, 2012, Mongolia—Deadline for Tavan Tolgoi negotiation: The Asia Miner, July. (Accessed September 20, 2013, at http://www.asiaminer.com/magazine/current-news/news-archive/148-july-2012/4418-mongolia-deadline-for-tavan-tolgoi-negotiations.html.)

Business-Mongolia.com, 2010, Rio Tinto increases ownership in Ivanhoe Mines to 22.4% with US$232 million purchase of shares: Business-Mongolia.com, March 1. (Accessed September 20, 2013, at http://www.business-mongolia.com/mongolia/2010/03/02/rio-tinto-increases-ownership-in-ivanhoe-mines-to-22-4-with-us232-million-purchase-of-shares.)

Coal Mongolia, 2012, Coal—Engine of Mongolian fast growing economy: Ulaanbaatar, Mongolia, Ministry of Mineral Resources and Energy, February, 373 p.

Commonwealth of Australia, 2011, AusAID annual report 2010–2011: Canberra, Australian Capital Territory, Australia, Commonwealth of Australia, 379 p.

Erdenetsogt, B.O., Lee, Insung, Delegiin, B.E., and Jargal Luvsanchultem, 2009, Mongolian coal-bearing basins—Geological settings, coal characteristics, distribution, and resources: International Journal of Coal Geology, v. 80, no. 2, p. 87–104.

Ernst & Young Global Ltd., 2012, Mongolia mining and tax guide: New York, New York, Ernst & Young Global Ltd., 36 p.

French, Cameron, and Ferreia-Marques, Clara, 2012, Ivanhoe cancels poison pill, paves way for Rio deal: Reuters.com, January 18. (Accessed September 20, 2013, at http://www.reuters.com/article/2012/01/18/us-ivanhoemines-idUSTRE80H1K620120118.)

Gotovsuren, Uguumur, Sodnomdorj, Amartuvshin, Batdorj, Khuukhnee, Nergui, Dashnvam, and Gochioco, L.M., 2012, Robust mining geophysics exploration in Mongolia: The Leading Edge, v. 31, no. 3, March, p. 304–306.

Ivanhoe Mines Ltd., 2001, Ivanhoe Mines expands discovery of large copper gold porphyry system at Turquoise Hill prospect in Mongolia: Vancouver, British Columbia, Canada, Ivanhoe Mines Ltd., February 15, 1 p.

Ivanhoe Mines Ltd., 2002a, Ivanhoe Mines acquires new exploration licenses in Mongolia covering an additional 6,900 square miles: Vancouver, British Columbia, Canada, Ivanhoe Mines Ltd., February 7, 1 p.

Ivanhoe Mines Ltd., 2002b, Ivanhoe Mines completes earn-in of 100% interest in Turquoise Hill gold and copper project in Mongolia: Vancouver, British Columbia, Canada, Ivanhoe Mines Ltd., February 4, 1 p.

Ivanhoe Mines Ltd., 2002c, Ivanhoe Mines important new copper and gold discovery expands Central Zone at Turquoise Hill gold and copper project in Mongolia: Vancouver, British Columbia, Canada, Ivanhoe Mines Ltd., May 21, 1 p.

Ivanhoe Mines Ltd., 2003a, Ivanhoe Mines granted long-term mining licenses for Turquoise Hill copper and gold project in Mongolia: Vancouver, British Columbia, Canada, Ivanhoe Mines Ltd., December 23, 1 p.

Ivanhoe Mines Ltd., 2003b, New independent resource estimate for Hugo Dummett deposit at Turquoise Hill project in Mongolia: Vancouver, British Columbia, Canada, Ivanhoe Mines Ltd., July 17, 1 p.

Ivanhoe Mines Ltd., 2003c, Review annual information form for the year ended December 31, 2002: Vancouver, British Columbia, Canada, Ivanhoe Mines Ltd., May 20, 67 p.

Ivanhoe Mines Ltd., 2005, Independent integrated development plan for Oyu Tolgoi highlights significant and long-lasting benefits for Mongolia: Vancouver, British Columbia, Canada, Ivanhoe Mines Ltd., September 29, 11 p.

Ivanhoe Mines Ltd., 2006a, Ivanhoe announces proven and probable copper and gold reserves for open-pit mine at Oyu Tolgoi project, Mongolia: Vancouver, British Columbia, Canada, Ivanhoe Mines Ltd. January 30, 4 p.

Ivanhoe Mines Ltd., 2006b, Ivanhoe Mines and Rio Tinto form strategic partnership to develop Mongolian copper-gold resources: Vancouver, British Columbia, Canada, Ivanhoe Mines Ltd., October 18, 7 p.

Ivanhoe Mines Ltd., 2006c, Ivanhoe Mines closes C$189 million bought-deal financing: Vancouver, British Columbia, Canada, Ivanhoe Mines Ltd., April 25, 1 p.

Ivanhoe Mines Ltd., 2006d, Mongolian Government establishes a working group to address investment agreement for Ivanhoe Oyu Tolgoi copper and gold project: Vancouver, British Columbia, Canada, Ivanhoe Mines Ltd., September 5, 1 p.

Ivanhoe Mines Ltd., 2010a, Ivanhoe Mines advances financing for Oyu Tolgoi copper-gold project in discussions with international financial institutions: Vancouver, British Columbia, Canada, Ivanhoe Mines Ltd., May 21, 3 p.

Ivanhoe Mines Ltd., 2010b, Ivanhoe Mines announces financial results and review of operations for the first quarter of 2010: Vancouver, British Columbia, Canada, Ivanhoe Mines Ltd., May 13, 17 p.

Ivanhoe Mines Ltd., 2011, Silver production at Ivanhoe Mine's Oyu Tolgoi project is expected to average more than 3 million ounces each year in first 10 years of operation: Vancouver, British Columbia, Canada, Ivanhoe Mines Ltd., March 3, 3 p.

Ivanhoe Mines Ltd., 2012a, Ivanhoe Mines negotiates an additional US$1.8 billion bridge financing as part of the comprehensive financing plan for Oyu Tolgoi: Vancouver, British Columbia, Canada, Ivanhoe Mines Ltd., January 18, 3 p.

Ivanhoe Mines Ltd., 2012b, Oyu Tolgoi project: Vancouver, British Columbia, Canada, Ivanhoe Mines Ltd., 490 p.

Kosich, Dorothy, 2007, Ivanhoe, Rio, Mongolia Govt. reach tentative agreement on Oyu Tolgoi copper/gold project: Mineweb.com, April 11. (Accessed September 20, 2013, at http://www.mineweb.com/mineweb/content/en/mineweb-base-metals?oid=19342&sn=Detail.)

Mineral Resources Authority of Mongolia, 2011, Mongolia—The land of opportunities: Ulaanbaatar, Mongolia, Mineral Resources Authority of Mongolia, 30 p.

National Statistical Office of Mongolia, 2013, Monthly bulletin of statistics: National Statistical Office of Mongolia, December, 151 p.

Politics.co.uk, 2010, Rio Tinto to exercise the Series A warrants in Ivanhoe Mines: Politics.co.uk, June 30. (Accessed September 20, 2013, at http://www.politics.co.uk/opinion-formers/rio-tinto/article/rio-tinto-to-exercise-the-series-a-warrants-in-ivanhoe-mines.)

Pravda.ru, 2006, Protesters from rival civic groups face off in Mongolia: Pravda.ru, April 11. (Accessed September 20, 2013, at http://english.pravda.ru/news/world/11-04-2006/79011-mongolia-0.)

Rio Tinto plc, 2010a, Rio Tinto investor seminar: Melbourne, Victoria, Australia, Rio Tinto plc, November 26, 3 p.

Rio Tinto plc, 2010b, Rio Tinto to manage Oyu Tolgoi project under new financial agreement with Ivanhoe Mines: Melbourne, Victoria, Australia, Rio Tinto plc, December 8, 3 p.

Rio Tinto plc, 2011a, Rio Tinto delivers record underlying earnings of $14 billion and announces $5 billion capital management programme together with 20 percent increase in dividend: Melbourne, Victoria, Australia, Rio Tinto plc, February 10, 44 p.

Rio Tinto plc, 2011b, Rio Tinto increases its stake in Ivanhoe Mines to 48.5%: Melbourne, Victoria, Australia, Rio Tinto plc, August 24, 2 p.

Rio Tinto plc, 2011c, Rio Tinto to exercise its remaining warrants in Ivanhoe Mines: Melbourne, Victoria, Australia, Rio Tinto plc, June 21, 2 p.

Thomson Reuters, 2010, Mongolia cancels $2 billion sale of Tavan Tolgoi stake—Interested parties included China Shenhua Energy Company Limited: Reuters, February 4. (Accessed September 20, 2013, at http://www.reuters.com/article/2010/02/04/mongolia-tavantolgoi-idUSHKU00014720100204.)

Turquoise Hill Resources Ltd., 2012a, Oyu Tolgoi celebrates commissioning of concentrator complex: Vancouver, British Columbia, Canada, Turquoise Hill Resources Ltd., October, 2 p.

Turquoise Hill Resources Ltd., 2012b, Turquoise Hill Resources rejects request from Mongolian Government to renegotiate Oyu Tolgoi investment agreement: Vancouver, British Columbia, Canada, Turquoise Hill Resources Ltd., October, 2 p.

Turquoise Hill Resources Ltd., 2013, Oyu Tolgoi (copper-gold), Mongolia—Overview: Turquoise Hill Resources. (Accessed September 18, 2013, at http://www.turquoisehill.com/s/Oyu_Tolgoi.asp.)

World Bank, The, 2013, Mongolia economic update: Washington, DC, The World Bank, April, 30 p.

Xinhuanet.com, 2013, Mongolia's economy grew 12.3% in 2012: Xinhuanet.com, January 16. (Accessed September 20, 2013, at http://news.xinhuanet.com/english/business/2013-01/16/c_132106840.htm.)

TABLE 1
MONGOLIA: PRODUCTION OF MINERAL COMMODITIES[1]

(Metric tons unless otherwise specified)

Commodity[2]		2008	2009	2010	2011	2012
Cement, hydraulic	thousand metric tons	270	235	323	426	350
Coal, unspecified	do.	9,692	13,164	25,246	30,940	28,561
Copper:						
Mine output, Cu content		126,796	129,800	124,985	121,590	121,660
Metal, refined		2,587	2,470	2,746	2,390	2,282
Fluorspar:						
Acid grade	thousand metric tons	116	115	141	116	116
Submetallurgical and other grade	do.	219	344	259	232	230
Total	do.	335	459	400	348 r	346
Gold, mine output, Au content	kilograms	15,184	9,803	6,037	5,703	5,995
Iron ore:						
Gross weight	thousand metric tons	1,387	1,380	3,203	5,678	7,561
Iron content	do.	800 r	800 r	1,900 r	3,400 r	4,537
Lime, hydrated and quicklime	do.	55	43	50	45	68
Molybdenum, mine output, Mo content		1,780	2,140	2,198	1,960	1,904
Petroleum, crude	thousand 42-gallon barrels	1,174	1,870	2,181	2,549	3,636
Salt, mine output		1,176	1,402	1,861	2,183	2,461
Silver, mine output, Ag content	kilograms	28,890	29,321	28,710	28,254	27,982
Steel, crude		81,400	50,100	64,200	60,000	68,100
Stone, crushed	thousand metric tons	103	123	101	94	233
Tungsten, mine output, W content		142	39	20	20	20
Zinc, mine output, Zn content		143,600	141,500	112,600	104,700	119,100

r Revised. do. Ditto.

[1] Table includes data available through November 20, 2013.

[2] In addition to the commodities listed, crude construction materials, such as gypsum, sand and gravel, and varieties of stone, such as limestone, are produced, but available information is inadequate to make reliable estimates of output.

TABLE 2
MONGOLIA: STRUCTURE OF THE MINERAL INDUSTRY IN 2012

(Thousand metric tons unless otherwise specified)

Commodity	Major operating companies and major equity owners	Location of main facilities[1]	Annual capacity[e]
Calcium oxide	Qinhua MAK Naryn Sukhait LLC (Mongolia-China joint venture)	316 km from Ulaanbaatar at the Olon Ovoot station of the Trans Mongolian Railway	50
Cement	Khutul Cement and Lime Factory	Darhan, Darhan-Uul Aymag	500
Coal	Baganuur Joint Stock Co. (Government, 51%, and public, 49%)	Baganuur Mine, Tov Aymag	3,000
Do.	Government, 95%, and public, 10%	Shivee Ovoo Mine, Dornogovi and Govisumber Aymag, 20 km from Choir City	2,000
Do.	SouthGobi Energy Resources Ltd. (Turquoise Hill Resources Ltd., 57.6%)	Ovoot Tolgoi Mine, Omnogovi [South Gobi] Aymag	4,600
Do.	do.	Tsagaan Tolgoi, Dornogovi Aymag, 95 km north of the Chinese border	3,000
Do.	Mongolian Mining Corp., 100%	Ukhaa Khudag Mine, Omnogovi Aymag, 61 km east of Dalanzadgad	8,600
Do.	MAK Mongolyn Alt Group, 100%	Naryn Sukhait mines, Gurvantes Soum, Omnogovi Aymag	3,000
Do.	Guilford Coal Ltd., 100%	South Gobi Mine, Omnogovi Aymag, 50 km east of Naryn Sukhait	3,000
Do.	Mongolian Mining Corp., 100%	Baruun Naran Mine, Omnogovi Aymag, 61 km east of Dalanzadgad	3,000
Copper, Cu in concentrates	Samsung Corp., 51%, and Erdenet Mining Corp. (Mongolia-Russia joint venture), 49%	Erdenet Ovoo open pit mine and processing plant, Bulgan Aymag, 180 km east of Darkhan city	140
Do.	Turquoise Hill Resources Ltd., 66%, and Government, 34%	Oyu Tolgoi Mine, Omnogovi Aymag, 80 km north of the Chinese border	420
Do.	Mongoyn Alt Corp., 100%	Tsagaan Suvarga Mine, Omnogovi Aymag, 560 km southeast of Ulaanbaatar	70
Copper, Cu in cathodes	Erdenet Mining Corp. (Mongolia-Russia joint venture), 51%, and Strand Holdings Ltd., 49%	Erdmin solvent extraction-electrowinning plant	3
Fluorspar	Mongolrostsvetmet LLC	Bor-Undur Mine and processing plant, Hentiy Aymag, 310 km southeast of Ulaanbaatar; 2 underground and 3 open pit mines	450 [2]
Do.	do.	Urgen Mine, Dornogovi Aymag, 535 km from Ulaanbaatar	100 [2]
Gold, Au in concentrates	Zinjin Mining Group Co. Ltd., 70%	Nari Tolgoi gold mine, Jierigron Sumu, Tov Aymag	90 [2]
Do.	North Asia Resources Holdings Ltd.	Khar Yamaat placer mine, 180 km north of Ulaanbaatar	NA
Do.	Mongolian Resource Corp. Ltd., 90%	Blue Eyes Mine, Bornuur Soum, Tov Aymag	36 [2]
Do. thousand cubic meters	Mongolrostsvetmet LLC	Zaamar placer gold operation, Tov Aymag, 240 km southwest of Ulaanbaatar	300
Do. do.	do.	Zeregtsee placer mine, 240 km southwest of Ulaanbaatar	500
Do.	Turquoise Hill Resources Ltd., 66%, and Government, 34%	Omnogovi Aymag, 80 km north of the Chinese border	420
Iron, Fe, in concentrates	Lung Ming Mining Co. Ltd., 66.7%, and China Investment Corp., 33.3%	Eruu Gol Mine	2,500
Lead	Shandong Xianglong Co Ltd	Tsav Mine, Dornod Aymag Ulaanbaatar	117 [2]
Limestone	MAK Mongolyn Alt Group, 100%	14 km from the Olon Ovoot station of the Trans Mongolia railway	NA
Molybdenum	Erdenet Mining Corp. (Mongolia-Russia joint venture) (Mongolia-Russia joint venture)	Erdenet Ovoo open pit mine and processing plant, Bulgan Aymag, 180 km east of Darkhan city	3,000
Do.	Turquoise Hill Resources Ltd., 66%, and Government, 34%	Omnogovi Aymag, 80 km north of the Chinese border	NA

See footnotes at end of table

TABLE 2—Continued
MONGOLIA: STRUCTURE OF THE MINERAL INDUSTRY IN 2012

(Thousand metric tons unless otherwise specified)

Commodity	Major operating companies and major equity owners	Location of main facilities[1]	Annual capacity[e]
Silver	Turquoise Hill Resources Ltd., 66%, and Government, 34%	Omnogovi Aymag, 80 km north of the Chinese border	93
Steel	Darkham metallurgy plant	Darhan, Darhan-Uul Aymag	100
Tungsten	Samsung Corp., 51%, and Erdenet Mining Corp. (Mongolia-Russia joint venture), 49%	Erdenet Ovoo open pit mine and processing plant, Bulgan Aymag, 180 km east of Darkhan city	140
Zinc	Tsairt Minerals Co. Ltd. (China-Mongolia joint venture)	Sukhe Bator, Suhbaatar Aymag	70
Do.	Shandong Xianglong Co. Ltd.	Tsav Mine, Dornod Aymag Ulaanbaatar	117 [2]
Do. metric tons	China Nonferrous Metals Group, 51%, and Government, 49%	Tumurtiin Ovoo Mine, Suhbaaater, 180 km southwest of Choibalsan	34,000

[e]Estimated. Do., do. Ditto. NA Not available.

[1]Abbreviations used for units of measure in this table include the following: km—kilometer.

[2]Mill capacity.

THE MINERAL INDUSTRY OF NEW CALEDONIA

By Susan Wacaster

New Caledonia is rich in a limited number of mineral resources. The country's economy remained heavily dependent upon nickel and byproduct cobalt production. Other metallic mineral resources included chromite, copper, gold, iron ore, manganese, and silver. The main island, Grande-Terre, is composed of greater than 80,000 square kilometers of late Eocene massive peridotite that represents one of the largest mantle rock complexes in the world. The peridotites, through which supergene alteration brought about widespread nickel concentrations, occupy one-third of the surface of Grande-Terre and include primarily harzburgite-dunite in the south with lesser amounts of lherzolite in the north. Mineral production within this structure is fundamental to the economy of New Caledonia (Direction de l'Industrie, des Mines et de l'Energie de la Nouvelle-Calédonie, 2009).

Interest among foreign mining companies in the development of nickel prospects in southern New Caledonia began in the early 1900s when subsidiaries of Inco Ltd. of Canada, which was created as the International Nickel Mining Co. in 1919 and ultimately acquired by Vale Ltd. (formerly Companhia Vale do Rio Doce [CVRD]) of Brazil in 2006, gained control of properties that had been sold or abandoned in the 1920s and 1930s. During World War II, Inco refined nickel matte from New Caledonia at its Sudbury operations on a not-for-profit basis. The Bureau de Recherches Géologiques et Minières of the New Caledonia Territory (BRGM NCT) (the New Caledonian branch of the French Geological Survey) had been active in New Caledonia since the 1950s. In the 1970s, the BRGM NCT began to inventory mining activities and develop plans for prospecting and exploiting mineral resources. The goal of the project was to diversify the mineral industry, which was based on the extraction primarily of nickel and lesser amounts of chromium and cobalt (Maurizot and Eberle, 1982; Vale S.A., 2013c).

Ore bodies were determined to be concentrated and aligned along tectonic structures in certain mineral provinces. The most significant metallic mineral concentrations were determined to include the central plutono-volcanic units, which contain copper and gold deposits in the form of volcanogenic massive sulfide (VMS) deposits; the volcano-sedimentary deposits of the Diahot province to the north, which contain copper, lead, and zinc (and lesser amounts of gold and silver); the East Coast Basalts province and West Coast Basalts province, which host VMS deposits of copper and lesser amounts of gold; mineral deposits related to major faults, which contain antimony, copper, lead, and tungsten; and mineral occurrences related to granodioritic intrusions, which contain minor deposits of molybdenum, tungsten, and lesser amounts of antimony (Direction de l'Industrie, des Mines et de l'Energie de la Nouvelle-Calédonie, 2009).

Production

In 2012, New Caledonia produced 102,400 metric tons (t) of nickel from 6.4 million metric tons (Mt) of saprolite ore and 29,300 t of nickel from 3.3 Mt of laterite ore compared with 100,400 t of nickel from 5.6 Mt of saprolite ore and 30,300 t of nickel from 3.2 Mt of laterite ore in 2011. In 2012, the amount of nickel derived from ferronickel production increased by about 7.5% to 43,030 t. Data on mineral production are in table 1 (Direction de l'Industrie, des Mines et de l'Energie de la Nouvelle-Calédonie, 2013).

Structure of the Mineral Industry

The country's most recent major mineral industry development was the commissioning in 2010 of Vale of Brazil's Goro nickel and byproduct cobalt operation, the startup of which had been delayed repeatedly, and the development of which involved large cost overruns. Ramping up to commercial production continued throughout early 2012. In 2012, Goro was jointly owned by Vale (74%), Japanese companies Sumitomo Metal Mining Co. Ltd. (11%) and Mitsui & Co. Ltd. (10%), and the Société de Participation Minière du Sud Calédonien S.A.S. (5%).

The Koniambo nickel project was a joint venture between Société Minière du Sud Pacifique (SMSP) (51%) and Xstrata Nickel of Switzerland (49%). In February 2012, Swiss companies Glencore International plc and Xstrata plc announced a potential merger (made final in May 2013) whereby Glencore would acquire the 65.92% of Xstrata that it did not already own, so a change of ownership was expected for the Koniambo project. SMSP continued to be a major nickel ore producer in New Caledonia at its SMSP Laterite Operation (also known as the Sud-Pacifique Mine), which is located northwest of Noumea. The Sud-Pacifique Mine was a joint venture between SMSP (51%) and POSCO of the Republic of Korea (49%) (table 2).

As of yearend 2012, 20 facilities were in place to handle the transport of mineral commodities, including conveyors (5), ports (3), or wharves (12), which had a combined annual gross tonnage capacity of 8.09 Mt. Of those facilities, 17 were used for exports. Six of the facilities were located in the South Province, and the rest were located in the North Province. The individual facilities ranged in capacity from 50,000 t (a wharf in Poum called Tanlé) to 786,000 t (a wharf in Koumac called Karembe). Of the 20 facilities, 17 were for the movement of ore. One port facility, Doniambo, was used for the export of ferronickel and matte, and another, Baie de Prony, was used for the export of nickel oxide (NiO), nickel hydroxide cake (NHC), and cobalt carbonate ($CoCO_3$). The Doniambo facility had the capacity to handle 206,000 t, and the Baie de Prony facility, 23,000 t. Another port, Vavouto, was ramping up for the export of ferronickel, but the capacity was not yet available. Ownership of the 20 facilities was distributed among eight companies (table 2).

Mineral Trade

In 2012, the total value of exports from New Caledonia decreased by 13.4% compared with that of 2011. Greater than 90% of that value was accounted for by mining and metallurgical products. The economy of New Caledonia was sensitive to the variations in world prices for mineral commodities, and the health of the economy was linked directly to the country's trade in, primarily, cobalt and nickel (Institut de la Statistique et des Études Économique, 2013, p. 17–18).

The value derived from exports of cobalt and nickel products from New Caledonia decreased compared with that of 2011 after having recovered somewhat in 2010 from the global economic slump and resultant price fluctuations that took place in 2008 and 2009. According to the United Nations Commodity Trade Statistics database, New Caledonia exported 4.3 Mt of nickel ore and concentrate valued at $243 million compared with 4.4 Mt of nickel ore and concentrate valued at about $302 million in 2011; 20,680 t of NiO and NHC valued at $79 million compared with 45,100 t and $127 million, respectively, in 2011; 186,319 t of ferronickel in granular or powder form valued at $674 million compared with 165,000 t and $847 million, respectively, in 2011; and about 19,523 t of nickel matte, NiO sinter, and other intermediate products of nickel metallurgy valued at $203 million compared with 19,700 t and $271 million, respectively, in 2011 (Direction de l'Industrie, des Mines et de l'Energie de la Nouvelle-Calédonie, 2013; United Nations Statistics Division, 2013).

About 50% of exports from New Caledonia were typically received by countries in Asia. On a year-on-year basis, exports to New Caledonia's Asian trade partners decreased by 10% in 2012 compared with a 21% increase in 2011 and a 42% increase in 2010. The decrease in 2012 was owing in part to decreased imports from Japan, which had typically been a leading recipient of New Caledonian exports but which was still recovering from the devastating effects of the 2011 Tohoku earthquake. About one-third of the volume of the products Japan typically received from New Caledonia was in the form of cobalt-nickel concentrates. The value of the trade deficit was 64% greater in 2012 than the average value of the trade deficit for the past 10 years. The increased trade deficit was mainly the result of decreased Japanese exports to France, as the financial crisis in the euro area continued to affect international markets negatively (Institut de la Statistique et des Études Économique, 2013, p. 17–18).

Commodity Review

Metals

Nickel and Cobalt.—Negotiations aimed at creating a major mining project in New Caledonia took place in the 1950s and 1960s and resulted in a joint proposal from Inco and Pechiney S.A. of France in 1966 that was rejected by the Government of France. In 1969, Inco and the BRGM joined a French consortium to develop a fully integrated nickel project in southeastern New Caledonia. In July 1970, Inco presented a proposal for a 45,000-t-capacity plant that would use Inco's carbonyl extraction and refining process—the Inco Carbonyl Process for Laterites—at a capital cost of what then would have been $500 million. The proposal, however, was rejected by the consortium. Another proposal was made in 1973 based on a reduction and acid-leaching process that would produce 18,000 t/yr of nickel and 1,300 t/yr of cobalt at an expected capital cost of $275 million. That proposal was based on the Goro ore body, which was then held in trust by the BRGM. The proposal was denied when the French Government split the rights to Goro, with some majority of the rights going to a French company. By that time, Inco had completed 85,000 meters (m) of exploratory drilling, sampled 11,000 t of ore for pilot testing, and spent $21 million on site investigations (Marcuson and others, 2009).

Subsequent attempts were made by other companies and the BRGM to develop Goro. By 1990, deadlocked negotiations between the BRGM and Dallhold Nickel Management (later known as Queensland Nickel Management) of Australia again caused the project to be put on hold. That same year, Inco acquired the mining licenses for the Goro ore body, completed the purchase from the BRGM, and created Goro Nickel Ltd. as a subsidiary to examine the feasibility of mining and processing laterite ore (Vale S.A., 2013c).

Inco projected that a future Goro mine operation could process about 2.6 million metric tons per year (Mt/yr) of ore to produce 40,000 metric tons per year (t/yr) of NiO, but the project was put on care-and-maintenance status in 1994 because it was considered small in size. The Goro deposit, however, offered a low-cost source of nickel and cobalt, so Inco investigated the use of acid leaching combined with solvent extraction-electrowinning (SX-EW) to process the laterite ore. In 1997, Inco and BRGM announced that they would construct a 12-metric-ton-per-day pilot plant. Inco completed a feasibility study that estimated open pit production of 27,000 t/yr of nickel and 2,720 t/yr of cobalt for 20 years. The company sought a partner to acquire up to 30% interest to help develop the project. Construction of the pilot plant took place in 1998, and it was commissioned in 1999 on the same site as the future commercial plant, located 58 kilometers east of Noumea in the South Province. Inco had expected to make a decision on building a commercial facility by late 2000, with initial production slated for 3 years after the start of construction (Bacon and Mihaylov, 2002; Vale S.A., 2013c).

By November 2000, Inco had applied for an operating permit and Goro was expected to produce 54,000 t/yr of nickel as a 78% ferronickel alloy and 5,400 t/yr of cobalt in cobalt carbonate. Reserve estimates included 47 Mt grading 1.59% nickel and 0.17% cobalt, with additional resources of 219 Mt grading 1.57% nickel and 0.18% cobalt. Production was expected to begin by late 2004 or early 2005. In 2001, Inco was still in discussions with companies interested in acquiring a stake in the project. Goro was expected to be operating at full capacity by yearend 2006. Inco owned 85% of the project and the BRGM held a 15% share. Inco wanted to sell the BRGM share plus another 15% of its own share to a third party because even with the sale of this 30% interest, Inco would still have needed to raise an estimated $600 million for project costs. The French Government had already contributed $350 million by midyear 2001 and it had provided a 15-year tax holiday to be followed by a rate of 50% of the prevailing tax rates for an additional 5 years. The construction permit for the commercial plant was received by Inco in April 2002 (SNL Metals Economics Group, 2013).

In July 2002, Inco reached an agreement in principle to sell a 25% share of Goro to a Japanese consortium led by Sumitomo Metal Mining. Inco would retain 70% of the project and the Government of New Caledonia would own 5%. By yearend 2002, the capital cost estimates for the development of Goro had increased to $1.45 billion, which represented a 30% to 40% cost overrun. In 2003, leaders of New Caledonia's indigenous Kanak group announced that they were seeking a role in Inco's development of the Goro project, and Inco announced that the cost overruns were going to delay the project for yet another year. In 2004, Inco completed a final review of the Goro project, which resulted in a $500 million reduction in the capital cost estimate, and in October approved the $1.9 billion mine development project. Construction was set to begin in early 2005, and production was projected to commence in September 2007. Goro was expected to reach about 75% of its expanded annual capacity of 60,000 t/yr of nickel within 12 months after commissioning and 90% within 2 years. The production capacity for cobalt was revised to about 4,650 t/yr (Northern Miner, 2002; ALTA Metallurgical Services, 2003; Guerriere, 2003; Inco Ltd., 2004; Stueck, 2004; Globe and Mail, The, 2005).

In 2005, Inco and the three Provinces of New Caledonia came to an agreement whereby the Provinces would acquire a total 10% interest in Goro. The Provinces formed a company, the Société de Participation Minière du Sud Calédonien (SPMSC) to hold the Territory's stake in the project, which would also include the 15% held by the BRGM. In April, Inco announced that Sumitomo Metal Mining and Mitsui had acquired a 21% interest in Goro for $150 million. By December, Inco reported that increased costs for commodities and construction materials, including fuel and lubricants, would again increase capital costs. In 2006, after incidents of environmental protests, legal proceedings, vandalism, a labor strike, revocation of the project's construction permit, and supply shortages, the estimated capital costs at Goro were increased to $3 billion and production was delayed until the end of 2008 (Globe and Mail, The, 2005; Mining Journal, 2005; PR Newswire, 2005).

In 2007, Vale [then Companhia Vale do Rio Doce (CVRD)] of Brazil completed a 100% takeover of Inco Ltd. to form CVRD Inco Ltd. The company announced a revised capital cost estimate of $3.2 billion. The project was still beset by protests and vandalism, including the complete destruction of a $1.5 million crane. Production was again delayed until the spring of 2009. In April 2009, commissioning of the acid plant was halted after an acid spill into a local waterway. By this time, total capital cost estimates had been increased to $4.3 billion. Goro was projected to commence operations in 2010 and to reach commercial production by 2013. The project did finally come online in the third quarter of 2010. In 2011, ramping up efforts continued, but the plant's high-pressure acid-leaching technology was difficult to get started. Output increased at Goro in the fourth quarter of 2011 (Companhia Vale do Rio Doce, 2007).

In the first 3 months of 2012, the operation produced 4,000 t of nickel and 385 t of cobalt in intermediate products compared with 5,100 t of nickel and 245 t of cobalt in all of 2011. At full production, Goro was expected to produce up to 60,000 t/yr of nickel and 4,650 t/yr of cobalt using high-pressure acid leaching to extract the large volumes of low-grade nickel. Proven and probable reserves at Goro included a revised 122.5 Mt of ore grading 1.44% nickel and 0.11% cobalt. By May 2012, however, Vale had suspended sales and purchases after declaring force majeure following an incident at the mine's sulfuric acid plant (Vale S.A., 2013a, p. 7, 9; 2013b, p. 64, 66).

As of December 31, 2011, proven and probable reserves at the Koniambo nickel project included 62.5 Mt grading 2.4% nickel. The project was expected to process 3 Mt/yr of limonite and saprolite ore during a 25-year mine life and to produce 60,000 t/yr of nickel at full production, which was expected to be reached sometime in 2014. In 2009, it was reported that ferronickel would be produced using an updated version of a process used at Xstrata's Falcondo operation called Nickel Smelting Technology. Initially, nickel would be extracted from the saprolite part of the ore body; plans were also in place to extract nickel from the limonite ore using a hydrometallurgical process (Xstrata plc, 2011, p. 36; SNL Metals Economics Group, 2013).

Outlook

New Caledonia is expected to remain a globally significant cobalt- and nickel-producing country, and output is expected to increase as production is ramped up at the Goro project and operations commence at the Koniambo project. An expansion is planned for Goro, in line with the original plans envisioned by Inco in 1999, for a phased commercial operation that would allow the initial capacity (then expected to be 30,000 t of nickel) to be doubled. The mine was expected to reach full operation in 2013. By the end of the second quarter of 2013, the plant at Goro was operating with two autoclaves. The operation produced a total of about 5,100 t of nickel in NHC and NiO in the first quarter of 2013 and about 3,400 t of nickel in the second quarter, which included 1,800 t contained in NHC and 1,600 t in NiO. Production slumped because of anticipated maintenance of the acid plant. Operations were reportedly normalized by June 2013 (Vale S.A., 2013a, p. 6).

References Cited

ALTA Metallurgical Services, 2003, Goro key to future of PAL: ALTA Metallurgical Services. (Accessed October 23, 2013, at http://www.altamet.com.au/about/altamet/contact.)

Bacon, G., and Mihaylov, I., 2002, Solvent extraction as an enabling technology in the nickel industry: Journal of the South African Institute of Mining and Metallurgy, November/December, p. 435-443. (Accessed September 21, 2013, at http://www.saimm.co.za/Journal/v102n08p435.pdf.)

Companhia Vale do Rio Doce, 2007, CVRD Inco announces completion of amalgamation and share redemption: Companhia Vale do Rio Doce, Rio de Janeiro, Brazil, January, 1 p.

Direction de l'Industrie, des Mines et de l'Energie de la Nouvelle-Calédonie, 2009, Geological map of New Caledonia—First edition: Direction de l'Industrie, des Mines et de l'Energie de la Nouvelle-Calédonie, 2 p.

Direction de l'Industrie, des Mines et de l'Energie de la Nouvelle-Calédonie, 2013, Productions et exportations minieres & metallurgiques—Mai 2013: Noumea, New Caledonia, Direction de l'Industrie, des Mines et de l'Energie de la Nouvelle-Calédonie, March 21, 3 p.

Globe and Mail, The, 2005, Higher material costs hitting Goro mine costs, Inco filing says: Metalspace.com, December 16. (Accessed October 22, 2013, at http://metalsplace.com/news/articles/3345/higher-material-costs-hitting-goro-mine-costs-inco-filing-says.)

Guerriere, Alison, 2003, Indigenous tribal leaders rip Inco over Goro nickel project: American Metal Market LLL, March 21. (Accessed October 22, 2013, at http://business.highbeam.com/436402/article-1G1-99129922/indigenous-tribal-leaders-rip-inco-over-goro-nickel.)

Inco Ltd., 2004, Updated capital cost estimate and schedule and progress on other key milestones provides basis for decision to proceed: Toronto, Ontario, Canada, Inco Ltd., October 19, 4 p.

Institut de la Statistique et des Études Économique, 2013, 2012 bilan economique et social: Noumea, New Caledonia, Institut de la Statistique et des Études Économique, 40 p.

Marcuson, S.W., Hooper, J., Osborne, R.C., Chow, K., and Burchell, J., 2009, Sustainability in nickel projects—50 years of experience at Vale Inco: Engineering and Mining Journal. (Accessed September 22, 2013, at http://www.e-mj.com/index.php/features/117-sustainability-in-nickel-projects-50-years-of-experience-at-vale-inco.html.)

Maurizot, Pierre, and Eberle, J.M., 1982, Preliminary metallogenic map of New Caledonia; Second part, Mineral deposits nonassociated with ultrabasic rocks: AAPG Bulletin, v. 66, p. 976.

Mining Journal, 2005, New Caledonia: Mining Journal Online. (Accessed October 23, 2013, at http://www.mining-journal.com/reports/new-caledonia---2005?SQ_DESIGN_NAME=print_friendly.)

Northern Miner, 2002, Inco reviews Goro development: Northern Miner. (Accessed October 23, 2013, at http://www.northernminer.com/news/inco-reviews-goro-development/1000149423.)

PR Newswire, 2005, INCO announces completion of acquisition by Sumitomo Metal Mining Co., Ltd. and Mitsui & Co., Ltd. of a 21% interest in Goro Nickel S.A.: PR Newswire (Accessed October 22, 2013, at http://www.prnewswire.com/news-releases/inco-announces-completion-of-aquisition-by sumitomo-metal-mining-co-ltd-and-mitsui--co-ltd-of-a-21-interest-in-goro-nickel-sa-54228312.html.)

SNL Metals Economics Group, 2013, MineSearch: SNL Metals Economics Group database. (Accessed October 23, 2013, via http://www.metalseconomics.com/database-services/minesearch.)

Stueck, Wendy, 2004, Inco's $1.9-billion (U.S.) Goro nickel project is finally a go: The Globe and Mail, [Toronto, Ontario, Canada]. (Accessed October 22, 2013, at http://www.theglobeandmail.com/reportonbusiness/incos-19-billion-us-goro-nickel-project-is-finally-a-go/article1005577.)

United Nations Statistics Division, 2013, United Nations commodity trade statistics database (UN Comtrade): United Nations Statistics Division database. (Accessed July 7, 2013, via http://comtrade.un.org/db.)

Vale S.A., 2013a, 4Q12 production report: Rio de Janeiro, Brazil, Vale S.A., February, 15 p.

Vale S.A., 2013b, Delivering value through capital efficiency—Annual report 2012: Vale S.A., 253 p. (Accessed August 22, 2013, at http://www.vale.com/EN/investors/Quarterly-results-reports/20F/20FDocs/20F_2012_i.pdf.)

Vale S.A., 2013c, Key dates for the Grand Sud industrial mining plant: Vale S.A., 1 p. (Accessed September 24, 2013, at http://www.vale.nc/sites/default/files/key_dates_v02.pdf.)

Xstrata plc, 2011, Mineral resources and ore reserves: Zug, Switzerland, Xstrata plc, December, 50 p.

TABLE 1

NEW CALEDONIA: PRODUCTION OF MINERAL COMMODITIES[1]

(Metric tons unless otherwise specified)

Commodity[2]		2008	2009	2010	2011	2012
Cement		139,498 r	140,173 r	161,236 r	147,761 r	123,668
Nickel:						
Ore:						
Gross weight	thousand metric tons	6,172	5,689	8,709	8,835	9,659
Co content		2,110 r	2,000 r	2,850 r	3,240 r	3,500
Ni content		102,700	92,500	129,800	130,700	131,700
Ferronickel:						
Gross weight[e]		123,600 r	126,100 r	131,300 r	132,100 r	142,000
Ni content		37,467	38,230	39,802	40,015	43,030
Nickel matte:						
Gross weight[e]		19,100 r	19,600 r	19,600 r	19,400 r	18,900
Ni content		13,564	13,902	13,917	13,780	13,417

[e]Estimated; estimated data are rounded to no more than three significant digits. [r]Revised.

[1]Table includes data available through October 28, 2013.

[2]In addition to the commodities listed, chromite, copper, crushed stone, gold, iron, manganese, silica sand, and silica are produced, but available information is inadequate to make reliable estimates of output.

TABLE 2
NEW CALEDONIA: STRUCTURE OF THE MINERAL INDUSTRY IN 2012

(Metric tons unless otherwise specified)

Commodity	Major operating companies and major equity owners	Location of main facilities	Annual capacity[e]
Cobalt, in ore and concentrate, Co content	Société Le Nickel (SLN) [Eramet Group, 56%; Société Territoriale Calédonienne de Participation Industrielle (STCPI), 34%; Nisshin Steel Co., 10%]	Kouaoua, Nepoui-Kopeto, Poum, Thio, and Tiebaghi mining centers	3,000
Do.	Vale S.A., 74%; Sumitomo Metal Mining Co. Ltd., 11%; Mitsui & Co. Ltd., 10%; Société de Participation Minière du Sud Calédonien SAS, 5 %	Goro, 58 kilometers east of Noumea in the South Province	4,650
Nickel			
In ore and concentrate, Ni content	Société Le Nickel (SLN) [Eramet Group, 56%; Société Territoriale Calédonienne de Participation Industrielle (STCPI), 34%; Nisshin Steel Co., 10%]	Kouaoua, Nepoui-Kopeto, Poum, Thio, and Tiebaghi mining centers	3,000
Do.	Société Minière du Sud Pacifique, 51%, and POSCO, 49%	SMSP Laterite Operation in the South Province	31,000
Do.	Société des Mines de la Tontouta, 100%	Moneo and Nakety mining centers	6,000
Do.	Other small nickel mining companies, which include Société Minière George Montagnat SA (SMGM)	Tontouta mining center	NA
Do.	Société Le Nickel (SLN) [Eramet Group, 56%; Société Territoriale Calédonienne de Participation Industrielle (STCPI), 34%; Nisshin Steel Co., 10%]	Goro, 58 kilometers east of Noumea in the South Province	60,000
Do.	Vale S.A., 74%; Sumitomo Metal Mining Co. Ltd., 11%; Mitsui & Co. Ltd., 10%; Société de Participation Minière du Sud Calédonien SAS, 5%	Goro, 58 kilometers east of Noumea in the South Province	1,000
In ferronickel, Ni content	Société Le Nickel (SLN) [Eramet Group, 56%; Société Territoriale Calédonienne de Participation Industrielle (STCPI), 34%; Nisshin Steel Co., 10%]	Doniambo, Noumea	60,000
In nickel matte, Ni content	do.	do.	15,000

[e]Estimated. Do., do. Ditto. NA Not available.

The Mineral Industry of New Zealand

By Pui-Kwan Tse

The economy of New Zealand continued to grow at a modest rate in 2012, and the real gross domestic product (GDP) increased by 2.7%. The agriculture, construction, and trade sectors were the main contributors to the economic growth. The Canterbury sequence of earthquakes in late 2010 and early 2011 caused substantial physical damage to the city of Christchurch, and the cost of reconstruction was estimated to be $30 billion. Reconstruction activities started slowly in late 2011 and continued in 2012; reconstruction of the city's infrastructure was expected to continue during the next several years. Residential and nonresidential construction increased in 2012 after a decrease in 2011 (Reserve Bank of New Zealand, 2013, p. 3–6; Statistics New Zealand, 2013a).

The output of the mineral industry of New Zealand was small compared with that of its neighboring country Australia. New Zealand has metallic mineral occurrences of antimony, bauxite, beryllium, chromium, copper, gallium, gold, iron, lead, lithium, magnesite, manganese, mercury, molybdenum, nickel, platinum-group metals, rare earths, silver, tin, titanium, tungsten, uranium, and zinc. Of these metallic minerals, only gold, iron, and silver were produced. Bentonite, clay, coal, diatomite, dolomite, limestone, perlite, phosphate rock, pumice, salt, silica, building and dimension stone, sulfur, and zeolites have also been discovered in the country (table 1).

New Zealand's total goods trade was NZ$93.3 billion (US$74.6 billion) in 2012. Exports were valued at NZ$46.1 billion (US$36.9 billion), which was a decrease of 3.4% from the value in 2011. Imports were valued at NZ$47.2 billion (US$37.8 billion), which was an increase of 0.7% from the value in 2011. Australia continued to be New Zealand's leading export destination, accounting for 21.5% of total exports. China remained New Zealand's second-ranked export market, receiving 15.0% of New Zealand's exports, followed by the United States, 9.2%; Japan, 7.0%; and the Republic of Korea, 3.4%. China continued to be New Zealand's leading source of imports, supplying 16.3% of New Zealand's imports, followed by Australia, 15.2%; the United States, 9.3%; Japan, 6.3%; and the Republic of Korea, 3.8%. Agricultural products were New Zealand's leading export commodity and accounted for more than 50% of total exports. Mineral fuels were New Zealand's leading nonagricultural commodity and accounted for 4.8% of the country's total export value; aluminum and its products accounted for 1.8%. Crude oil and oil products were New Zealand's most valuable imported commodities, accounting for 17.7% of the country's total import value. Iron and steel products were the most valuable metallic imports and they accounted for 2.0% of the country's total import value. Because it had no alumina refinery, New Zealand depended on imported alumina from Australia for its aluminum production (Statistics New Zealand, 2013b, p. 12, p. 82–87).

Government Policies and Programs

The laws that were in effect in 2012—The Crown Minerals Act 1991 (which was amended by the Crown Minerals Amendment Act 2013) and the Crown Minerals Amendment Act 2003—set forth the broad legislative framework for the prospecting for, exploring for, and mining of Crown-owned (meaning Government-owned on behalf of all New Zealanders) minerals within New Zealand's territorial area, which extends to 12 nautical miles off the New Zealand coast. The Ministry of Economic Development, through the Crown Minerals Group, is responsible for the overall management of all state-owned minerals in New Zealand. Crown-owned minerals include gold, petroleum, silver, uranium, and all minerals on or under Crown-owned land. In some cases, the Government also has rights to certain minerals on some private land. The Crown Minerals Group also advises on policy and regulations and promotes investment in the mineral sector. The royalty regimes for coal, nonfuel minerals, and petroleum are defined in the Government mineral program that is reviewed every 10 years.

In 2009, the Government announced that it would review the legislative, regulatory, royalty, and taxation arrangements for nonfuel minerals and petroleum. The changes that the Government proposed to consider would allow more flexibility on permit duration (to deal with operating challenges, such as the limit of 5 years for an exploration permit), set up a new permit class, and ensure that the regime is able to include new technologies and resources. The Government also would evaluate Schedule 4 of the Crown Minerals Act 1991, which restricts mineral-related activity in specified public conservation areas. Schedule 4 lands accounted for about 40% of public conservation land, or 13% of New Zealand's total land area.

The Government's review process was completed in 2010 and a proposed bill to revise the Crown Minerals Act 1991 was drafted. Under the draft bill, the Government would maintain the existing Schedule 4 areas. The Government and the Regional Council would perform joint technical studies on mineral prospective areas on the North Island and the South Island. The Ministry of Energy and Resources and the land-holding minister would approve jointly the mineral-related access to Crown land based on the economic, mineral, and national significance of the proposal. The bill would introduce a 2-tiered system for permit management. The Tier 1 permits relate to gold [other than alluvial gold, unless the royalty payment in the fifth and subsequent permit years exceeded NZ$50,000 (US$40,000)] and petroleum. The permits for coal, iron sand, silver, and other metallic minerals depend on whether or not specified royalty thresholds are reached. The Tier 1 permits require a hands-on coordination management and regulatory regime. The Tier 2 permits are for industrial rocks, and for small business and hobby mineral operations. The proposed amended bill would improve coordination between the Crown Minerals permitting regime and health and safety and environmental regulatory functions for Tier 1 activities. The Crown Minerals Amendment Act 2013, which amends the Crown Minerals Act 1991, was enacted on May 24, 2013 (Parliament, The, 2013, p. 9–30).

Minerals in the National Economy

New Zealand's mineral resources were dominated by aggregate and gold, which together accounted for 80% of the total value of New Zealand's mineral resources. Gold, iron sand, and silver were major metallic commodities that made a notable contribution to New Zealand's economy. Production of other metallic minerals, such as bauxite, copper, lead, and zinc, could potentially be economically feasible if technologies and prices become favorable. Excluding the petroleum industry, the value of New Zealand's mineral sector accounted for less than 1% of the GDP. The total value of New Zealand's minerals and mineral fuel production accounted for about 2% of the GDP (Statistics New Zealand, 2013c, p. 2).

Production

Production of such mineral commodities as diatomaceous earth, dimension stone, dolomite, and perlite increased by more than 10% compared with that of 2011. Mineral commodities for which production decreased significantly included serpentine, silica sand, silver, and zeolite. Data on mineral production are in table 1.

Structure of the Mineral Industry

Table 2 is a list of major mineral industry facilities in New Zealand.

Commodity Review

Metals

Aluminum.—New Zealand Aluminium Smelters Ltd. was the sole primary aluminum producer in New Zealand. In 2011, New Zealand Aluminium completed the replacement of all transformers at its reduction line at the Tiwai Point smelter to increase the power delivery level. Owing to unfavorable market conditions, the company decided to shut down reduction line No. 4 in April. As a result, aluminum production decreased to 326,963 metric tons (t) in 2012 from 354,029 t in 2011. The company signed an 18-year electricity supply contract with state-owned Meridian Energy Ltd. in 2007 (before the global financial crisis that began in 2008), which was to come into force in 2013. The demand for aluminum decreased and price of aluminum also decreased during the past several years. As a result, New Zealand Aluminium faced a financial loss of $50 million in 2012. The increase in electricity prices under the new contract placed a significant burden on the production costs of the smelter. Pacific Aluminium of Australia (a major shareholder of New Zealand Aluminium) held discussions with Meridian Energy to reduce the price of electricity supplied to the smelter in 2012. No agreement had been reached at yearend 2012. The smelter employed 750 people and accounted for an estimated 10.5% of the economy of the Southland region of the South Island. If no new agreement is reached, Pacific Aluminium might consider shutting down the smelter in New Zealand (Stuff.co.nz, 2013).

Gold.—New Zealand's gold production was mainly from the Waihi area in the North Island and from the Otago region and along the west coast on the South Island. Hard rock gold mines were mined by Newmont Mining Corp. of the United States on the North Island and OceanaGold Corp. of Australia on the South Island. Newmont Waihi Gold, which was a subsidiary of Newmont Mining, mined the Favona, the Martha, and the Trio Mines in and around Waihi. The mine at Favona had been scheduled to close in 2011; however, the operation continued in 2012. Newmont Waihi Gold had received approval from the Government to mine the Trio deposit at the end of 2010. Construction of two development drifts that were 510 meters (m) and 790 m in length, respectively, began in 2010. The Trio ore bodies were situated between the Martha and the Favona Mines and would be accessed from the Favona portal. Waihi Gold Co. Ltd. (trading as Newmont Waihi Gold) planned to mine at the deepest level (350 m) and to work up towards the surface. Ore production at Trio began in mid-2012, and construction at Trio was scheduled to be completed in 2014. Once completed, the mine was projected to produce about 1 million metric tons per year (Mt/yr) of ore containing 6.2 t (200,000 troy ounces) of gold at an average grade of 6 to 7 grams per metric ton gold (Newmont Mining Corp., 2012, p. 2–10).

In 2011, Newmont Waihi Gold introduced a new underground exploration project in Waihi East, the "Golden Link" project, which was composed of the Correnso exploration project and the Martha exploration project. The Correnso project had the potential to be an underground mine that would replace the existing underground Favona and Trio Mines. As proposed, mining at Correnso would take place at a depth of 350 m, which is considerably deeper than the other mines in Waihi East. Mining at Correnso would start at the bottom of the ore body and progress up to the top of ore body, which would be about 130 m below the surface. The Golden Link project would extend Newmont Waihi Gold's mining operation to 2020. Newmont Waihi Gold submitted the Golden Link project for Government approval in 2012 (Waihi Gold Co. Ltd., 2012, p. 1–5).

Owing to a fire accident in the Trio gold mine in July and a shutdown of the mine for several days after the accident, gold production in the area of Waihi decreased in 2012 by about 37% to 1,864 kilograms. The production of silver in the area of Waihi also decreased significantly in 2012. The accident was caused by a truck engine fire and resulted in 28 workers being trapped in the underground mine (Ministry of Economic Development, 2013b).

Iron Ore and Iron and Steel.—New Zealand's iron ore deposits are iron sands, which are placer deposits formed from the erosion of andesitic and rhyolitic volcanic rocks. These iron sands occur in onshore dunes and beaches and in offshore marine sands along the coastline from Kaipara Harbor south to Wanganui on the west coast of the North Island. Iron sand concentrate from Taharoa, which contained about 57% iron, was exported to other countries in the Asia and the Pacific region. Iron sand from the Waikato North Head site was pumped to the Glenbrook steel plant of New Zealand Steel Ltd. (a subsidiary of BlueScope Steel Ltd. of Australia) by way of an

18-kilometer (km)-long underground pipe. The Glenbrook steel plant, which was the sole integrated steel producer in the country, had an output capacity of 650,000 metric tons per year (t/yr).

The Government granted Trans-Tasman Resources Ltd. an exploration license to explore for iron ore deposits off the west coast of the North Island from the Waikato River in the north to the Rangitikei River in the south. Iron sand in the area was vanadium-bearing titanomagnetite. The company submitted a mining permit application to the Government for extracting iron sand in the South Taranaki Bight. The application covered an area of 65.76 square kilometers within the existing prospecting license zone (Trans-Tasman Resources Ltd., 2013).

Industrial Minerals

Cement.—New Zealand's cement industry was dominated by two producers—Golden Bay Cement on the South Island and Holcim New Zealand Ltd., which was a subsidiary of Holcim Ltd. of Switzerland, on the North Island. The two companies had a combined output capacity of 1.4 Mt/yr; however, domestic cement demand was about 1.43 Mt/yr. Holcim planned to build a 2-Mt/yr plant at Weston (near Oamatu) to replace the existing wet kilns cement plant. Owing to the uncertainty of the international financial situation, Holcim postponed a decision on building a new plant until 2013. The new $200 million plant would increase production capacity and reduce carbon dioxide emissions by 25% below its 1990 benchmark by 2015 (Global Cement, 2013).

Mineral Fuels

Coal.—New Zealand's coal resources were estimated to be 15 billion metric tons (Gt), of which about 8.6 Gt was economically recoverable. Coal accounted for about 4% of the country's total energy consumption. Bituminous coal resources are located in the West Coast region of the South Island; subbituminous coal resources are found mainly in the Waikato region of the North Island, as well as in the Otago, the Southland, and the West Coast regions of the South Island. Lignite resources are found in the Otago and the Southland regions of the South Island. The South Island lignite deposits accounted for 80% of the country's coal resources.

Pike River Coal Ltd. completed the construction of its Pike River Mine, which is located about 50 km northeast of Greymouth on the west coast of the South Island, in 2010. The company planned to produce about 800,000 t/yr of coking coal for 18 years. In November 2010, an explosion in the mine killed 29 people; the company was unable to continue normal operations thereafter and went into receivership. PricewaterhouseCoopers International Ltd. was appointed as the receiver under the terms of a General Security Deed dated May 21, 2010, and planned to put the assets of Pike River Coal up for sale. The sale of Pike River Coal to state-owned Solid Energy New Zealand Ltd. was completed in 2012. Solid Energy planned to secure the mine site and review the exploration plan to determine the coal resource. It might take several years before the company is able to determine whether the mine could be operated commercially.

The Government accepted the Royal Commission's recommendations on addressing systemic failures in the country's health and safety regulatory regime. The implementation of the recommendations would be in place by the end of 2013 (Berry, 2012; Ministry of Business, Innovation and Employment, 2012).

Bathurst Resources Ltd. had two coal operating mines on the South Island—the Cascade Mine, which is located near Westport, and the Takitimu Mine, which is located at Nightcaps in the Southland region. The Cascade open pit mine, which was part of the Buller coal projects, was operated by Bathurst Resources' subsidiary, Buller Coal Ltd.; the mine produced high-quality, low-sulfur coking coal. Bathurst Resources acquired the Takitimu coal mine in 2011 when it acquired the assets of Eastern Resources Group. The two mines (Cascade and Takitimu) had a total combined output capacity of 350,000 t/yr. In 2011, the Government granted Buller Coal the right to develop the Escarpment Block, which is located next to the Cascade Mine. The mine was estimated to contain about 3 Mt of high-quality coking coal and had an estimated life of 5 years. Three local environmental groups appealed to the Environment Court to block the development of this coal mine at Denniston Plateau in 2012. The Environment Court would announce the decision to permit coal mining at Denniston Plateau in 2013 (Bathurst Resources Ltd., 2013, p. 3–6).

Natural Gas and Oil.—New Zealand's natural gas and oil were produced from 19 fields, all of which are located in the Taranaki basin. In 2012, New Zealand's production of natural gas increased by about 14%, whereas production of oil decreased by about 15% compared with that of 2011. Natural gas production increased at the Mangahewa and the Maui fields, and the owner, Todd Energy Ltd., invested $800 million to drill five wells to expand the output capacity at the Mangahewa field. Natural gas production in New Zealand was expected to increase during the next several years.

New Zealand was a net importer of oil. Nearly all domestically produced New Zealand oil was exported because New Zealand crude oil was low density, had low sulfur content, and attracted a premium price on the international market. Cheaper foreign oil was imported to refine at the Marsden Point refinery. About 59% of the imported oil was from Middle Eastern countries, and about 28% was from Asian countries, mainly Brunei and Indonesia. The Government extended the tax exemption for exploration companies until December 31, 2014, to encourage exploration for offshore hydrocarbons in New Zealand territory (Ministry of Economic Development, 2013a).

Outlook

Most mineral production in New Zealand is consumed locally, with the exception of aluminum, coal, gold, and amorphous silica. Coal and gold are the leading exported mineral commodities. Under the Crown Minerals Amendment Act 2013, some mineral exploration restrictions have been redefined on public areas where the mineral potential is significant and mineral production could contribute significantly to the economy of New Zealand. The development of the mining sector in New Zealand, however, is constrained by the population's concerns about the environmental issues related to mining, the ecological sensitivity of the country, and New Zealand's location far from major industrial markets. Consistent with these trends, New Zealand's mineral development is expected to continue to increase only gradually.

References Cited

Bathurst Resources Ltd., 2013, Quarterly activities and cash flow report 31 December 2012: Wellington, New Zealand, Bathurst Resources Ltd., January 31, 12 p.

Berry, Michael, 2012, Pike River Coal sale completed: Stuff.co.nz, July 17. (Accessed October 4, 2012, at http://www.stuff.co.nz/business/industries/7294998/Pike-River-Coal-salw-completed.)

Global Cement, 2013, Weston uncertainty ends in New Zealand: Global Cement, August 7. (Accessed October 25, 2013, at http://globalcement.com/news/item/1859-weston-tension-ends-in-new-zealand?tmp.)

Ministry of Business, Innovation and Employment, 2012, Pike River implementation plan: Wellington, New Zealand, Ministry of Business, Innovation and Employment, December, 13 p.

Ministry of Economic Development, 2013a, New Zealand energy quarterly: Wellington, New Zealand, Ministry of Economic Development, 8 p.

Ministry of Economic Development, 2013b, New Zealand metal production: Wellington, New Zealand, Ministry of Economic Development, 1 p.

Newmont Mining Corp., 2012, Newmont Waihi's gold: Denver, Colorado, Newmont Mining Corp., 28 p.

Parliament, The, 2013, The Crown Minerals Amendment Act 2013: Wellington, New Zealand, The Parliament, May 24 reprint, 117 p.

Reserve Bank of New Zealand, 2013, Monetary policy statement: Wellington, New Zealand, Reserve Bank of New Zealand, September, 35 p.

Statistics New Zealand, 2013a, A good year for our economy: Statistics New Zealand. (Accessed September 30, 2013, at http://www.stats.govt.nz/browse_for_stats/snapshots-of-nz/yearbook/economy/national/gross-domestic.aspx.)

Statistics New Zealand, 2013b, Analysis of NZ merchandise trade—Year to December 2012: Wellington, New Zealand, Statistics New Zealand, 208 p.

Statistics New Zealand, 2013c, Gross domestic product—December 2012 quarter: Wellington, New Zealand, Statistics New Zealand, 22 p.

Stuff.co.nz, 2013, Smelter still in talks over power price deal: Stuff.co.nz, April 13, 2013. (Accessed April 24, 2013, at http://stuff.co.nz/southland-times/business/8545556/Smelter-still-in-talks-over-power-price-deal.)

Trans-Tasman Resources Ltd., 2013, Trans-Tasman Resources submits mining permit application: Wellington, New Zealand, Trans-Tasman Resources Ltd. press release, July 26, 1 p.

Waihi Gold Co. Ltd., 2012, Golden Link project including the Correnso underground—Application to HDC for resource consent and AEE: Waihi, New Zealand, Waihi Gold Co. Ltd., June, 86 p.

TABLE 1
NEW ZEALAND: PRODUCTION OF MINERAL COMMODITIES[1]

(Metric tons unless otherwise specified)

Commodity		2008	2009	2010	2011	2012
METALS						
Aluminum metal, smelter, primary		315,500	271,902	343,335	354,029	326,963
Gold, mine output, Au content	kilograms	13,403	13,442	13,494	11,761 r	10,164
Iron and steel:						
Iron sand, titaniferous magnetite, gross weight	thousand metric tons	2,020	2,092	2,439	2,357	2,395
Pig iron[e]	do.	622	608	667	659	669
Steel, crude[e]	do.	799	765	853	844	912
Lead, refinery output, secondary[e]		9,000	13,000	9,000	9,000	9,000
Silver, mine output, Ag content	kilograms	18,269	14,264	17,136	14,324	5,629
INDUSTRIAL MINERALS						
Cement, hydraulic[e]	thousand metric tons	1,200	1,200	1,100	1,200	1,200
Clays:						
Bentonite		753	880	1,216	--	--
Kaolin, pottery		12,761	9,016	10,700	21,545	11,578
For brick and tile		34,650	40,740	30,192	10,911	71,487
Diatomaceous earth		14	10	95	--	--
Lime[e]		180,000 r	175,000 r	170,000 r	175,000 r	175,000
Marble[e]		15,000	15,000	14,000	14,000	14,000
Nitrogen, N content of ammonia[e]		125,000	125,000	120,000	120,000	120,000
Perlite		--	8,848	5,088	--	3,598
Pumice		174,729	159,357	118,249	229,268	72,414
Salt[e]		100,000	100,000	95,000	95,000	95,000
Sand and gravel:						
Silica sand, glass sand		48,575	43,458	113,231	109,346	73,064
Other industrial sand		1,160,543	1,453,793	1,726,236	1,203,103	1,517,308
For roads and ballast	thousand metric tons	20,889	15,471	13,257	15,258 r	15,439
For building aggregate	do.	9,743	8,064	7,528	6,183 r	6,561
Stone:						
Dolomite		16,962	52,000	86,399	59,782	86,040
Limestone and marl:						
For agriculture	thousand metric tons	1,918	2,020	1,686	1,387 r	1,020
For cement	do.	2,018	1,888	1,800	1,705 r	1,797
For other industrial uses	do.	874	664	1,054	185	319
Serpentine		4,494	14,197	43	41,201 r	36,731
Dimension		16,998	17,795	18,911	140	8,614
Zeolites		25,800	21,750	--	3,523 r	--
MINERAL FUELS AND RELATED MATERIALS						
Coal, all grades	thousand metric tons	4,909	4,563	5,330	4,944	4,926
Liquefied petroleum gas[e]	thousand 42-gallon barrels	979 [2]	857 [2]	1,200	1,200	1,200
Natural gas:						
Gross production	million cubic meters	4,484	4,644	5,052	4,678	5,188
Marketed production	do.	3,994	4,097	4,432	4,003	4,559
Petroleum:						
Crude	thousand 42-gallon barrels	21,436	20,026	19,302	16,591	14,149
Refinery products[e]	do.	34,000	35,000	34,000	33,000	39,000

[e]Estimated; estimated data are rounded to no more than three significant digits; may not add to totals shown. [r]Revised. do. Ditto. -- Zero.

[1]Table includes data available through August 10, 2013.

[2]Reported figure.

TABLE 2
NEW ZEALAND: STRUCTURE OF THE MINERAL INDUSTRY IN 2012

(Thousand metric tons unless otherwise specified)

Commodity		Facilities, major operating companies, and major equity owners	Location of main facilities	Annual capacity[e]
Aluminum		Tiwai Point smelter [New Zealand Aluminium Smelters Ltd. (Pacific Aluminium, 79.36%, and Sumitomo Chemical Co., 20.64%)]	Southland, Invercargill	350
Cement		Golden Bay Cement (Fletcher Building Ltd.)	Portland	900
Do.		Holcim New Zealand Ltd.	Cape Foulwind, Westport	500
Coal		Stockton open pit mine (Solid Energy New Zealand Ltd., 51%, and Cargill Inc., 49%)	Buller, 35 kilometers northeast of Westport	2,500
Do.		Pike River underground mine (Pike River Coal Ltd.)	50 kilometers northeast of Greymouth	1,000
Do.		Spring Creek underground mine (Solid Energy New Zealand Ltd.)	Greymouth	1,000
Do.		Rotowaro open pit mine (Solid Energy New Zealand Ltd.)	Huntly	1,500
Do.		Huntly East underground mine (Solid Energy New Zealand Ltd.)	do.	500
Do.		New Vale open pit mine (Solid Energy New Zealand Ltd.)	50 kilometers northeast of Invercargill	300
Do.		Ohai open pit mine (Solid Energy New Zealand Ltd.)	Ohai	200
Do.		Terrace underground mine (Solid Energy New Zealand Ltd.)	Reefton	100
Gold	metric tons	Newmont Waihi Gold (subsidiary of Newmont Mining Corp.)	Waihi	5
Do.	do.	Macraes gold project (OceanaGold Corp.)	Otago	6
Do.	do.	Reefton gold project (OceanaGold Corp.)	Reefton	10
Iron and steel:				
Iron ore		New Zealand Steel Ltd. (BlueScope Steel Ltd. of Australia)	Taharoa, 150 kilometers south of Auckland	1,300
Do.		do.	Waikato North Head, 30 kilometers south of Auckland	1,000
Steel		do.	Glenbrook	650
Do.		Otahuhu Mill [Pacific Steel Group (Fletcher Building Ltd.)]	Auckland	300
Kaolin		Imerys Tableware New Zealand Ltd.	80 kilometers northwest of Whangarei	25
Petroleum, refinery	barrels per day	Marsden Point Oil Refinery (New Zealand Refinery Co., operator)	Marsden Point	95,000
Salt		Dominion Salt Ltd.	South of Blenheim	70
Silver	metric tons	Newmont Waihi Gold (Newmont Mining Corp.)	Waihi	30
Do.	do.	OceanaGold Corp.	Otago	1

[e]Estimated. Do., do. Ditto.

The Mineral Industry of Pakistan

By Chin S. Kuo

Pakistan is rich in such mineral resources as barite, coal, copper, iron ore, limestone, and salt, and the identified resources of copper and iron ore are large. The country also produced a variety of industrial minerals and some other metallic minerals. Pakistan also has extensive energy resources and is known to have moderate oil reserves, sizable gas reserves, large coal resources, and large hydropower potential. In the past several years, gas and oil production met only about one-half of the country's energy needs, but exploitation of energy resources continued to be slow owing to a shortage of capital and to political instability. In particular, the supply of natural gas from domestic sources was expected to decline in the next 2 to 3 years, as the existing fields were nearing depletion. The Government was reviewing its gas development policy and is encouraging foreign companies to participate in gas production.

The Competition Commission of Pakistan (CCP) conducted searches and inspections at the offices of the All Pakistan Cement Manufacturers Association (APCMA) and the facilities of Kohat Cement Co. Ltd. in Lahore under Section 34 of the Competition Act 2010 to look for proof of suspected cartelization in the cement sector. It was found that the cement producers had devised a vigilance plan by which the cement dispatches were monitored. If the CCP finds the APCMA and Kohat Cement guilty of collusion, it could impose a penalty of nearly $700 million on the APCMA and its members. The president of the APCMA was also the chief executive of Kohat Cement (Global Cement News, 2012a).

Minerals in the National Economy

In 2012, Pakistan's economy was dominated by the services, industrial, and agriculture sectors, which accounted for 53%, 25%, and 22% of the gross domestic product (GDP), respectively. Industrial output increased by 11%, and production from mining and quarrying accounted for 15% of industrial production and increased by 20% compared with that of 2011. The value of output from the mineral industry accounted for 3.3% of the GDP, which posted a growth rate of 11% in 2012. Cement and jewelry were the country's major export items in 2012. Aluminum, crude petroleum, iron and steel, and petroleum products were the major import commodities (State Bank of Pakistan, 2013, p. 238).

Production

In 2012, the metallic minerals mined in Pakistan included bauxite, chromite, copper, iron ore, lead, and zinc. Pakistan produced a variety of industrial minerals, including aragonite, barite, clays, dolomite, gypsum, limestone, and salt. Output of iron ore decreased by 12% compared with that of 2011. Production of pig iron increased significantly after a several years' decline, probably owing to better capacity utilization. The production pattern for fuller's earth was similar to that of pig iron. The output of secondary lead fluctuated from year to year, which might have to do with the scrap availability. Production of kaolin also tended to fluctuate from year to year. Steady increases followed by a modest drop in 2012 were noted in the production of feldspar, gypsum, magnesite, marine salt, and phosphate rock (table 1).

Structure of the Mineral Industry

State-owned companies controlled the production and marketing of chromite, coal, copper, iron ore, and steel. Private-sector companies were allowed to own and produce nonfuel minerals—mainly industrial minerals, including cement. Despite the Government's efforts to privatize large-scale state-owned companies, the public sector companies continued to account for a significant percentage of the country's mineral production.

The Mineral Department of the Ministry of Petroleum and Natural Resources is responsible for the exploration, planning, development, and operation of mining ventures that are controlled by the state-owned companies. The Ministry's Petroleum Department is responsible for the exploration and production of hydrocarbons and for the transmission and distribution of natural gas. Table 2 is a list of major mineral-producing facilities in the country.

Commodity Review

Metals

Copper and Gold.—The mining lease application case of 2011 between Tethyan Copper Co. (TCC) (a joint venture between Barrick Gold Corp. of Canada and Antofagasta plc of Chile) and the Balochistan Provincial government was under review by the International Court of Arbitration. If the court decides the case in favor of TCC, then the government of Balochistan would be obligated to pay for its investments in exploration and construction of the Reko Diq gold and copper mine project. TCC held a 75% interest in the project and the Provincial government had a 25% stake, which was a disputed point. Mine production was initially scheduled to begin in 2014, and the mine was estimated to have reserves of 11.65 million metric tons (Mt) of contained copper and 659,000 kilograms of contained gold. The dispute regarding the project included two issues—(a) the Provincial government's refusal to take financial responsibility for its 25% stake in the project, and (b) the purported involvement of Metallurgical Construction Corp. (MCC) of China in the project. The Government proposed that the Provincial government re-negotiate with TCC to find an amicable settlement to the dispute (Bhutta, 2012).

ME Resource Corp. of Canada entered into an agreement to acquire a 70% interest in two exploration licenses in the Chagi mineral belt in Balochistan Province that are contiguous to the Reko Diq copper-gold deposit, which had resources of 5,900 Mt at grades of 0.4% copper and 0.22 gram per metric ton gold. The company's objective was to explore and develop the mineral claims and pursue additional acquisitions (ME Resource Corp., 2012).

Iron and Steel.—Tuwairqi Steel Mills planned to commission its 1.28-million-metric-ton-per-year (Mt/yr) direct-reduced iron (DRI) plant at Port Qasim in Karachi in July. A wastewater treatment plant and a powerplant were completed in early 2012. Kobe Steel of Japan's Midrex unit supplied the DRI plant, which required iron ore pellets of a minimum of 56.5% iron content that Tuwairqi sourced from Bahrain and Oman. Output of DRI would be sold to the domestic market, as well as to India and Malaysia. An electric arc furnace-based billet plant and an iron ore mine were also planned. A new 2-Mt/yr steel plant was expected to be constructed in 2015. Pakistan's steel consumption of 7 Mt/yr was more than its production capacity of between 4 and 5 Mt/yr (ArabSteel, 2012).

Industrial Minerals

Cement.—Cement consumption in Pakistan had stagnated, and cement producers operated at 73% of installed capacity. The total installed capacity for cement production was 45 Mt/yr. The country's cement demand was estimated to be about 31 Mt, 70% of which was for domestic consumption and 30% of which was exports that went mostly to Afghanistan and India. Cement exports to India had decreased steadily in recent years to 590,104 t in fiscal year 2011 from 786,672 t in fiscal year 2008. They went through India's Gujarat Port to benefit the southern States. Although Pakistan had granted "most favored nation" status to India, Pakistan's cement exports were not likely to increase owing to nontariff barriers to trade, which the Government of India said were uniform for all countries and not specific to Pakistan. The volume and price of Pakistan's cement exports to Afghanistan, however, were expected to increase in 2012 (Aggregate Research, 2012).

Pakistan's cement exports to Afghanistan represented 50% of the total cement exports, by value, from Pakistan because the majority of the cement was shipped from companies located close to the border between the countries. Pakistan-supplied cement dominated cement consumption in the central and northern regions of Afghanistan where major reconstruction activities were underway. The Pakistani cement companies with the most exposure to the Afghan market included Bestway Cement Co. Ltd., Cherat Cement Co. Ltd., DG Khan Cement Co. Ltd., Fauji Cement Co. Ltd., Lafarge Pakistan Cement Ltd., and Lucky Cement Ltd. (Global Cement News, 2012b).

Arif Habib Corp. sold about a 61% share in Thatta Cement Co. Ltd. (one of Pakistan's smallest cement producers) to a consortium of four companies. Sky Pak Holding and Al-Miftah Holdings each bought 22.7% of the shares and Golden Global Holding and Rising Star Holding acquired 8.6% and 7%, respectively. Thatta Cement, which is located at Makli in the District of Thatta in Sindh Province, had the capacity to produce 450,000 metric tons per year (t/yr) of cement. Arif Habib also owned a 75% stake in Al-Abbas Cement Industries Ltd., which would have an installed capacity of 900,000 t/yr after an expansion of the plant was completed in 2013 (Express Tribune, The, 2013).

To expand its plant, Al-Abbas Cement planned to invest in new machinery and increase its capacity utilization to become one of the most efficient cement producers in Pakistan. The company previously had a small production capacity of 750,000 t/yr compared with the Nation's installed capacity of 45 Mt/yr. Current capacity utilization rates were between 80% and 85%. The company invested in a vertical cement-grinding mill, which increased the plant's capacity by 20% to 900,000 t/yr. Al-Abbas Cement received 30% of its revenues from cement exports but planned to focus on supplying the surging demand in the domestic market that was the result of the Government's increased spending on construction and development projects (Zaheer, 2012).

Fauji Cement planned to acquire Askari Cement Co. Ltd., which had a 1.1-Mt/yr-capacity cement plant at Wah and a 1.6-Mt/yr-capacity cement plant at Nizampur. Askari Cement's capacity utilization rate was only 45%. If the deal moves forward, Fauji Cement would become the second-ranked cement producer in Pakistan, and its cement production capacity would increase to 6.1 Mt/yr from 3.4 Mt/yr (International Cement Research, 2012).

Talc.—Microcrystalline talc associated with magnesite and dolomite is found at Sherwan, which is located west of Abbottabad. Mines in the four areas, including Bandi Sadique and three others where talc was mined, had a total output capacity of 165,000 t/yr. CapriCorn Minerals produced 20,000 t/yr from its Bandi Sadique deposit. The company operated a 5,000-t/yr plant in Lahore to process high-quality white pure talc, all of which was exported (Industrial Minerals, 2012).

Mineral Fuels

Coal.—Oracle Coalfields plc of the United Kingdom, which explored for and developed a lignite property located in Sindh Province, performed a feasibility study on Block VI of the Thar coalfield. The exploration license was for a 66.1-square-kilometer (km^2) area that was held by Oracle Coalfields' 80%-owned subsidiary Sindh Carbon Energy Ltd. Coal resources (measured and indicated) in the 20-km^2 mining area were estimated to be 459 Mt (wet basis) with ash content of 5.89% and sulfur content of 0.91%, and were suitable for power generation. Inferred resources for the license area were 70 Mt. The total capital expenditure for mine development was estimated to be $610 million. Lignite production of 5 Mt/yr by the opencast method would extend the mine life to 23 years (Oracle Coalfields plc, 2012).

Outlook

Full production of copper and gold from the mining operation at Reko Diq is expected to be delayed until after 2014 because of the dispute involving the government of Balochistan Province and MCC. Pakistan's cement industry is expected to increase its capacity utilization rate to nearly 90% to meet the increased cement demand as rising domestic cement prices reflect a tightening market in the near future. Cement production is expected to increase at an average of 3.2% per year. Cement demand from Afghanistan and India is also expected to be on the rise. Development and mining of coal resources, including lignite, in the Thar District in Sindh Province is expected to proceed as planned.

References Cited

Aggregate Research, 2012, MFN won't help spur cement exports: Aggregate Research, April 17. (Accessed April 23, 2012, at http://www.aggregateresearch.com/article.aspx?id=24912.)

ArabSteel, 2012, Tuwairqi Steel Mills to commission DRI in Q2 2012: ArabSteel, January 2. (Accessed February 3, 2012, at http://www.arabsteel.info/total/long_news_total_e.esp?id=1035.)

Bhutta, Zafar, 2012, Centre distances itself from Reko Diq dispute: The Express Tribune [Karachi, Pakistan], April 8. (Accessed April 16, 2012, at http://tribune.com.pk/story/361384/centre-distances-itself-from-reko-diq-dispute.)

Express Tribune, The, 2013, Arif Habib Corp. agrees to sell Thatta Cement: The Express Tribune [Karachi, Pakistan], March 8. (Accessed March 11, 2013, at http://tribune.com.pk/story/51740/mergers-and-acquisitions-arif-habib-corporation-agrees-to-sell-thatta-cement.)

Global Cement News, 2012a, CCP inspects APCMA over cartelization claims: Global Cement News, January 18. (Accessed January 18, 2012, at http://www.propubs.com/resources/global-cement/news/itemlist/tag/GCW32?utm.)

Global Cement News, 2012b, Pakistani export price to Afghanistan rises by 25%: Global Cement News, January 27. (Accessed February 13, 2012, at http://www.globalcement.com/news/itemlist/tag/gcw34?utm_source=newsletter&utm_medium=email&utm_campaign+gcw34.)

Industrial Minerals, 2012, The emergence of Afghanistan as a significant talc supplier: Industrial Minerals, no. 540, September, p. 60.

International Cement Research, 2012, Fauji Cement seeks Askari merger: International Cement Research, April, p. 9.

ME Resource Corp., 2012, ME Resource Corp. announces acquisition of mineral licenses in the prolific Tethyan belt of Pakistan: Vancouver, British Columbia, Canada, ME Resource Corp. press release, January 21, 1 p.

Oracle Coalfields plc, 2012, Technical feasibility study: Oracle Coalfields plc, February 6. (Accessed February 6, 2012, at http://www.londonstockexchange.com/exchange/news/market-news-detail.html?announcementid=11106161.)

State Bank of Pakistan, 2013, Statistical bulletin: State Bank of Pakistan, June, 260 p.

Zaheer, Farhan, 2012, Timely investments bring Al-Abbas back to profitability: Aggregate Research, November 19. (Accessed November 21, 2012, at http://www.aggregateresearch.com/print.aspx?ID=26622.)

TABLE 1
PAKISTAN: PRODUCTION OF MINERAL COMMODITIES[1]

(Metric tons unless otherwise specified)

Commodity		2008	2009	2010	2011	2012[e]
METALS						
Bauxite, gross weight		25,000 [e]	11,300	9,576	12,997 [r]	12,000
Chromium ore:						
Gross weight		104,000	133,000	252,000	240,000 [r]	260,000
Cr_2O_3 content		46,800	59,900	113,400	112,000	114,000
Copper, mine output, Cu content[e]		18,700	18,500	18,000	19,000	19,000
Gold, mine output, Au content[e]	kilograms	1,600	1,600	1,600	1,600	1,600
Iron and steel:[e]						
Iron ore, gross weight	thousand metric tons	250	333 [2]	418 [2]	430 [r, 2]	380
Pig iron	do.	1,000	700	483 [2]	232 [r, 2]	450
Steel, crude	do.	1,100	1,100	1,100	1,200	1,200
Lead:[e]						
Pb content in concentrate		--	26,000	26,000	27,000	27,000
Refined, secondary		3,000	85 [2]	2,889 [2]	919 [r, 2]	2,900
Silver, mine output, Ag content[e]	kilograms	2,800	2,800	2,800	2,800	2,800
Zinc, Zn content in concentrate		--	1,000	10,000	15,000	12,000
INDUSTRIAL MINERALS						
Abrasives, natural, emery[e]		150	150	150	150	150
Barite		56,500	56,333	49,038	56,202 [r]	52,000
Cement, hydraulic[e]	thousand metric tons	30,800 [r, 2]	32,800 [r, 2]	30,000	32,000	33,000
Chalk		5,000 [e]	8,343	1,322	1,422 [r]	1,500
Clays:						
Bentonite		31,500	33,300	42,100	44,500 [r]	41,000
Fire clay		359,500	359,200	307,300	333,900 [r]	340,000
Fuller's earth		10,500	11,055	6,370	4,761 [r]	9,000
Kaolin, china clay		24,500	15,318	27,265	16,481 [r]	25,000
Other[e]		220,000	250,000	240,000	260,000	250,000

See footnotes at end of table.

TABLE 1—Continued
PAKISTAN: PRODUCTION OF MINERAL COMMODITIES[1]

(Metric tons unless otherwise specified)

Commodity		2008	2009	2010	2011	2012[e]
INDUSTRIAL MATERIALS—Continued						
Feldspar		28,300[r]	46,000[r]	102,000[r]	107,000[r]	62,000
Fluorspar[e]		1,700	1,400	1,500	1,600	1,700
Gypsum, crude		730,000	856,000	946,000	1,215,000[r]	952,000
Magnesite, crude		3,500	3,918	8,330	16,826[r]	7,500
Nitrogen, N content of ammonia[e]		2,300,000	2,350,000	2,400,000	2,450,000	2,500,000
Phosphate rock:						
Gross weight		3,900	30,467	87,807	126,194[r]	82,000
P_2O_5 content		700	5,480	15,800	22,700[r]	14,800
Pigments, mineral, natural, ocher		51,417[r]	55,985[r]	50,220[r]	40,932[r]	40,000
Salt:						
Rock	thousand metric tons	1,883	1,941	2,058	2,028[r]	1,900
Marine	do.	50[e]	93	190	315[r]	180
Total	do.	1,930[e]	2,034	2,248	2,343[r]	2,080
Sodium compounds, n.e.s.:[e, 3]						
Caustic soda		240,000	250,000	172,000[r, 2]	162,000[r, 2]	150,000
Soda ash, manufactured		250,000	260,000	378,000[r, 2]	335,000[r, 2]	410,000
Stone:						
Aragonite and marble		1,341,000	1,223,387	1,471,014	1,816,254[r]	1,800,000
Dolomite		305,000	150,619	306,940	283,768[r]	300,000
Limestone	thousand metric tons	32,488	35,375	17,984	33,285[r]	30,000
Other, as "ordinary stone"[e]	do.	6	7	7	8	8
Strontium minerals, celestite		1,000	--	--	--	--
Sulfur, native[e]		27,400[r]	26,200[r]	27,100[r]	26,600[r]	26,000
Talc and related materials, soapstone		26,000	40,792	121,800	114,100[r]	110,000
MINERAL FUELS AND RELATED MATERIALS						
Coal, all grades	thousand metric tons	3,691	3,292	3,429	4,026[r]	4,000
Coke	do.	310	320	302[r]	175[r]	180
Gas, natural:						
Gross production	million cubic meters	41,261	41,658	42,000[e]	43,000[e]	43,000
Marketed production, sales[e]	do.	38,000	39,000	40,000	41,000	41,000
Natural gas liquids[e]	thousand 42-gallon barrels	750	750	760	760	770
Petroleum:						
Crude	do.	24,818	23,870	70,800[r]	61,300[r]	65,000
Refinery products:[e]						
Gasoline	do.	11,152[2]	11,161[2]	11,000	12,000	12,000
Jet fuel	do.	7,868[2]	7,584[2]	6,631[r, 2]	5,204[r, 2]	5,500
Kerosene	do.	1,527[2]	1,217[2]	903[r, 2]	912[r, 2]	1,000
Distillate fuel oil	do.	32,000	31,000	32,000	31,000	32,000
Residual fuel oil	do.	21,369[2]	18,615[2]	20,000	21,000	22,000
Lubricants	do.	3,759[2]	3,689[2]	1,393[r, 2]	1,358[r, 2]	1,500
Other	do.	15,000	16,000	17,000	18,000	19,000
Total	do.	92,700	89,300	88,900[r]	89,500[r]	93,000

[e]Estimated; estimated data are rounded to no more than three significant digits; may not add to totals shown. [r]Revised. do. Ditto. -- Zero.
[1]Table includes data available through August 12, 2013.
[2]Reported figure.
[3]Not elsewhere specified.

TABLE 2
PAKISTAN: STRUCTURE OF THE MINERAL INDUSTRY IN 2012

(Thousand metric tons unless otherwise specified)

Commodity		Major operating companies and major equity owners	Location of main facilities	Annual capacity^e
Barite		Bolan Mining Enterprises	Khuzdar, Balochistan Province	24
Do.		Razvi Mining (Private) Ltd.	Gandori, Kalan, and Retri	30
Cement		Al-Abbas Cement Industries Ltd.	Karachi	900
Do.		Askari Cement Co. Ltd.	Nizampur and Wah	2,700
Do.		Attock Cement Pakistan Ltd.	Hub Chowki	800
Do.		Bestway Cement Co. Ltd.	Chakwal and Hattar	3,000
Do.		Cherat Cement Co. Ltd.	Nowshera	750
Do.		Dandot Cement Co. Ltd.	Dandot	500
Do.		Fauji Cement Co. Ltd.	Jhang Bahtar	1,170
Do.		do.	do.	2,200
Do.		Gharibwal Cement Ltd.	Jhelum	540
Do.		Javedan Cement Ltd.	Karachi	600
Do.		D.G. Khan Cement Co. Ltd.	Chakwal and Dera Ghazi Khan	1,650
Do.		Kohat Cement Co. Ltd.	Kohat	700
Do.		Lafarge Pakistan Cement Ltd.	Chakwal	2,500
Do.		Lucky Cement Ltd.	Karachi	3,750
Do.		do.	Pezu	4,000
Do.		Maple Leaf Cement Factory Ltd.	Daudkhel	1,500
Do.		Mustehkam Cement Ltd.	Haripur	600
Do.		Pakistan Cement Co.	Between Islamabad and Lahore, Punjab Province	2,200
Do.		Pioneer Cement Ltd.	Chenki	1,300
Do.		Thatta Cement Co. Ltd.	Thatta	450
Do.		Zeal Pak Cement Factory Ltd.	Hyderabad	1,080
Chromite		Pakistan Chrome Mines Ltd.	Gwal, Khanozai, Muslim Bagh, and Nisai, Balochistan Province	20
Coal		Sindh Coal Authority	Dadu, Sindh Province	4,000
Do.		do.	Tharparkar, Sindh Province	NA
Copper, mine		Saindak Metals Ltd. [Metallurgical Construction Corp. (MCC), operator]	Chaghi, Balochistan Province	22
Gas, natural	million cubic meters per day	Pakistan Petroleum Ltd. (PPL)	Adhi, Punjab Province; Kandhkot and Mazarani, Sindh Province; and Sui, Balochistan Province	24
Do.	do.	Oil and Gas Development Co. Ltd. (OGDC)	37 oilfields and gasfields, including Mari, Sindh Province	31
Lead and zinc, ore		MCC Duddar Minerals Development Co. Pvt.	Duddar, Balochistan Province	660
Petroleum:				
Crude	42-gallon barrels per day	Pakistan Petroleum Ltd. (PPL)	Adhi, Punjab Province	1,600
Do.	do.	Oil and Gas Development Co. Ltd. (OGDC)	37 oilfields and gasfields	46,000
Refined	do.	Bosicor Pakistan Ltd.	Karachi	30,000
Do.	do.	Pak-Arab Refinery Co. Ltd. (joint venture of the Governments of Pakistan and the Emirate of Abu Dhabi)	Mahmood Kot, Punjab Province	100,000
Phosphate rock		Pakistan Mining Co. Ltd.	NA	90
Steel, crude		Pakistan Steel Mills Corp. (Pvt) Ltd. (PSM)	Karachi	1,100
Talc		CapriCorn Minerals	Bandi Sadique	20

^{e}Estimated. Do., do. Ditto. NA Not available.

The Mineral Industry of Papua New Guinea

By Susan Wacaster

Copper, gold, and silver were the major mineral commodities produced in Papua New Guinea in 2012. The country was highly prospective for mineralization in epithermal and porphyry-related high- and low-sulfidation systems; skarns; volcanic massive sulfides; exhalative manganese deposits; lateritic nickel, chromite, and cobalt deposits; and seafloor massive sulfides. Other metallic mineral resources that may occur in subeconomic deposits, that had not yet been extensively explored, or that were under development included iron ore, manganese, molybdenum, platinum-group elements, and zinc. Occurrences of industrial minerals included economically important limestone and phosphate rock deposits and minor deposits of asbestos, diatomite, graphite, pozzolan, silica, and sulfur. Papua New Guinea also produced crude petroleum. The country's petroleum production had been in decline, but petroleum exploration was at an all-time high in 2012. Papua New Guinea also had large quantities of natural gas resources. The Papua New Guinea Liquefied Natural Gas Project (PNG LNG) was scheduled to begin in 2014 and was expected to supply liquid natural gas (LNG) to companies in Asia, including the Chinese Petroleum Corp. of Taiwan; Osaka Gas Co.; Tokyo Electric Power Company of Japan; and Unipec Asia Co. (a subsidiary of China Petroleum and Chemical Corp.) (Mineral Resources Authority of Papua New Guinea, 2013a, p. 1).

Minerals in the National Economy

Comprehensive official economic data for Papua New Guinea were not available. In terms of gross domestic product (GDP), Papua New Guinea was the leading developing economy among the Pacific island nations. The country's real GDP growth rate was an estimated 7.7% in 2012 compared with 8.9% in 2011. The growth was driven by domestic demand created by ongoing construction of the $19 billion PNG LNG project, which was jointly financed by Exxon Mobil Corp. of the United States [with Esso Highlands Ltd. as subsidiary operator (33.2%)], Oil Search Ltd. of Australia (29%), National Petroleum Co. of Papua New Guinea (16.6%), Santos Ltd. of Australia (13.5%), JX Nippon Oil and Gas Exploration of Japan (4.7%), Mineral Resources Development Co. [Papua New Guinea landowners (2.8%)]; and Petromin PNG Holdings Ltd. (0.2%). The value derived from the construction sector was estimated to have increased by about 25% in 2012 compared with that of 2011 followed closely by the wholesale and retail trade and transport and communications sectors. All sectors had experienced a surge in demand associated with the PNG LNG project. That demand was expected to diminish as the project was completed, although the increased physical capacity installed as part of the project could continue to contribute to some level of sustained increased economic activity (Bulman, 2012, p. 1; Global Finance, 2013).

Government Policies and Programs

Mineral resources in Papua New Guinea are owned by the state. Laws that regulate the mining industry in the country include the Mining Safety Act of 1977, the Mining Act of 1992, and the Mineral Resources Authority Act of 2005. The Mineral Resources Authority Act of 2005 provides for the establishment of the Mineral Resources Authority (MRA) and defines the agency's powers and functions. Those functions include (a) promoting the development and exploitation of the country's mineral resources; (b) overseeing the administration and enforcement of the Mining Development Act of 1955, the Mining Act (Bougainville Copper Agreement) of 1967, the Mining Safety Act of 1977, the Mining Act of 1992, the Ok Tedi Agreement, and any other laws related to the development of Papua New Guinea's mineral resources; and (c) acting as the agent for the state for international mining agreements. The Geological Survey Division (GSD) is one of four divisions of the MRA. A few of the core functions of the GSD include compiling geoscience data and information aimed at encouraging mineral sector investment; conducting regional geoscience and geologic resource mapping; and archiving statutory reports submitted by mineral tenement holders (Government of Papua New Guinea, 2005, p. 10; Mineral Resources Authority of Papua New Guinea, 2013b).

Production

Copper production decreased in 2012 by 3.9% compared with that of 2011, which was a typical decrease in production related to the mining of lower grade ores at the aging Ok Tedi Mine. A greater production decrease of about 18% in 2011 compared with that of 2010 was related to two shutdowns of the mining operation in 2011. Silver production in the country decreased by 10.3% and gold production decreased by 16.2% owing to decreased production from the Hidden Valley Mine. Data on mineral production are in table 1.

Structure of the Mineral Industry

In 2012, several private international mining companies were majority owners or shareholders in producing metals operations in Papua New Guinea, including Newcrest Mining's wholly owned Lihir Island Mine and Harmony Gold and Newcrest Mining's Hidden Valley Mine, from which the Government received a royalty of 2%. Barrick Gold Corp. of Canada operated and held a 95% interest in the Porgera Mine, and the Government held the remaining 5% share through the state-owned Mineral Resources Development Corp. New Guinea Gold Corp. of Canada was the operator of and held a 92% interest in the Sinivit Gold Mine, and Gold Mines of Niugini Holdings Pty Ltd. of Papua New Guinea held the remaining 8% interest. The Ok Tedi Mine was operated by the PNG Sustainable Development Program Ltd., which held a 70%

share in the operation; the Government held the remaining 30% share. The Simberi Mine was wholly owned by St Barbara Ltd. of Australia, and the Tolukuma Hill Mine was wholly owned by Petromin PNG Holdings. Table 2 is a list of major mineral facilities (SNL Metals Economics Group, 2013).

Mineral Trade

According to the Central Bank of Papua New Guinea, the value of total goods exported in 2012 from the country was $5.96 billion. Of this amount, the value of mineral exports was $4.34 billion, or 74.3% of the total, which was a 19.5% decrease from that of 2011. The value of exported gold decreased by 13.9% in 2012 compared with that of 2011 to $2.4 billion. The gold export volume was 46,300 kilograms (kg), which was a decrease of 10.8% compared with that of 2011. The decreased export volume was owing to lower volumes of ore extracted at the Ok Tedi Mine and downtime because of maintenance work on the mill at the Simberi Mine, which offset increased production at the Porgera, the Toukuma, the Lihir, and the Hidden Valley Mines. The average gold price on the London Metal Exchange increased by 6.4% in 2012 compared with that of 2011; however, Papua New Guinea's decreased export volume was too large to offset increased prices received for the country's gold (Bank of Papua New Guinea, 2013, p. 14; United Nations Statistics Division, 2014).

According to the Central Bank of Papua New Guinea, the copper export volume was about 125,300 metric tons (t), which was a 12.7% decrease compared with that of 2011. The decrease in copper exports was also owing to extraction of lower grade ore at the Ok Tedi Mine. The total value of copper exports decreased by 32% in 2012 compared with that of 2011 to about $966.6 million. Data from the United Nations Commodity Trade Statistics Database indicated that about 48,600 t of copper exports, including copper scrap and waste, copper wire, and refined copper, was exported in 2012 (Bank of Papua New Guinea, 2013, p. 14).

The crude oil export volume was 8.2 million barrels, which was a decrease of 6.1% compared with that of 2011. The decrease in the export volume of crude oil was owing to decreased extraction rates brought on by the decline in reserves at the Gobe Main, the Kutubu, the Moran, and the South East Gobe oilfields. The average export price of crude oil from Papua New Guinea was $111 per barrel in 2012, which was a 13.7% decrease compared with that of 2011. The decreased price per barrel was attributed to higher production by member nations of the Organization of the Petroleum Exporting Countries (OPEC) and to lower demand from China and Europe. The value of crude oil exports was about $9.2 million (Bank of Papua New Guinea, 2013, p. 14).

Commodity Review

Metals

Cobalt and Nickel.—The Ramu nickel and cobalt laterite project [a joint venture among Ramu NiCo Ltd. (85%), Highlands Pacific Ltd. of Australia (8.56%), Mineral Resources Development Corp. (3.94%), and unnamed landowners (2.5%)], also known as the Kurumbukari, the Kurbukan, and (or) the Ramu River project, is located on the Kurumbukari plateau near the Ramu River in Madang Province. China's Metallurgical Corporation of China Ltd. (MCC) held a 61% interest in Ramu NiCo, and a number of other Chinese entities held the remaining 39%. Highlands announced that its share in Ramu would increase to 11.3% after an unspecified amount of internal project debt had been repaid from operating revenue, and that it had an option to increase its interest to 20.55% (Highlands Pacific Ltd., 2013, p. 17).

The Ramu open pit mine is connected by a 135-kilometer (km) slurry pipeline to the Basamuk processing plant, which is located 75 km west of the Provincial capital city of Madang in the Rai Coast District. Parts of the slurry pipeline are buried, and the pipeline drops about 680 m in elevation from the mine to the processing plant. The Ramu Mine is a low-strip-ratio, free-digging open pit mine. Face shovels and backhoe configured excavators were to mine the 12-meter (m)-thick (on average) ore body and load the ore into trucks for delivery to the washing plant and then to the beneficiation plant. Treatment would remove the chromite and create a slurry feed for overland pipeline transport to the Basamuk processing plant. The expected output of the mine as of April 2010 included an annual mining capacity of 5.7 million metric tons (Mt) of ore; an annual processing capacity of 4.28 Mt of wet ore, or 3.21 Mt of dry ore; and an annual output of 79,300 t of nickel-cobalt mixed hydroxides, or 33,000 t of nickel metal and 3,300 t of cobalt metal. When fully operational, the Ramu operation was expected to account for about one-third of all China's nickel imports (Metallurgical Corporation of China Ltd., 2010, p. 23; Highlands Pacific Ltd., 2013, p. 18).

Commissioning of the Ramu Mine took place in March 2012, and the rampup to full production was expected to take 12 to 18 months. In 2012, a total of 1.54 Mt of wet ore was delivered to the beneficiation plant, of which 735,000 t of wet ore was mined and delivered in the fourth (December) quarter. The total amount of ore treated and transported through the pipeline in 2012 was 647,000 t (dry weight); of that amount, 298,000 t (dry weight) was transported in the fourth (December) quarter. At full capacity, the operation was expected to process 3.3 Mt (dry weight) of ore with 1.04% nickel and 0.11% cobalt (Highlands Pacific Ltd., 2013, p. 18).

Copper.—The Ok Tedi Mine, which was also known as the Mt. Fubilan Mine, is located at Mt. Fubilan in a region of steep cliffs known as the Hindenberg Wall within the rainforest of Papua New Guinea's Western Province approximately 16 km east of the border with Indonesia. The Hindenberg Wall is a limestone terrain that is underlain by easily eroded siltstone. The location and geology surrounding the mine contributed to the collapse of a tailings dam in 1984 that was part of the site's initial construction. Since then, from 20 to 60 Mt/yr of tailings from the operation had been released into the upper Ok Tedi River, resulting in massive environmental degradation of the Fly River and Ok Tedi River systems (Garrett, 2013).

In 2012, negotiations continued within the country regarding a mine life extension plan. Ok Tedi's Mine Life Extension proposal would extend the life of the mine from its original scheduled closing date in 2013 to 2022, during which time the mine would produce at one-half of its original capacity from two underground mines and one open pit mine. By yearend, seven of nine affected community groups had agreed to continue production for 11 more years. In 2012, the Ok Tedi Mine produced about 125,350 t of copper contained in about 482,900 t of concentrate compared with 130,500 t of copper contained in 494,900 t of concentrate in 2011.

Gold.—In 2012, the Hidden Valley Mine produced about 5,524 kg of gold and about 53,345 kg of silver compared with 3,118 kg of gold and 20,934 kg of silver in 2011. The Porgera Mine produced about 14,276 kg of gold in 2012 compared with about 15,600 kg in 2011. The Ok Tedi Mine produced about 12,286 kg of gold in 2012 compared with about 12,977 kg in 2011 (SNL Metals Economics Group, 2013).

Mineral Fuels

Natural Gas.—The PNG LNG project was 80% complete and on track for its first deliveries in 2014. The project was being developed with two processing trains, but additional resources had been discovered that would allow for an expansion to three trains. The consortium was deciding whether to expand the project rather than build a second plant because expansion would be less expensive and less labor intensive as the essential infrastructure (including roads and pipelines) would already be in place. A decision to expand the plant could cause conflict with the Government, however, which reportedly was seeking as much infrastructure investment as possible in order to stimulate continued economic growth (Kelly, 2013).

Outlook

GDP growth in Papua New Guinea is expected to decrease in 2013 and 2014 as construction of the PNG LNG and related private sector investments are completed; the decrease could be offset in part by commissioning of the Ramu nickel cobalt mine in 2012. In late 2014 and 2015, however, the GDP is predicted to increase by 20% to 25% as a result of the first production from the PNG LNG project. Construction activity is expected to decrease by between 8% and 9% in 2014, which is nonetheless about 20% more activity than in 2011 and greater than twice that of 2008 (Bulman, 2012, p. 16).

References Cited

Bank of Papua New Guinea, 2013, December 2010 economic bulletin: Port Moresby, Papua New Guinea, Bank of Papua New Guinea, 34 p.

Bulman, Tim, 2012, Papua New Guinea economic briefing: The World Bank. (Accessed November 11, 2013, at http://documents.worldbank.org/curated/en/2013/01/17364817/papua-new-guinea-economic-briefing-last-days-boom-lasting-improvements-living-standards.)

Garrett, Jemima, 2013, PNG's Ok Tedi—From disaster to dividends: Australian Network News, January 6. (Accessed February 19, 2014, at http://www.abc.net.au/news/2013-01-07/an-radio-doco3a-ok-tedi/4455092.)

Global Finance, 2013, Papua New Guinea country report: Global Finance. (Accessed November 12, 2013, at http://www.gfmag.com/gdp-data-country-reports/201-papua-new-guinea-gdp-country-report.html#axzz2kj7zTVCv.)

Government of Papua New Guinea, 2005, Mineral Resources Authority Act of 2005: Government of Papua New Guinea, December 23, 76 p.

Highlands Pacific Ltd., 2013, 2012 financial statements: Brisbane, Queensland, Australia, Highlands Pacific Ltd.

Kelly, Ross, 2013, Exxon seeks to expand PNG LNG plant's capacity: The Wall Street Journal, May 27. (Accessed November 15, 2013, at http://online.wsj.com/news/articles/SB10001424127887323855804578508512752180272.)

Metallurgical Corporation of China Ltd., 2010, Annual report 2009: Beijing, China, Metallurgical Corporation of China Ltd., 250 p.

Mineral Resources Authority of Papua New Guinea, 2013a, Geological framework and mineralization of Papua New Guinea—An update: Port Moresby, Papua New Guinea, Mineral Resources Authority of Papua New Guinea, 62 p.

Mineral Resources Authority of Papua New Guinea, 2013b, Papua New Guinea Mining Information: Port Moresby, Papua New Guinea, Mineral Resources Authority of Papua New Guinea. (Accessed November 13, 2013, at http://www.mra.gov.pg/Investors/PNGMiningInformation.aspx.)

SNL Metals Economics Group, 2013, MineSearch: SNL Metals Economics Group database. (Accessed November 8, 2013, via http://www.metalseconomics.com/database-services/minesearch.)

United Nations Statistics Division, 2014, United Nations commodity trade statistics database (UN Comtrade): United Nations Statistics Division database. (Accessed July 7, 2013, via http://comtrade.un.org/db.)

TABLE 1
PAPUA NEW GUINEA: PRODUCTION OF MINERAL COMMODITIES[1]

Commodity[2]		2008	2009	2010	2011	2012
Copper, mine output, Cu content	metric tons	159,650	166,700	159,821	130,473 r	125,348
Gas, natural, marketed	million cubic meters	141 r	113 r	113 r	113 r	100 e
Gold, mine output, Au content	kilograms	67,463	63,600	58,983 r	62,200	52,100
Nickel hydroxide:						
Gross weight	metric tons	--	--	--	--	13,777
Co content	do.	--	--	--	--	469
Ni content	do.	--	--	--	--	5,283
Petroleum, crude	thousand 42-gallon barrels	13,993	12,806	11,100 r	11,000 r	10,000
Silver, mine output, Ag content	kilograms	51,300	50,000	74,000 r	90,700 r	81,300

eEstimated; estimated data are rounded to no more than three significant digits. rRevised. do. Ditto. -- Zero.

[1]Table includes data available through November 15, 2013.

[2]In addition to the commodities listed, cement, construction materials (common clays, sand and gravel, and stone), and refined petroleum products are produced, but available information is either inadequate to make a reliable estimate of output or output is insignificant.

TABLE 2
PAPUA NEW GUINEA: STRUCTURE OF THE MINERAL INDUSTRY IN 2012

(Metric tons unless otherwise specified)

Commodity		Major operating companies and major equity owners	Location of main facilities[1]	Annual capacity[e]
Cement	thousand metric tons	Papua New Guinea-Halla Cement Pty. Ltd. (Halla Cement Corp., 50%, and Government, 50%)	Lae, Morobe Province	200
Cobalt	do.	MCC Ramu NiCo, Ltd., 85%; Highlands Pacific Ltd., 8.56%; Mineral Resources Development Corp., 3.94%	Open pit mine facility, Basamuk beneficiation plant, Madang Province	3
Copper	do.	Ok Tedi Mining Ltd., operator (PNG Sustainable Development Program Ltd., 70%, and Government, 30%)	Ok Tedi open cut, Western Province, 20 km northwest of Tabubil and 390 km southwest of Wewak	170
Gold		Newcrest Mining Ltd., operator, 100%	Lihir open cut, Lihir Island, New Ireland Province, 700 km northeast of Port Moresby	26
Do.		St Barbara Ltd., 100%	Simberi Island open cut, New Ireland Province	3
Do.		Newcrest Mining Ltd., 50%, and Harmony Gold Mining Co. Ltd., 50%	Hidden Valley open cut, Morobe Province, 90 km southwest of Lae	8
Do.		Ok Tedi Mining Ltd. (PNG Sustainable Development Program Ltd., 70%, and Government, 30%)	Ok Tedi open cut, Western Province, 20 km northwest of Tabubil and 390 km southwest of Wewak	17
Do.		Porgera Joint Venture (Barrick Gold Corp., 95%, and Mineral Resources Development Corp., 5%)	Porgera open cut and underground mines, Enga Province, 620 km northwest of Port Moresby	22
Do.		New Guinea Gold Corp., 92%	Sinivit open pit, East New Britain Province, about 50 km south southwest of Rabaul	1
Do.		Petromin PNG Holdings, 100%	Tolukuma Hill open pit, 100 km north of Port Moresby	216
Nickel	thousand metric tons	MCC Ramu NiCo, Ltd., 85%	Open pit mine facility, Basamuk beneficiation plant, Madang Province	33
Petroleum, crude	thousand 42-gallon barrels per day	Petroleum development license 2: Chevron Niugini Ltd. (operator and manager), 19.37%; Oil Search (Kutubu) Ltd., 27.14%; Orogen Minerals Ltd., 25.44%; Exxon Mobil Corp., 14.52%; Petroleum Resources (Kutubu) Ltd., 6.75%; Merlin Petroleum Co., 6.78% Petroleum development license 5: Exxon Mobil Corp. (operator and manager), 47.5%, and Oil Search Ltd., 52.5%	Central Moran oilfield, Southern Highlands Province (includes Agogo and Iaqufi-Hedinia fields). Onshore Papuan Basin, petroleum development licenses 2 and 5	20
Do.	do.	Chevron Niugini Ltd. (operator and manager), 19.37%; Oil Search Ltd., 27.14%; Orogen Minerals Ltd., 30.19%; Exxon Mobil Corp., 14.52%; Merlin Petroleum Co., 6.78%; Petroleum Resources Ltd. (Gobe), 2.0%	Gobe Main oilfield, Southern Highlands Province. Onshore Papuan Basin, petroleum development license 4	11
Do.	do.	Chevron Niugini Ltd. (operator and manager), 19.37%; Oil Search Ltd., 27.14%; Orogen Minerals Ltd., 25.44%; Exxon Mobil Corp., 14.52%; Petroleum Resources (Kutubu) Ltd., 6.75%; Merlin Petroleum Co., 6.78%	Kutubu oilfield, Southern Highlands Province. Onshore Papuan Basin, petroleum development license 2	16
Do.	do.	Santos Ltd. (operator and manager), 15.5%; Southern Highlands Petroleum Ltd., 39.14%; Orogen Minerals Ltd., 20.5%; Oil Search Ltd., 15.50%; Cue PNG Oil Co. Ltd., 5.42%; Petroleum Resources (Gobe) Ltd., 2.0%; Mountains West Exploration, Inc., 1.94%	South East Gobe oilfield, Gulf and Southern Highlands Provinces. Onshore Papuan Basin, petroleum development licenses 3 and 4	11
Silver		Porgera Joint Venture (Barrick Gold Corp., 95%, and Mineral Resources Development Corp., 5%)	Porgera open cut and underground mines, Enga Province, 620 km northwest of Port Moresby	4
Do.		Newcrest Mining Ltd., 50%, and Harmony Gold Mining Co. Ltd., 50%	Hidden Valley open cut, Morobe Province, 90 km southwest of Lae	124
Do.		Ok Tedi Mining Ltd. (operator) (PNG Sustainable Development Program Ltd., 70%, and Government, 30%)	Ok Tedi open cut, Western Province, 20 km northwest of Tabulio and 390 km southwest of Wewak	30

[e]Estimated. Do., do. Ditto.
[1]Abbreviations used for units of measure in this table include the following: km, kilometer.

THE MINERAL INDUSTRY OF THE PHILIPPINES

By Yolanda Fong-Sam

In 2012, nickel was one of the most valuable mineral commodities produced in the Philippines. Other mineral commodities for which domestic production was significant to the national economy included cement, chromium, copper, gold, marine salt, and silver (table 1).

Minerals in the National Economy

The mining and quarrying sector contributed about 1.14% of the Philippines' gross domestic product (GDP) (at constant 2000 prices) in 2012 compared with 1.19% in 2011. The construction sector contributed 5.4% of the GDP compared with 5% in 2011. The value of metallic mineral production was $2.386 billion,[1] which was a decrease of 16% from the value of $2.839 billion in 2011. The value of gold produced in 2012 amounted to $831.9 million (0.6% of the GDP) (Bangko Sentral ng Pilipinas, 2013; Mines and Geosciences Bureau, 2013c).

Government Policies and Programs

The 1995 Philippine Mining Act regulates mineral resource development, requires the Government to monitor mineral activity (production, trade, and value) and maintain a database of mineral reserves, and encourages direct investment by the private and public sectors in mineral exploration and development activities in the Philippines. The Government grants exploration permits to qualified applicants to explore for mineral resources. Exploration permits are valid for a period of 2 years and are renewable for not more than 4 years for exploration of nonmetallic minerals and 6 years for exploration of metallic minerals (Mines and Geosciences Bureau, 2013d).

The 1995 Philippine Mining Act allows for three types of mining agreements. Each type of mining agreement is valid for 25 years and is renewable for an additional 25 years. The first is a mineral agreement in which the Government grants a domestic investor (a Filipino individual or corporation) an exclusive right to conduct mining operations in the contracted area. The second type of mining agreement is a Financial or Technical Assistance Agreement (FTAA), which is available to domestic and foreign corporations for a maximum area of 81,000 hectares (ha) onshore or 324,000 ha offshore. The third type of agreement is a mineral production-sharing agreement for properties with a maximum area of 16,200 ha; it is open to domestic and foreign corporations (Mines and Geosciences Bureau, 2013d).

The Philippine Government approved Executive Order 79 on July 6, 2012. The Executive order imposed a moratorium on the approval of new mining projects until Congress passes a law to increase royalty fees. The Executive order proposes an increase in royalty fees to 5% from 2%. The new policy bans mining in agricultural lands, fishery zones, island ecosystems, and tourist destinations. The Executive order also changes the rules for the acquisition of new mining permits by establishing a bidding process rather than the previously used "first come, first served" approach. Small-scale mining activities are limited to certain areas of the country, and small-scale mining of metallic minerals is prohibited except for mining of chromite, gold, and silver. The use of mercury is prohibited in small-scale mining operations. Existing mining agreements, concessions, contracts, and projects that were approved before the Executive order was signed are exempted from the new rules; however, the Executive order authorizes the Department of Environmental and Natural Resources of the Philippines (DENR) to review all existing mining contracts and renegotiate the contract terms with the contractors. The DENR is also in charge of implementing the Extractive Industries Transparency Initiative, which is an international effort designed to set the standards for transparency in the management of revenue derived from extractive industries, such as mining and oil and natural gas extraction. Executive Order 79 also proposes the creation of a Mining Industry Coordinating Council, which would be an interagency body in charge of conducting dialogue with stakeholders, reviewing all existing mining-related laws and regulations, conducting public auctions for mining tenements, ensuring that contractors have the proper environmental insurance coverage and perpetual liability, and implementing other industry reforms (Mines and Geosciences Bureau, 2012; Mining Weekly, 2012; Office of the President of the Philippines, 2012).

In September, the DENR announced a temporary suspension of Executive Order 79 because of ambiguity in the provisions related to the renewal of expiring mining contracts. The temporary suspension was put in place while new amendments were issued to address the ambiguous provisions. According to the Chamber of Mines of the Philippines, Executive Order 79 mandates renegotiation of the terms of all mining contracts (even those eligible for renewal), which effectively shortens the total maximum agreement period to 25 years from 50 years (Andrade, 2012).

In December, the Mines and Geosciences Bureau (MGB) announced that it planned to lift the ban that prohibits the issuance of new mining permits by March 2013. The MGB also added that it would increase mining permit fees, although the amount of the increases was not specified. The MGB was waiting for the release of a revised map indicating the specific areas banned from mining activity before lifting the ban. The updated map was being created by the National Mapping and Resources Information Agency (GMA News Online, 2012).

Production

In 2012, the Philippines produced 424,000 metric tons (t) of nickel, which was an increase of 93.6% compared with

---
[1]Where necessary, values have been converted from Philippine pesos (PHP) to U.S. dollars (US$) at the annual rate of PHP42.23=US$1.00 for 2012 and PHP43.31=US$1.00 for 2011.

that of 2011. Other metals for which production increased significantly compared with that of 2011 included silver (48.2%) and chromite (43.7%). Production data for industrial minerals and mineral fuels were not available; hence the production numbers in table 1 are estimated. The metals for which production decreased in 2012 included smelted copper (53%), gold (49.4%), refined copper (40%), and iron ore (26.1%) (table 1).

Structure of the Mineral Industry

In 2012, an estimated 252,000 people, or 0.6% of the total number of people employed in the country, worked in the mining and quarrying industry (Mines and Geosciences Bureau, 2013a). Some of the main producers of mineral commodities in the Philippines were CGA Mining Ltd. of Australia (gold and silver); Lafayette Mining Ltd. of Australia (copper, gold, and silver); Lepanto Consolidated Mining Co. of the Philippines (copper, gold, and silver); Nickel Asia Corp. of the Philippines (nickel); Philex Mining Corp. of Canada (copper, gold, and silver); and TVI Resources Development Philippine Inc. (gold and silver), which was the Philippines affiliate of TVI Pacific Inc. of Canada. The country's major mineral industry facilities are listed in table 2.

Mineral Trade

In 2012, total trade between the Philippines and other countries increased by 5% to $114.2 billion from $108.8 billion in 2011. The country's total exports were valued at $52.1 billion compared with $48.3 billion in 2011, which was an increase of 7.9%; also, the total value of imports in 2012 increased by 2.7% to $62.1 billion from $60.5 billion in 2011. Exports of refined copper were valued at $505 million, which was a decrease of 58.3% compared with the value in 2011; exports of gold were valued at $480.6 million, which was an increase of 10.2% compared with the value in 2011; and petroleum products exports were valued at $465 million, which was a decrease of 28.2% compared with the value in 2011. Imports of iron and steel and mineral fuels and related materials were valued at $15.1 billion in 2012, which was about 24% of the country's total import value (National Statistics Office of the Philippines, 2013).

In 2012, the Philippines' leading trading partner was Japan, which accounted for 14.3% ($16.4 billion) of the country's total trade, broken down as $9.9 billion in exports to Japan, and $6.5 billion in imports from Japan. The second-ranked trading partner was the United States, which accounted for 12.7% of total trade; exports to and imports from the United States were valued at $7.417 billion and $7.124 billion, respectively. The country's third- and fourth-ranked trading partners were China and Singapore, which accounted for 11.2% and 8.1% of the Philippines' total trade, respectively. Exports to China were valued at $6.17 billion, and imports from China were valued at $6.68 billion. Exports to Singapore were valued at $4.867 billion, and imports from Singapore were valued at $4.405 billion. Total trade with the European Union accounted for $10.571 billion, which was equivalent to 9.3% of total trade (National Statistics Office of the Philippines, 2013).

Commodity Review

Metals

Copper, Gold, Silver, and Zinc.—In 2012, the mined copper production in the country totaled 65,444 t of metal content, which was an increase of 2.5% compared with the 63,835 t produced in 2011. Copper mine output at Carmen Copper Corp.'s Toledo Copper Complex was a reported 40,783 t compared with 32,206 t in 2011 (an increase of 27%), and at Rapu-Rapu Processing Inc.'s polymetallic project, production reached 7,760 t compared with 6,810 t in 2011 (an increase of 14%). At Philex Mining Corp.'s Padcal copper project, however, production decreased by 41.2% to a reported 10,118 t in 2012 from 17,216 t in 2011. Copper production also decreased at TVI Resources' Canatuan mining project by 11% (Mines and Geosciences Bureau, 2013b).

Gold production decreased by 49.4% to 15,762 kilograms (kg) in 2012 from 31,120 kg in 2011 (table 1). Based on gold purchases made by the Bangko Sentral ng Pilipinas (Philippines Central Bank) in 2012, small-scale mines produced 709 kg of gold compared with 17,389 kg of gold in 2011. At the Padcal copper project, gold production dropped by 49% to 2,218 kg in 2012 from 4,358 kg in 2011. Some gold production increases were reported at polymetallic projects that produced gold as a byproduct, such as at Rapu-Rapu's polymetallic project, where production increased by 131% to 2,016 kg of gold in 2012 from 874 kg of gold in 2011; at the Toledo Copper Complex, where production increased by 100% to 426 kg of gold from 213 kg of gold in 2011; at Canatuan, where production increased by 64% to 350 kg of gold from 214 kg of gold in 2011; and at the Victoria gold project (which was owned by Lepanto Consolidated Mining Co.), where production increased by 61% to 910 kg of gold from 564 kg of gold in 2011 (Mines and Geosciences Bureau, 2013b).

On January 27, Indophil Resources NL of Australia (37.5% interest) received notice from Xstrata plc of the United Kingdom (62.5% interest) about the revised estimates on the mineral resources at the Tampakan project. The reestimation increased Tampakan's mineral resources (measured, indicated, and inferred) to 2.94 billion metric tons at a grade of 0.51% copper at a cutoff grade of 0.2% copper and 0.19 gram per metric ton (g/t) gold, which represented 15 million metric tons (Mt) of copper and 547,000 kg of gold (reported as 17.6 million troy ounces). The feasibility study concluded that the project had the potential to produce an average of 450,000 metric tons per year (t/yr) of copper during the first 5 years of production, and an average of 375,000 t/yr of copper and about 11,200 kilograms per year (kg/yr) of gold (reported as 360,000 troy ounces per year) for a minimum of 17 years. The project was projected to cost a total of $5.9 billion. On December 12, Sagittarius Mines, Inc. (SMI) [a joint venture between GlencoreXstrata (62.5%) and Indophil Resources NL (37.5%)], which was the operator of the Tampakan project, announced that it had provided the Government with a revised development plan, which estimated that production would begin in 2019 rather than in 2016 (the previous target date). The Tampakan project is located in South Cotabato Province near the town of Tampakan on the southern island of Mindanao. In June 2010, South Cotabato Province banned the use of open pit mining as a method to extract mineral resources in the Province after passing an environmental code that was implemented in October of that year. The ban in South Cotabato Province was likely to affect mostly the Tampakan copper and gold mining project (Indophil Resources NL, 2013, p. 3, 14).

In August, SMI entered into a joint-development agreement with Alsons Energy Development Corp. (AEDC), a local powerplant developer, to determine the feasibility of constructing a coal-fired powerplant to meet the Tampakan project's power demand. The proposed powerplant was to be located in the town of Maasim, Saragani Province (Indophil Resources NL, 2013, p. 3).

In April, Crazy Horse Resources Inc. of Canada reported the completion of a prefeasibility study based on a 15-million-metric-ton-per-year (Mt/yr) operation at the Taysan property. The Taysan property covers a total area of 11,254 ha and is located in Batangas Province, Luzon Island, about 100 kilometers (km) south of Manila and 20 km east of the Provincial capital of Batangas City. The property consists of a copper and gold porphyry deposit with an estimated measured and indicated resource of about 460 Mt at a cutoff rate of 0.1% copper; contained copper was estimated to be about 1.18 Mt; contained gold, about 47,000 kg (reported as 1.5 million troy ounces); contained silver, about 383,000 kg (reported as 12.3 million troy ounces); and contained magnetite, 14.9 Mt. The cost of developing the property was estimated to be $521 million, and the life of the mine was expected to be 24 years (Crazy Horse Resources Inc., 2012).

On September 25, Mindoro Resources Ltd. of Canada and TVI Pacific signed the following four joint-venture agreements: the Agata mining joint-venture, the Agata processing joint-venture, the Pan de Azucar joint-venture, and the Pan de Azucar processing joint-venture. Under the agreements, TVI would earn a 60% interest in each joint venture by developing and operating the projects and complying with requirements, such as funding the developments with $2 million during the first year, and $500,000 thereafter, and performing and completing feasibility studies. Feasibility studies for each of the developments were planned for 2013, and additional exploratory drilling was scheduled for the Pan de Azucar area to define copper and gold mineral resources (Mindoro Resources Ltd., 2013, p. 4, 7).

On July 24, Red Mountain Mining Ltd. of Australia signed a share-sale agreement with Mindoro Resources to acquire a 100% interest in Mindoro Resources's Batangas gold project, which is located 120 km south of Manila in Batangas Province, Luzon Island. The project, which covers an area of 270 square kilometers (km^2), includes the Archangel and the Lobo areas (which are under a mineral production-sharing agreement), and El Paso and the Talahib areas, which are under a copper-gold exploration permit. In addition, the company acquired a 75% interest in the Tapian San Francisco joint-venture copper-gold project, which is located in Surigao Province in Northern Mindanao; this transaction was completed on October 30. After the acquisition of the projects, Red Mountain completed a 1,800-meter diamond-drilling program at Archangel. The company also determined the indicated and inferred mineral resources at the Batangas gold project to be 5.78 Mt at an average grade of 2.2 g/t gold and 3.3 g/t silver for about 12,700 kg of contained gold (reported as 408,000 contained troy ounces of gold) and about 18,800 kg of contained silver (reported as 606,000 contained troy ounces of silver) at an 0.85 g/t gold cutoff grade (Red Mountain Mining Ltd., 2013).

Throughout 2012, Metals Exploration plc of the United Kingdom focused on preparing and readying the mining site of the Runruno copper, gold, and molybdenum project, which would include such infrastructure features as road access, the main processing plant, and other buildings. Exploration on site continued during the year, mainly in the Magnetite Creek and Malilibeg South areas. The company estimated the gold resources for the Malilibeg South area to be 7.55 Mt at an average grade of 1.4 g/t gold for approximately 10,600 kg of gold (reported as 340,000 troy ounces). The company also carried out other exploration activities in the area, such as geochemical and geophysical studies, mapping, and soil sampling, and a diamond drilling program to determine additional gold mineralization and the potential for porphyry copper mineralization. The Runruno polymetallic project, which is located in Nueva Viscaya Province about 322 km north of Manila, was planned to be an open pit mining operation. As of March 2011 (the latest date for which a resource estimate was available), the Runruno resources were estimated to be 43,200 kg of gold (reported as 1.39 million troy ounces) and 11,600 t of molybdenum (reported as 25.6 million pounds) (Metals Exploration plc., 2013).

In June 2011, OceanaGold Corp. of Australia (92% interest) announced that it had begun the construction of its Didipio gold-copper project, which is located on the north of the island of Luzon in northern Philippines, approximately 270 km north of the capital city of Manila. Production from the Didipio Mine commenced during the fourth quarter of 2012, and the first copper concentrate was produced in December 2012. The project, which was developed at a cost of $250 million, was expected to start commercial production during the second quarter of 2013. The initial throughput was projected to be 2.5 Mt/yr, and the company expected production to reach full capacity of 3.5 Mt/yr by 2014. The project was under an FTAA for an area of about 158 km^2 and had an estimated mine life of 16 years as an open pit mining operation, and approximately 8 additional years as an underground operation. The FTAA agreement allowed OceanaGold to share its net revenue with the Government at a basis of 60/40, of which 60% of the revenue would be the Government's portion. The company forecasted annual production of more than 3,100 kg of gold (reported as 100,000 troy ounces), and 14,000 t of copper in concentrate. The measured and indicated resources of the Didipio Mine were approximately 66,600 kg of gold (reported as 2.14 million troy ounces), and 290,000 t of copper; reserves were estimated to be about 52,300 kg of gold (reported as 1.68 million troy ounces) and 230,000 t of copper. Also in 2012, OceanaGold announced the completion of the construction of a mineral processing facility located to the north of the Didipio Mine. In July 2012, OceanaGold announced that it had signed an agreement with Trafigura Beheer BV of the Netherlands for the sale and purchase of 100% of the copper and gold concentrate output from the Didipio Mine. The agreement was valid for a minimum of 5 years. OceanaGold planned to conduct a feasibility study in 2012 to increase the processing facility's capacity to 5 Mt/yr from 3.5 Mt/yr, which was expected to increase production to about 4,700 kg/yr of gold (reported as 150,000 troy ounces per year) and 20,000 t/yr of copper (OceanaGold Corp., 2013a; 2013b, p. 23; 2013c).

Nickel.—During 2012, a drilling program was conducted in the Zambales Chromite Mining Corp. tenement, which is located approximately 5 km to the north of the Acoje nickel project on Luzon Island. ENK plc of the United Kingdom (40%) and local partner Montemina Resources Corp. (60%), which owned Zambales, estimated that the inferred resources at the Zambales property were 23.5 Mt at average grades of 1.18% nickel and 0.05% cobalt (ENK plc, 2012, p. 10).

In 2012, the production of nickel increased by about 93.6% to 424,000 t from 219,000 t (revised) in 2011 (table 1). In January, Intex Resources ASA of Norway announced its partnership, through a memorandum of understanding, with MCC8 Group Co. Ltd. of China for the construction of the Philippines' first refined nickel plant. The plant, which would be located on the island of Mindoro, was to process low-grade lateritic nickel ore to produce nickel metal (Lucas, 2012).

Nickel Asia Corp. announced that construction of the Taganito high-pressure acid-leach (HPAL) hydrometallurgical processing plant was on schedule during 2012. The project, which was managed by Taganito HPAL Nickel Corp., was expected to start commercial operations by the last quarter of 2013. Sumitomo Metal Mining Co. Ltd. of Japan (the majority shareholder) reported the total costs of the project to be approximately $1.7 billion. When completed and in full production, the Taganito plant would have a production capacity of 30,000 t/yr of contained nickel in the form of nickel and cobalt sulfide; an expansion project was also planned for 2016 to increase the plant's capacity to 36,000 t/yr of contained nickel (Nickel Asia Corp., 2012, p. 13).

In October 2011, ENK produced its first 50 kg of nickel hydroxide during the trial phase of the pilot plant at its Acoje nickel project on Luzon Island. The company expected to produce about 2.4 t/yr of nickel hydroxide when the plant is fully operational. Production was expected to commence formally in mid-2013. During 2012, the company finalized a feasibility study and completed a drilling program to try to identify additional resources within the tenement. The Acoje project covers an area of 3,765 ha and had estimated indicated and inferred resources of 50.14 Mt at grades of 1.08% nickel and 0.05% cobalt (ENK plc, 2012, p. 8, 9).

Industrial Minerals

Cement.—In 2012, the production of cement in the Philippines increased by about 18% to 18.9 Mt from 16 Mt (table 1). APC Group Inc. of the United States announced in July that it planned to resume the construction of a $200 million cement plant in Cebu. The cement plant, which had a design capacity of 1.5 Mt/yr, was expected to be completed by 2014 (International Cement Review, 2012).

In September, Cemex Corp. Philippines announced a plan to invest $65 million in a capacity expansion project at the company's APO plant. The expansion, which was planned to begin in early 2014, would increase the plant's annual capacity by 1.5 Mt/yr (Global Cement, 2012).

Outlook

The DENR has forecasted that investments in the mineral industry may reach $14 billion in the next 5 to 6 years, according to the Philippines Chamber of Mines (Mining Weekly, 2012). During 2012, the adoption of Executive Order 79, which extended the moratorium on the approval of new mining projects and proposed an increase in royalty fees, among other regulatory changes, caused an atmosphere of uncertainty among the foreign investors. In the wake of the approval of the order, it remains to be seen how the changes will affect the mineral industry as a whole.

The Philippines expect several mining investment projects that started in 2011 and 2012 to be commissioned in the next several years. Among the projects are the Cemex Philippines APO plant expansion project (2014), the Taganito nickel and cobalt plant expansion (2016), and the Tampakan copper-gold project (2019).

Production in the Philippines' industrial minerals and construction materials sector increased steadily during 2012, partly as a result of increased demand in the Southeast Asia region. The Philippines is likely to be a key supplier and producer of a number of mineral commodities, as well as a significant producer of such metals as chromium, nickel, and zinc. The production of metallic minerals will continue to be important to the economy of the Philippines as many established projects have started to expand their facilities and production capacities. Similarly, the production of copper, gold, and nickel is expected to increase as major exploration activities result in new discoveries and increases in resources and as proposed developments are commissioned in the near future.

References Cited

Andrade, J.I., 2012, DENR suspends implementation of new mining policy guidelines: Inquirer News [Makati City, Philippines], September 28. (Accessed December 18, 2013, at http://newsinfo.inquirer.net/279042/denr-suspends-implementation-of-new-mining-policy-guidelines.)

Bangko Sentral ng Pilipinas, 2013, Gross national income (GNI) by industrial origin—Table 43: Bangko Sentral ng Pilipinas, 1 p. (Accessed November 4, 2013, at http://www.bsp.gov.ph/statistics/spei_pub/november 2013.zip.)

Crazy Horse Resources Inc., 2012, Copper & gold development in the Philippines: Vancouver, British Columbia, Canada, Crazy Horse Resources Inc. (Accessed December 20, 2013, at http://crazyhorseresources.com/s/Home.asp.)

ENK plc, 2012, Annual report 2012: London, United Kingdom, ENK plc, 76 p. (Accessed November 22, 2013, at http://www.enickel.co.uk/files/5313/4433/8684/ENK_2012AR_Final_LR.pdf.)

Global Cement, 2012, Cemex to expand in Philippines: Global Cement. (Accessed September 20, 2012, at http://www.globalcement.com/news/item/1152-cemex-to-expand-in-philippines.)

GMA News Online, 2012, PHL to lift ban on new mining permits in 2013, eyes 100% increase in fees: Gmanetwork.com [Quezon City, Philippines], December 4. (Accessed December 18, 2013, at http://www.gmanetwork.com/news/story/284958/economy/agricultureandmining/phl-to-lift-ban-on-new-mining-permits-in-2013-eyes-100-increase-in-fees.)

Indophil Resources NL, 2013, Annual report 2012: Melbourne, Victoria, Australia, Indophil Resources NL, May 23, 90 p.

International Cement Review, 2012, Project considered—Philippines: London, United Kingdom, International Cement Review, July, p. 10.

Lucas, D.L., 2012, Intex, Chinese partner to build Philippines' first nickel plant: Philippine Daily Inquirer [Makati City, Philippines], January 13. (Accessed September 6, 2013, at http://business.inquirer.net/39545/intex-chinese-partner-to-build-philippines%E2%80%99-first-nickel-plant.)

Metals Exploration plc., 2013, Final results for the year ended 31 December 2012: London, United Kingdom, Metals Exploration plc., May 23, 18 p. (Accessed November 22, 2013, at http://www.metalsexploration.com/pdf/13_05_23_Annual_Report_RNS_2013.pdf.)

Mindoro Resources Ltd., 2013, Consolidated financial statements & management's discussion and analysis for the year ended December 31, 2012: Edmonton, Alberta, Canada, Mindoro Resources Ltd., 51 p. (Accessed September 20, 2013, at http://www.mindoro.com/pdf/2012-Annual-Financials.pdf.)

Mines and Geosciences Bureau, 2012, Mining issues: Quezon City, Philippines, Mines and Geosciences Bureau. (Accessed December 18, 2013, at http://www.mgb.gov.ph/art.aspx?artid=385.)

Mines and Geosciences Bureau, 2013a, Employment—Mining industry statistics: Quezon City, Philippines, Mines and Geosciences Bureau, 1 p. (Accessed November 22, 2013, at http://www.mgb.gov.ph/Files/Statistics/MineralIndustryStatistics.pdf.)

Mines and Geosciences Bureau, 2013b, Table 1—Philippine metallic mineral production—As of February 2013: Quezon City, Philippines, Mines and Geosciences Bureau, 1 p.. (Accessed July 15, 2013, at http://www.mgb.gov.ph/Files/Statistics/MetallicProduction.pdf.)

Mines and Geosciences Bureau, 2013c, The Philippine minerals industry at a glance: Quezon City, Philippines, Mines and Geosciences Bureau. (Accessed December 17, 2013, at http://www.mgb.gov.ph/Files/ItemLinks/ThePhilippineMineralsIndustryAtAGlance.jpg.)

Mines and Geosciences Bureau, 2013d, Types of mining rights and contracts: Quezon City, Philippines, Mines and Geosciences Bureau. (Accessed December 17, 2013, at http://www.mgb.gov.ph/pgs.aspx?pgsid=32.)

Mining Weekly, 2012, Eyeing more—Philippines Aquino signs mining order seeking higher royalty: Mining Weekly, v. 18, no. 27, p. 23.

National Statistics Office of the Philippines, 2013, Foreign trade statistics of the Philippines 2012: Manila, Philippines, National Statistics Office press release no. 2013–082, November 7. (Accessed December 17, 2013, at http://www.census.gov.ph/content/foreign-trade-statistics-philippines-2012.)

Nickel Asia Corp., 2012, Annual report 2012: Makati City, Philippines, Nickel Asia Corp., 40 p. (Accessed December 4, 2013, at http://www.nickelasia.com/2012_NAC_Annual_report_FINAL.pdf.)

OceanaGold Corp., 2013a, Didipio Mine: Melbourne, Victoria, Australia, OceanaGold Corp. (Accessed December 20, 2013, at http://www.oceanagold.com/our-business/philippines/didipio-mine/.)

OceanaGold Corp., 2013b, Fact book 2013: Melbourne, Victoria, Australia, OceanaGold Corp., 40 p. (Accessed December 4, 2013, http://www.oceanagold.com/assets/documents/annual-reports/OGC-2013-Fact-Book-2013-FINAL.pdf.)

OceanaGold Corp., 2013c, OceanaGold announces offtake heads of agreement for Didipio copper concentrate with Trafigura: Melbourne, Victoria, Australia, OceanaGold Corp., 1 p. (Accessed November 22, 2013, http://www.oceanagold.com/assets/documents/filings/2012-Press-Releases/120710-Didipio-Copper-Concentrate-Heads-of-Agreement-Final.pdf.)

Office of the President of the Philippines, 2012, Executive Order No. 79: Manila, Philippines, Official Gazette, July 6. (Accessed December 27, 2013, at http://www.gov.ph/2012/07/06/executive-order-no-79-s-2012/.)

Red Mountain Mining Ltd., 2013, Annual financial report—30 June 2013: Perth, Western Australia, Australia, Red Mountain Mining Ltd., 92 p. (Accessed September 13, 2013, at http://www.redmm.com.au/files/nrteUploadFiles/202F092F201363A533A03PM.pdf.)

TABLE 1
PHILIPPINES: PRODUCTION OF MINERAL COMMODITIES[1]

(Metric tons unless otherwise specified)

Commodity[2]		2008	2009	2010	2011	2012
METALS						
Chromium, chromite, gross weight		15,268	14,322	14,807	25,483	36,628
Cobalt, mine output, Co content[3]		1,200 e	1,500 e	2,200	2,200 e	2,600
Copper:						
Mine output, Cu content		21,235	49,060	58,412	63,835	65,444
Metal:						
Smelter		239,700	230,100	216,200	205,000	97,000
Refined		174,600	178,000	171,900	164,000	98,400
Gold, mine output, Au content	kilograms	35,726	37,047	40,847	31,120	15,762
Iron and steel:						
Iron ore, gross weight		--	--	--	468,000	346,000
Iron ore, Fe content (62.5%)		--	--	--	292,608	216,176
Metal, steel, crude	thousand metric tons	711	824	1,050	1,200	1,200 e
Lead, metal, secondary, refined		34,000	32,000	30,000	34,000	32,000
Manganese:						
Gross weight		12,800	8,500	11,300	4,300	5,000
Mn content (43%)		5,500	3,600	4,900	1,900	2,000
Nickel, mine output, Ni content[4, 5]		90,000 r	128,500 r	207,000 r	219,000 r	424,000
Silver, mine output, Ag content	kilograms	14,224	33,808	41,004	45,530	67,477
Zinc, mine output, Zn content		1,619	10,035	9,268	18,170	19,559

See footnotes at end of table.

TABLE 1—Continued
PHILIPPINES: PRODUCTION OF MINERAL COMMODITIES[1]

(Metric tons unless otherwise specified)

Commodity[2]		2008	2009	2010	2011	2012
INDUSTRIAL MINERALS						
Cement, hydraulic	thousand metric tons	13,369	14,865	15,900	16,063	18,907
Clays:						
Bentonite		1,422	1,413	1,475	2,087	2,000 e
Red		7,181	7,357	7,050	8,243	8,300 e
White		8,745	8,519	8,857	12,246	12,300 e
Other		5,601	5,599	5,878	8,143	8,200 e
Feldspar		15,838	16,394	15,882	22,050	22,000 e
Lime		4,299	4,327	4,524	5,934	6,000 e
Perlite		4,593	4,606	4,756	6,272	6,300 e
Phosphate rock		2,271	2,257	2,308	2,778	2,800 e
Salt, marine		510,059	516,066	557,644	720,146	720,000 e
Sand and gravel:						
Silica sand	thousand metric tons	270	284	296	352	350 e
Other[6]	thousand cubic meters	46,659	46,602	49,009	58,815	59,000 e
Stone:						
Crushed, broken, other[7]	do.	3,077	3,069	3,258	4,259	4,300 e
Dolomite		1,150,035	1,176,991	1,259,152	1,431,118	1,430,000 e
Limestone[8]	thousand metric tons	31,528	33,090	35,540	42,526	42,500 e
Marble, dimension, unfinished	cubic meters	5,410	5,629	6,001	8,043	8,000 e
Pumice		2,063	2,064	2,274	2,797	2,800 e
Tuff		17,570	18,830	19,166	22,106	22,100 e
Volcanic cinder[9]	cubic meters	6,519	6,686	7,325	9,219	9,200 e
MINERAL FUELS AND RELATED MATERIALS						
Coal, all grades	thousand metric tons	3,610	4,687	6,650	6,881	7,000 e
Gas, natural, gross	million cubic meters	3,881	3,909	3,681	3,975	4,000 e
Petroleum:						
Crude	thousand 42-gallon barrels	965	2,920	3,059	2,326	2,500 e
Refinery products:e						
Liquefied petroleum gas	do.	3,556 [10]	3,286 [10]	3,500	3,500	3,500
Gasoline	do.	11,988 [10]	9,153 [10]	9,000	9,000	9,000
Jet fuel	do.	46,000	46,000	46,000	46,000	46,000
Kerosene	do.	6,596 [10]	1,002 [10]	1,000	1,000	1,000
Distillate fuel oil	do.	23,871 [10]	17,541 [10]	17,500	17,500	17,500
Residual fuel oil	do.	15,975 [10]	10,776 [10]	10,000	10,000	10,000
Refinery fuel and losses	do.	2,307 [10]	2,068 [10]	2,000	2,000	2,000
Other	do.	2,882 [10]	4,635 [10]	4,500	4,500	4,500
Total	do.	113,000	94,500	93,500	93,500	93,500

eEstimated; estimated data are rounded to no more than three significant digits; may not add to totals shown. rRevised. do. Ditto. -- Zero.

[1]Table includes data available through December 23, 2013.

[2]In addition to the commodities listed, the Philippines produces platinum-group metals as byproducts of other metal production, quartz, and sulfur, but available information is inadequate to make reliable estimates of output.

[3]The majority of the nickel laterite produced in the Philippines is exported to China, but whether cobalt content is recovered is not known.

[4]Production of mined nickel (Ni content) was reported by the Government as follows, in metric tons: 2008—80,644; 2009—139,744; 2010—184,330; 2011—319,363 (revised); and 2012—317,621. The numbers in the table have been adjusted to take into account data received from individual company sources as well as trade statistics (see footnote 5).

[5]Data compiled using trade data from the United Nations Comtrade database for nickel ores and concentrates exported from the Philippines to Australia, China, and Japan.

[6]Includes "pebbles" and "soil" not further described.

[7]Includes materials described as rock, crushed or broken/blasted; stones, cobbles, and boulders; pebbles; rock aggregates; and broken adobe.

[8]Includes limestone for agriculture, cement manufacturing, industrial use, and other.

[9]Reported as "black cinder" for the years 2008–11 by the Philippines Mines and Geosciences Bureau.

[10]Reported figure.

TABLE 2
PHILIPPINES: STRUCTURE OF THE MINERAL INDUSTRY IN 2012

(Metric tons unless otherwise specified)

Commodity		Major operating companies and major equity owners	Location of main facilities	Annual capacity
Cement		Eagle Cement Co.	Plant located in Akle, San Ildefonso, Bulacan	1,500,000.
Do.		Fortune Cement Corp.	Bulacan plant at Norzagaray, Bulacan Province; Batangas plant at Taysan, Batangas Province	2,100,000.
Do.		Holcim Philippines, Inc.	Bulacan plant at Norzagaray, Bulacan Province; Davao plant at Barrio Ilang, Davao City; La Union plant at Bacnotan, La Union Province; Lugait plant at Lugait, Misamis Oriental Province	7,200,000.
Do.		Solid Cement Corp., APO Cement Corp., and Cemex Philippines Group of Companies	Cement plants at three locations— Naga, Cebu Province (APO Cement Corp.); Antipolo City, Rizal Province (Solid Cement Corp.); Binangonan, Rizal Province (Rizal Cement Corp.)	4,300,000.
Chromite, Cr content		Consolidated Mines Inc. (owner) and Benguet Corp. (operator)	Masinloc chromite mine (Coto chromite deposit) located in Coto 27 kilometers east of the Port of Mansiloc in Zambales Province	5,000.
Do.		Heritage Resources Mining Corp.	Homonhon chromite project	17,000.
Do.		Krominco Inc.	Dinagat chromite project—Redondo Mine (Mt. Redondo deposit) located in the Municipality of Loreto, Dinagat Island	26,000.
Copper, Cu content of concentrate		Carmen Copper Corp.	Toledo Copper Complex, located in the Central Highlands of Cebu, Cebu Island	20,000.
Do.		Lafayette Mining Ltd., 75%, and LG International and Korean Resources Corp., 25%	Rapu-Rapu Mine under the Rapu-Rapu polymetallic project, located in Albay Province	36,000.
Do.		Lepanto Consolidated Mining Co.	Victoria and Teresa Mines located in Mankayan, Benguet Province	200.
Do.		Philex Mining Corp. (through its subsidiary Philex Gold Inc.), 81%	Padcal copper project located in Tuba, Benguet Province, Island of Luzon	21,000.
Do.		TVI Resources Development Philippine Inc., 100%	Canatuan project, located east of Siocon, Province of Zamboanga del Norte, Mindanao Island	10,000.
Do.		Glencore International plc.	Philippine Associated Smelting and Refining Corp. (PASAR), located in Isabel, Leyte Province	250,000 smelter; 173,000 refinery.
Gold, Au content	kilograms	APEX Mining Company Inc.	APEX Maco operation	100.
Do.	do.	CGA Mining Ltd.	Masbate gold project, located 350 kilometers south of Manila, Masbate Island	6,000.
Do.	do.	Lafayette Mining Ltd., 75%, and LG International and Korean Resources Corp., 25%	Rapu-Rapu Mine under the Rapu-Rapu polymetallic project, located in Albay Province	1,500.
Do.	do.	Lepanto Consolidated Mining Co.	Victoria and Teresa Mines located in Mankayan, Benguet Province	2,000.
Do.	do.	Philex Mining Corp. (through its subsidiary Philex Gold Inc.), 81%	Padcal Mine (Sto. Tomas II deposit) located in Tuba, Benguet Province, Island of Luzon	5,000.
Do.	do.	Philippine Mining Development Corp.	Diwalwal Direct State Development Project at Mount Diwalwal in Davao del Norte Province	100.
Do.	do.	Philsaga Mining Corp.	Banahaw Gold Project	NA.
Do.	do.	TVI Resources Development Philippine Inc., 100%	Canatuan project, located east of Siocon, Province of Zamboanga del Norte, Mindanao Island	500.

See footnotes at end of table.

TABLE 2—Continued
PHILIPPINES: STRUCTURE OF THE MINERAL INDUSTRY IN 2012

(Metric tons unless otherwise specified)

Commodity	Major operating companies and major equity owners	Location of main facilities	Annual capacity
Nickel, Ni content	CRAU Mineral Resources Corp.	Santa Cruz-Candelaria nickel project located in Zambales Province	1,000.
Do.	CTP Construction & Mining Corp.	Adlay-Cagdianao-Tandawa (ACT) nickel project, located in Barangay Adlay, Municipality of Carrascal, Province of Surigao del Sur	10,000.
Do.	Hinatuan Mining Corp.	South Dinagat project located on Nonoc Island	4,000.
Do.	Nickel Asia Corp., 100%	Cagdianao nickel project located near Barangay Valencia on Dinagat Island	10,000.
Do.	do.	Tagana-an nickel project located on Hinatuan Island	30,000.
Do.	Nickel Asia Corp., 65%; Pacific Metals Co. Ltd., 33.5%; Sojitz Philippines, 1.5%	Claver nickel project (Taganito) located in Surigao del Norte Province, Mindanao Island	16,000.
Do.	Nickel Asia Corp., 60%; Pacific Metals Co. Ltd., 36%; Sojitz Philippines, 4%	Rio Tuba nickel project, located in Barrio Rio Tuba, Municipality of Bataraza in Palawan Province.	5,000.
Do.	SR Metals, Inc.	SR Nickel project, Tubay Mine, located in Tubay, Agusan del Norte Province	25,000.
Do.	Toledo Mining Corporation Plc., 56.1%	Berong nickel project located on Palawan Island	10,000.
Nickel, mine and plant	Coral Bay Nickel Corp. (Sumitomo Metal Mining Co. Ltd., 54%; Mitsui & Co. Ltd. 18%; Rio Tuba Nickel Mining Corp., 10%; Nickel Asia Corp., 6%)	Coral Bay nickel high-pressure acid-leach (HPAL) plant located on Palawan Island	24,000 nickel; 1,800 cobalt.
Silver, kilograms Ag content	Lafayette Mining Ltd., 75%, and LG International and Korean Resources Corp., 25%	Rapu-Rapu Mine under the Rapu-Rapu polymetallic project, located in Albay Province	18,000.
Do. do.	Lepanto Consolidated Mining Co.	Victoria and Teresa Mines located in Mankayan, Benguet Province	4,000.
Do. do.	Philex Mining Corp. (through its subsidiary Philex Gold Inc.), 81%	Padcal Mine (Santo Tomas II deposit) located in Tuba, Benguet Province, Island of Luzon	5,000.
Do. do.	TVI Resources Development Philippine Inc., 100%	Canatuan project, located east of Siocon, Province of Zamboanga del Norte, Mindanao Island	17,000.
Zinc, Zn content	Lafayette Mining Ltd., 75%, and LG International and Korean Resources Corp., 25%	Rapu-Rapu Mine under the Rapu-Rapu polymetallic project, located in Albay Province	8,000.

Do., do. Ditto. NA Not available.

The Mineral Industry of Sri Lanka

By Chin S. Kuo

Sri Lanka has modest resources of industrial and precious minerals. It produced cement, clays, feldspar, graphite, mica, mineral sands, phosphate rock, salt, stone, and such gemstones as ruby and sapphire. The country produced no metals or crude oil, but it imported petroleum for refining. In 2012, the Government was offering tax incentives and generous regulations to lure investors to its mining sector. Mining companies would be charged a royalty of 5% on revenues and no other duties.

In 2012, Sri Lanka's gross domestic product (GDP) was $128.4 billion, and the country's per capita income was $6,200 based on purchasing power parity. The country recorded an increase in the GDP of 6.4% for the year owing to ongoing reconstruction projects and infrastructure development. The industrial sector registered growth of 10.3% and contributed 31.5% to the GDP. Mining and quarrying accounted for 2% of the GDP (U.S. Central Intelligence Agency, 2012).

In 2012, Sri Lanka exported minerals valued at $42.7 million, of which mineral sands accounted for 65%; silica and quartz, 20%; graphite, 10%; and mica, 3%. Exports of gemstones mainly went to China and Thailand. The export value of minerals to China increased to $15.14 million in 2012. The Government invited Chinese companies to invest in mining and processing mineral sands and graphite for export. Australia and Germany were also interested in developing Sri Lanka's rare-earth minerals and mineral sands (Lanka Business Online, 2013). Crude petroleum was a major import item for Sri Lanka; the country imported 41,000 barrels per day in 2012.

Production

Production of graphite and zircon was estimated to have increased by 43% and 17%, respectively, in 2012 compared with that of 2011 owing to expansions in production capacity that were completed in 2011. The value of Sri Lanka's gemstone production was estimated to have increased by 11% compared with the value in 2011 owing mainly to the high prices of gemstones on the world market; although the output of ruby was estimated to have increased by 8.6%, the output of sapphire decreased by 6.3%. Production of steel manufactures and residual fuel oil was estimated to have decreased slightly. Production of cement, clay for cement, kaolin, limestone, and salt each was estimated to have increased by between 5% and 10% (table 1).

Structure of the Mineral Industry

The mining of graphite, mineral sands, phosphate rock, and salt was performed by state-owned companies. The Ministry of State Resources and Enterprise Development (SRED) owned companies that mined mineral sands and phosphate rock. State-owned Ceylon Petroleum Corp. operated a crude oil refinery. The Government had no plans to privatize any state-owned enterprises but planned instead to retain ownership and management of these enterprises and to make them profitable. The private sector produced all other mineral output, with the exception of cement, which was manufactured and sold by a combination of the private sector and foreign investors and by state-owned Sri Lanka Cement Corp. Foreign companies in partnership with local investors operated several cement plants in the country. AMG Mining AG (formerly Graphit Kropfmühl AG) of Germany owned an 89.68% stake in Bogala Graphite Lanka Ltd., and other investors owned the remaining 10.32% (table 2).

Commodity Review

Metals

Titanium.—Lanka Mineral Sands Ltd., which was under the SRED, was engaged in extracting mineral sands along Pulmoddai beach north of Trincomalee. The mineral sands contained garnet, ilmenite, monazite, rutile, and zircon. The mineral sands deposit was estimated to have a resource of about 12.5 million metric tons (Mt), and ilmenite accounted for 60% of the deposit. The company operated a mineral processing plant for separating different minerals from the deposit. These minerals were the raw materials for rare-earth elements, thorium, titanium, and zirconium (Ministry of State Resources and Enterprise Development, 2013a).

Lanka Mineral Sands planned to construct a new mineral sands processing plant at Kokkilai in northern Sri Lanka. The black mineral sands of Pulmoddai beach have heavy-mineral concentrations of 50% to 60% and were considered to be some of the richest mineral sands in the world. The company also planned to increase mine production of ilmenite and zircon (Ollett, 2012, p. 30).

Mirama Minerals was also engaged in mining and processing mineral sands at its plant at Dambulla in central Sri Lanka. Mirama produced garnet, ilmenite, mica, quartz, rutile, and zircon for export. The company planned to set up a synthetic rutile plant with the intention of expanding it eventually to include production of titanium dioxide (Mirama Minerals, 2013).

Industrial Minerals

Graphite.—Sri Lanka's vein graphite is the highest quality form of natural graphite in the world; it contains more than 90% carbon content. Graphite is used in such traditional industries as refractories and steelmaking, and it is also in demand for use in such emerging technologies as lithium-ion batteries. State-owned Kahatagaha Graphite Lanka Ltd. planned to increase its production of vein graphite by 50% in 2012. The company, which operated one of the three graphite mines in the country, intended to increase its processing capacity of vein graphite to 150 metric tons per month (t/mo) from 100 t/mo to meet market demand. The company mined seams of graphite 610 meters underground in the Maduragoda-Dodangaslanda area in the Kurunegala District. Its graphite resource was estimated to be 500,000 metric tons (t) (Moores, 2012).

The country's Board of Investment (BoI) signed a deal with Plumbago Lanka (Pvt) Ltd. in which the BoI agreed to invest $78 million during a 4-year period to process and export graphite. The Esna business advisory group would own a 25% stake in the project. Plumbago Lanka would work with the Geological Survey and Mines Bureau in operating the project. Earlier, the BoI had approved a proposal by Sarcon Development (Pvt) Ltd. to set up a $15.2 million graphite processing plant to produce graphite for export (Syrett, 2012, p. 62).

Sakura Pvt. Ltd., in collaboration with the Government, restarted the old Ragedara underground vein graphite mine, which is located in Hiriyala in northwestern Sri Lanka. The mine became Sri Lanka's third-ranked vein graphite producer in 2012. Sri Lanka exported 95% of its graphite production. Germany received a shipment of 10 t of the mine's high-purity graphite, which graded 90% to 99% carbon, in July 2012. Canada and France also were the possible destinations of Sri Lanka's exports of vein graphite. Bora Bora Resources Ltd. of Australia acquired a Sri Lankan graphite project for $500,000 in late 2012 and raised $1.25 million to fund the acquisition and project development (Australia's Paydirt, 2013).

Phosphate Rock.—Lanka Phosphate Ltd., which was also under the SRED, operated the Eppawala phosphate project, which covered an area of approximately 324 hectares in the Anuradhapura District, North Central Province. The resources of the deposit were estimated to be 60 Mt containing 33% to 40% phosphorus pentoxide. The company produced two types of phosphate rock—Eppawala and high-grade Eppawala. The company planned to manufacture single superphosphate (SSP) to substitute for triple superphosphate (TSP), which was currently imported. With the production of SSP, the Government was expected to reduce its imports of TSP and, consequently, to reduce the amount of foreign exchange spent on imports (Ministry of State Resources and Enterprise Development, 2013b).

Outlook

Slower economic growth (6.3%) is forecast for 2013 for the country (International Monetary Fund, 2013). Sri Lanka's production of gemstones is expected to contribute $250 million to the country's GDP in 2013. Graphite production is expected to increase owing to planned expansions and investment. The country is expected to expand its output of downstream steel products to meet domestic demand, particularly that of the construction industry. Production of mineral sands is expected to increase as a result of Lanka Mineral Sands's construction of a new mineral processing plant and the Government's investment in the mineral sands industry. Mirama Minerals is expected to start up a value-added business, such as production of synthetic rutile and titanium dioxide.

References Cited

Australia's Paydirt, 2013, ASX 2012 IPOs: Australia's Paydirt, February, p. 33.

International Monetary Fund, 2013, World economic outlook: International Monetary Fund, April. (Accessed August 19, 2013, at http://www.imf.org/external/pubs/ft/weo/2013/01/index.htm.)

Lanka Business Online, 2013, Sri Lanka invites China to invest in mineral sector: Lanka Business Online, April 7. (Accessed July 10, 2013, at http://www.lankabusinessonline.com/news/print/1875474855.)

Ministry of State Resources and Enterprise Development, 2013a, Institutions, Lanka Mineral Sands Ltd.: Ministry of State Resources and Enterprise Development. (Accessed July 10, 2013, at http://www.sredmin.gov.lk/index.php?opinion=com_content&view=article&id=49.)

Ministry of State Resources and Enterprise Development, 2013b, Institutions, Lanka Phosphate Ltd.: Ministry of State Resources and Enterprise Development. (Accessed July 10, 2013, at http://www.sredmin.gov.lk/index.php?opinion=com_content&view=article&id=48.)

Mirama Minerals, 2013, Home page: Mirama Minerals. (Accessed July 11, 2013, at http://www.miramaminerals.com/index.php?id=1.)

Moores, Simon, 2012, Canada surges as flake tops graphite wish list: Industrial Minerals, no. 540, September, p. 31.

Ollett, John, 2012, Zircon buoys as prices rise: Industrial Minerals, no. 536, May, p. 28–30.

Syrett, Laura, 2012, The grade unknowns—Graphite mining projects "under the radar": Industrial Minerals, no. 543, December, p. 51–63.

U.S. Central Intelligence Agency, 2012, Sri Lanka, in The world factbook: U.S. Central Intelligence Agency. (Accessed July 18, 2013, at https://www.cia.gov/library/publications/the-world-factbook/geos/ce.html.)

TABLE 1
SRI LANKA: ESTIMATED PRODUCTION OF MINERAL COMMODITIES[1,2]

(Metric tons unless otherwise specified)

Commodity[3]		2008	2009	2010	2011	2012
Cement, hydraulic	thousand metric tons	1,800	1,900	2,600 [r]	2,200	2,400
Clays:						
Ball clay		52,966 [4]	54,873 [4]	47,826 [4]	50,000	52,000
Clays for cement manufacture		950	950	1,000	1,100	1,200
Kaolin		10,039 [4]	9,538 [4]	8,207 [4]	8,000	8,500
Feldspar, crude and ground		32,586 [4]	73,365 [4]	75,405 [4]	70,000	72,000
Gemstones:						
Precious and semiprecious, other than diamond, value	thousands	$108,000	$110,000	$150,000	$180,000	$200,000
Cat's eye	carats	50,000	51,000	54,000	55,000	56,000
Ruby	do.	47,900 [4]	20,300 [4]	31,336 [4]	35,000	38,000
Sapphire	do.	541,900 [4]	986,500 [4]	1,491,698 [4]	1,600,000	1,500,000
Other	do.	2,300,000	2,400,000	2,500,000	2,400,000	2,500,000
Graphite, all grades		6,615 [4]	3,171 [4]	3,437 [4]	3,500	5,000
Iron and steel, metal, semimanufactures		66,809 [4]	72,000	75,000	76,000	75,000
Mica, scrap		2,364 [4]	2,347 [4]	2,095 [4]	2,100	2,200
Petroleum refinery products:						
Gasoline	thousand 42-gallon barrels	2,300	2,400	2,600	2,700	2,800
Jet fuel	do.	750	750	800	800	850
Kerosene	do.	1,500	1,500	1,500	1,500	1,500
Distillate fuel oil	do.	5,500	5,600	5,700	5,800	5,900
Residual fuel oil	do.	4,800	4,800	4,500	4,600	4,500
Refinery fuel and losses	do.	740	750	760	800	820
Other	do.	2,500	2,600	2,700	2,800	2,900
Total	do.	18,100	18,400	18,600	19,000	19,300
Phosphate rock, gross weight		41,947 [4]	36,347 [4]	47,778 [4]	48,000	49,000
Salt		110,856 [4]	10,500	10,400	11,000	12,000
Stone:						
Limestone	thousand metric tons	1,091 [4]	1,145 [4]	1,192 [4]	1,200	1,300
Quartzite		37,196 [4]	30,409 [4]	34,437 [4]	36,000	37,000
Titanium mineral concentrates, gross weight:						
Ilmenite		22,159 [4]	122,424 [4]	52,637 [4]	52,000	53,000
Rutile		11,335 [4]	2,276 [4]	2,568 [4]	2,700	2,800
Zirconium, zircon, gross weight		41,000	9,000	11,000	30,000	35,000

[r]Revised. do. Ditto.

[1]Estimated data are rounded to no more than three significant digits; may not add to totals shown.

[2]Table includes data available through July 15, 2013.

[3]In addition to the commodities listed, crude construction materials, such as calcite, clay for brick and tile, dolomite, sand and gravel, sulfur, and varieties of stone presumably are produced, but available information is inadequate to make reliable estimates of output.

[4]Reported figure.

TABLE 2
SRI LANKA: STRUCTURE OF THE MINERAL INDUSTRY IN 2012

(Thousand metric tons unless otherwise specified)

Commodity		Major operating companies and major equity owners	Location of main facilities	Annual capacity[e]
Cement		Holcim (Lanka) Ltd. (part of Holcim Ltd.)	Puttalam	1,000
Do.		Sri Lanka Cement Corp. (Ministry of Industry and Commerce)	Kankesanthurai	1,000
Do.		do.	Puttalam	400
Do.		Tokyo Cement Co. (Lanka) Ltd.	Trincomalee	300
Clay, ball		Lanka Ceramic Ltd.	Dediyawala	NA
Graphite		Kahatagaha Graphite Lanka Ltd. (Ministry of Industry and Commerce)	Kahatagaha Mine	6
Do.		Bogala Graphite Lanka Ltd. (AMG Mining AG, 89.68%, and others, 10.32%)	Bogala Mine	7
Do.		Sakura Pvt. Ltd.	Ragedara Mine	NA
Petroleum, refined	42-gallon barrels per day	Ceylon Petroleum Corp. (Ministry of Petroleum and Petroleum Resources Development)	Sapugaskanda	51,000
Phosphate rock		Lanka Phosphate Ltd. (Ministry of State Resources and Enterprise Development)	Eppawala	40
Titanium, mineral sands		Lanka Mineral Sands Ltd. (Ministry of State Resources and Enterprise Development)	Pulmoddai	150
Do.		Mirama Minerals	Dambulla	NA

[e]Estimated. Do., do. Ditto. NA Not available.

The Mineral Industry of Taiwan

By Pui-Kwan Tse

Taiwan is an island with few mineral resources. Taiwan's economy was oriented towards exports, and the health of the economy was highly susceptible to conditions in the world markets. In 2012, the global economic conditions remained weak, especially in major industrialized countries in the Western World. Demand for Taiwan's goods from the Asian markets was weaker in 2012 than in 2011; however, these markets still accounted for 65% of Taiwan's total exports. China was the leading destination of Taiwan's goods and received 25.8% of the total value of exports; Hong Kong, 13.0%; the United States, 11.0%; Singapore, 6.7%; and Japan, 6.3%. About 36% of Taiwan's exports were electronic products in 2012, which was a decrease of about 4% from that of 2011. Metal products accounted for 17% of the total value of exports. Domestic demand for goods was flat, and, in response the output of the manufacturing sector decreased by 3.8%. Taiwan's gross domestic product (GDP) increased by 1.3% in 2012 compared with an increase of 4.0% in 2011. The output value of the mining and quarrying sector increased by 10.4% in 2012 but accounted for only a small share of Taiwan's GDP. The service sector grew by a modest 1.0%; however, it accounted for 69.0% of Taiwan's economy. Both Government and private-sector consumption increased by a modest 1.3%, and the spending contributed to growth of 0.84% (Ministry of Economic Affairs, 2013, p. 2; Ministry of Finance, 2013a, b; Taiwan Statistical Bureau, 2013, p. 3, 41).

Minerals in the National Economy

Taiwan's significant identified mineral resources included clay, coal, copper, dolomite, feldspar, gold, gypsum, natural gas, petroleum, serpentine, and talc. After several decades of mining, nearly all the recoverable coal, metallic minerals, and talc had been depleted. The output of the mining industry, which had a very small effect on Taiwan's economy, was less than 1% of total industrial production. In 2011, the employment in the mining sector decreased to about 4,000 people (Taiwan Statistical Bureau, 2013, p. 13).

Production

The major mining activities in Taiwan were the production of dolomite, limestone, marble, natural gas, and petroleum. Natural gas and petroleum were produced on the western part of the island, and limestone and marble were mined on the eastern part of the island. In 2012, the production value of the major mineral commodities was $457 million, of which $283 million was from natural gas and $97 million was from marble. Besides natural gas and marble, sulfur was Taiwan's most valuable mineral commodity. Because Taiwan had no domestic primary aluminum, copper, lead, or zinc smelting capacity, downstream metal producers relied on imports of ingots and scrap to produce products from these metals. Owing to high labor costs, environmental problems, and weak domestic demand, the output of these industries had gradually decreased during the past several years, and companies had moved their manufacturing facilities to mainland China and Southeast Asian countries. In 2012, the production of natural gas, mica, and sulfur increased significantly, but the output of silica sand decreased sharply (table 1; Taiwan Bureau of Mines, 2013).

Structure of the Mineral Industry

Table 2 is a list of major mineral industry facilities.

Commodity Review

Metals

Aluminum.—Without any primary aluminum production in Taiwan, aluminum product producers depended on imports of aluminum ingot and scrap to meet their needs. In 2012, Taiwan imported 337,411 metric tons (t) of unwrought aluminum, 430,464 t of aluminum alloys, and 126,851 t of scrap to meet domestic demand. Unwrought aluminum was imported from, in descending order of volume of imports, Australia, Iran, Tajikistan, the United Arab Emirates, Russia, and Bahrain. Aluminum alloy was from the United Arab Emirates, Bahrain, Russia, Australia, and Qatar, and aluminum scrap was from Mexico, Australia, Belgium, Colombia, Israel, the United States, and South Africa. Owing to increased demand for aluminum alloy on the island, CS Aluminum Corp., which was a subsidiary of China Steel Corp. (CSC), expanded its aluminum products output capacity to 167,000 metric tons per year (t/yr) in 2010 from 122,000 t/yr in 1998. The company started the second phase of its aluminum alloy production capacity expansion in 2011. The company's output capacity increased to 173,000 t in 2011. Once the expansion project is completed in 2014, the production capacity would be 319,000 t/yr (CS Aluminum Corp., 2013; Ministry of Finance, 2013b, p. 3-645–3-646).

Copper.—Without any refined copper production, Taiwan relied on imported copper to meet its demand. In 2012, Taiwan imported 435,534 t of refined copper, 458,242 t of copper alloys, and 93,767 t of scrap. Refined copper was mainly from, in descending order of volume of imports, Chile, Japan, Australia, Peru, and the Democratic Republic of the Congo [Congo (Kinshasa)], and copper alloy was from Japan, Russia, the Republic of Korea, and Ukraine. Because of surging demand for copper from the electronics sector, Taiwan's copper consumption was estimated to be about 1 million metric tons (Mt) in 2012 (Ministry of Finance, 2013b, p. 3-635–3-636).

Iron and Steel and Iron Ore.—Taiwan was the fifth-ranked iron and steel producer in Asia behind China, Japan, India, and the Republic of Korea. Owing to the decreased demand for steel products and the sluggish global steel market, especially in Europe, in 2012, Taiwan's steel producers reduced their steel output in the first half of the year. Taiwan's downstream producers of steel products reduced their new orders, and CSC, which was the only integrated iron and steel producer in Taiwan, decided to halt the production of its plate mill for 20 days for annual maintenance (World Steel Association, 2013).

In 2012, Taiwan imported 18.4 Mt of iron ore, which was a decrease of 10.2% from that of 2011. Iron ore was mainly from Australia, 69.5%; Brazil, 26.6%; and Canada, 3.6%. Without any iron ore resources, CSC imported all its iron ore from overseas markets. During the past several years, the company tried to secure iron ore resources in Australia and other countries. CSC's subsidiary CSC Steel Australia Holding Pty Ltd. invested $270 million to acquire a 3.68% share in ArcelorMittal Mines Canada's (ArcelorMittal Mines) Labrador Trough iron ore project in Canada. CSC and Pohang Iron and Steel Corp. of the Republic of Korea together held a 15% stake in ArcelorMittal Mines' iron ore project. CSC would receive 1 million metric tons per year (Mt/yr) of iron ore from ArcelorMittal Mines when it starts producing in 2013. This transaction was subject to Government approval in Canada and Taiwan. CSC sought eventually to increase its iron ore self-sufficiency to 15% during the next several years from its current level of 11.6% self-sufficiency (Ministry of Finance, 2013b, p. 3-115; Taiwan Iron and Steel Industries Association, 2013).

CSC invested $6.7 billion to increase the company's output capacity to 20 Mt/yr in the next several years. The company's subsidiary Dragon Steel Corp. completed the construction of a 2.5-Mt/yr blast furnace in 2010, and the installation of a second 2.5-Mt/yr blast furnace was scheduled to be completed in early 2013. After completing the second phase expansion in 2014, Dragon Steel would have a crude steel output capacity of 6 Mt/yr. CSC and Dragon Steel together would have a total crude steel output capacity of more than 16 Mt/yr (South East Asia Iron and Steel Institute, 2013).

Industrial Minerals

Cement.—Owing to a lack of limestone resources and a limited market on the island, many of Taiwan's cement producers had gradually moved their production bases to China in the late 1990s and expanded their cement output capacities there. Most of Taiwan's cement producers were located in the eastern part of the island, which accounted for more than 80% of Taiwan's total output capacity. In 2012, Taiwan produced less than 16 Mt of cement and consumed about 12 Mt. Taiwan exported 4.8 Mt of cement to, in descending order of volume, Ghana, Malaysia, Indonesia, Mauritius, and Australia. Taiwan's cement consumption had decreased gradually to about 12 Mt in recent years from 28 Mt in the 1990s, and cement exports had increased. The Taiwan authorities planned to cap cement exports at 30% of the total output by 2015 to encourage cement producers to phase out old, inefficient plants (Ministry of Finance, 2013a, p. 3-121; Taiwan Cement Corp., 2013, p. 40).

Mineral Fuels

Coal.—Taiwan had no domestic coal production and depended on imported coal to meet its demand for coal. Taiwan Power Co. was the leading coal consumer followed by the cement and iron and steel sectors. In 2012, Taiwan imported 64.6 Mt of coal, which was a decrease of 3.0% from the amount imported in 2011; of that amount, 46.7 Mt was for power generation and 8.1 Mt was for iron and steel production. Thermal coal was imported mainly from, in descending order of volume, Australia, Indonesia, South Africa, Russia, and China. Taiwan consumed 64.1 Mt of coal in 2012 (Taiwan Bureau of Energy, 2013a).

Natural Gas and Petroleum.—With its limited mineral fuel resources, Taiwan produced only about 0.02% of its petroleum requirements and relied on imports to meet the remaining demand—mainly through long-term contracts with, in descending order of supply, Saudi Arabia, Kuwait, Angola, Oman, the United Arab Emirates, and Iran. The island produced about 6.7% of its liquefied natural gas (LNG) in 2012. Taiwan imported a total of 16.7 billion cubic meters of LNG from, in descending order of volume of imports, Qatar, Malaysia, and Indonesia, which accounted for 76% of the island's total LNG imports. Taiwan consumed 15.3 billion cubic meters of LNG, of which power generation accounted for 84.3% (Taiwan Bureau of Energy, 2013b).

Outlook

Taiwan's economic growth is heavily dependent on external trade. The slow economic recovery in the United States and other developed countries is expected to affect its exports. Trade and investments between Taiwan and China are expected to continue to increase in the years to come. Economic growth in Taiwan is expected to increase slowly during the next 2 years and to be more dependent on the economic growth in the Asia and the Pacific region. The service sector accounts for more than 70% of the GDP and, given Taiwan's limited mineral resources, the mining sector is expected to have only a minimal effect on its economy in the future. The growth of the manufacturing sector is likely to be led by the computer, electronics components, and telecommunication products sectors. Taiwan relies on imports of raw materials to support its iron and steel and nonferrous metals sectors. The rising prices of these raw materials could affect producers' profit margins, and tightened environmental regulations may force nonferrous metal and steel producers to relocate their production facilities to mineral-rich countries with lower labor costs. Taiwan's economy has gradually been transforming from one that is led by a labor-intensive manufacturing sector to one that is led by a knowledge-intensive service sector. Taiwan authorities continue their efforts to promote Taiwan as a green island and to ease restrictions on economic ties with China, primarily in the areas of investment, tourism, trade, and transportation. Such changes would likely stimulate growth in the service sector.

References Cited

CS Aluminum Corp., 2013, Company profile: CS Aluminum Corp. (Accessed July 25, 2013, at http://www.csalu.com.tw/c/company.htm.)

Ministry of Economic Affairs, 2013, 2012 yearbook of industrial production statistics: Taipei, Taiwan, Department of Statistics, March, 233 p.

Ministry of Finance, 2013a, Monthly statistics of exports: Taipei, Taiwan, Directorate General of Customs, Statistical Department, December, variously paginated.

Ministry of Finance, 2013b, Monthly statistics of imports: Taipei, Taiwan, Directorate General of Customs, Statistical Department, December, variously paginated.

South East Asia Iron and Steel Institute, 2013, Taiwan's steelmaking industry produces NT302.4B in Q4: South East Asia Iron and Steel Institute Newsletter, May, p. 7.

Taiwan Bureau of Energy, 2013a, Energy statistics monthly—Coal: Taipei, Taiwan, Bureau of Energy, July, 3 p.

Taiwan Bureau of Energy, 2013b, Energy statistics monthly—Natural gas and petroleum: Taipei, Taiwan, Bureau of Energy, July, 6 p.

Taiwan Bureau of Mines, 2013, Annual statistical report of minerals 2012: Taipei, Taiwan, Bureau of Mines, 1 p.

Taiwan Cement Corp., 2013, 2012 annual report: Taipei, Taiwan, Taiwan Cement Corp. 239 p.

Taiwan Iron and Steel Industries Association, 2013, CSC increases self-supply iron ore from overseas: Taiwan Iron and Steel Industries Association. (Accessed August 5, 2013, at http://www.tsiia.org.tw/frontend message.aspx?message_id=21375.)

Taiwan Statistical Bureau, 2013, Monthly statistics of the Republic of China: Taipei, Taiwan, Directorate General of Budget, Accounting, and Statistics, no. 569, 66 p.

World Steel Association, 2013, Crude steel production—December 2012: Brussels, Belgium, World Steel Association press release, January 23, 1 p.

TABLE 1
TAIWAN: PRODUCTION OF MINERAL COMMODITIES[1]

(Metric tons unless otherwise specified)

Commodity		2008	2009	2010	2011	2012
METALS						
Iron and steel:						
Pig iron	thousand metric tons	9,750	7,939	9,358	12,940	11,800
Steel, crude	do.	19,222	15,566	20,498	22,879	21,083
Nickel, refined[e]		11,000	11,000	11,000	11,000	11,000
INDUSTRIAL MINERALS						
Cement, hydraulic	thousand metric tons	17,330	15,918	16,301	16,852	15,806
Fire clay		746	9	--	--	--
Lime[e]		450,000	450,000	460,000	460,000	450,000
Mica		3,179	557	--	1,455	6,844
Nitrogen, liquid		414,000 [r]	346,000 [r]	334,000 [r]	328,000 [r]	316,000
Silica sand		249,824	328,153	305,882	173,354	58,157
Sodium compounds, caustic soda		1,808,174 [r]	1,662,793 [r]	1,782,680 [r]	1,693,241 [r]	1,727,597
Stone:						
Dolomite	thousand metric tons	104	70	117	67	47
Limestone	do.	227	232	45	7	6
Marble	do.	25,811	24,146	25,118	24,351	22,524
Serpentine	do.	264	242	97	64	53
Sulfur		211,869	252,392	231,700	219,975	231,296
MINERAL FUELS AND RELATED MATERIALS						
Gas, natural:						
Gross	million cubic meters	357	351	290	330	442
Marketed[e]	do.	310	310	250	280	390
Petroleum:						
Crude	thousand 42-gallon barrels	101	101	91	71	72
Refinery products	do.	443,200	410,000	445,000	405,000	425,000

[e]Estimated; estimated data are rounded to no more than three significant digits. [r]Revised. do. Ditto. -- Zero.

[1]Table includes data available through July 18, 2013.

TABLE 2
TAIWAN: STRUCTURE OF THE MINERAL INDUSTRY IN 2012

(Thousand metric tons unless otherwise specified)

Commodity		Major operating companies	Location of main facilitites	Annual capacity[e]
Cement		Asia Cement Corp.	Hsinchu	1,800
Do.		do.	Hualien	4,020
Do.		Chia Hsin Cement Corp.	Kaohsiung	1,860
Do.		Chien Tai Cement Co. Ltd.	do.	1,720
Do.		Lucky Cement Corp.	Tungao	2,000
Do.		Southeast Cement Corp.	Kaohsiung	1,090
Do.		do.	Chutung	1,400
Do.		Taiwan Cement Corp.	Hualien	1,600
Do.		do.	Hualien Hsien	5,600
Do.		do.	Suao	3,400
Do.		Universal Cement Corp.	Kaohsiung	1,550
Marble		Taiwan Marble Co., Ltd.	Panchiao	15
Nickel		Taiwan Nickel Refinery	Kaohsiung	14
Petroleum:				
Crude	thousand 42-gallon barrels per year	Chinese Petroleum Corp.	Chuhuangkeng and Tungtzuchiao	850
Refinery products	do.	do.	Kaohsiung	570
Do.	do.	do.	Taoyuan	200
Do.	do.	Formosa Plastics Group	Yunlin	450
Steel		An Feng Steel Co. Ltd.	Kaohsiung	2,000
Do.		China Steel Corp.	do.	9,900
Do.		Dragon Steel Corp. (China Steel Corp.)	Taichung	3,500
Do.		Hai Kwang Enterprise Corp	Kaohsiung	600
Do.		Tang Eng Stainless Steel Plant	do.	300
Do.		Yieh Hsing Enterprise Co. Ltd.	do.	450
Do.		Yieh Phui Enterprise Co. Ltd.	do.	1,300
Do.		Yieh United Steel Co.	do.	1,000
Do.		Feng Hsin Iron and Steel Co. Ltd.	do.	1,200
Sulfur		China Petrochemical Development Corp.	Taipei	280

[e]Estimated; estimated data are rounded to no more than three significant digits. Do., do. Ditto.

THE MINERAL INDUSTRY OF THAILAND

By Lin Shi

In 2012, Thailand's mineral production increased by 8.2%, its manufacturing production increased by 6.9%, and its gross domestic product (GDP) increased by 6.5% compared with that of 2011, respectively. In late 2011, Thai economic expansion was interrupted suddenly by the worst flood that the country had encountered in 70 years—the industrial areas in Bangkok and its surrounding Provinces were seriously affected. The economic recovery began in the second quarter of 2012 following the previous year's decline (Bank of Thailand, 2013a).

Minerals in the National Economy

Thailand is one of the world's leading producers of cement, feldspar, gypsum, and tin. The country's identified mineral resources were being produced for domestic consumption and export. Thailand's manufacturing sector produced automobiles and automotive parts, cement, integrated circuits, and jewelry (table 1; Bank of Thailand, 2013b, p. 9; Carlin, 2013; Crangle, 2013; Tanner, 2013; U.S. Central Intelligence Agency, 2013; van Oss, 2013).

Government Policies and Programs

Thailand's economy in recent years was affected by both domestic and international financial circumstances. The Government policies that followed the world financial crisis in 2008 helped move the country's economy to positive growth in 2010, and the Thai economy continued to have strong growth in the first three quarters of 2011 until the flooding in the fourth quarter. In 2012, the Government initiated $11.7 billion worth of infrastructure projects to modify the manufacturing infrastructure after the flood and prevent similar economic damage in the future. The Government also applied a nationwide minimum daily wage and implemented policies to increase the country's exports and lower business and employee taxes (Bank of Thailand, 2013b, p. 51–59; U.S. Central Intelligence Agency, 2013).

The Ministry of Industry is the principal Government agency that oversees the country's mineral sector. The Ministry's Department of Primary Industry and Mines (DPIM) administers the Minerals Act and issues mining regulations. The DPIM also provides technical assistance to the metallurgical, mineral processing, and mining industries. The Department of Mineral Resources (DMR) drafts national mineral policies and provides technical assistance for geologic prospecting and mineral exploration. The DMR conducts geologic mapping, manages mineral resources, performs mineral analyses, and administers the country's mineral resources information center (Department of Mineral Resources, 2012).

Structure of the Mineral Industry

Table 2 is a list of major mineral industry facilities in Thailand. Most of the nonfuel mineral mining and mineral processing companies in Thailand were privately owned and operated. The Electricity Generating Authority of Thailand (EGAT) and several coal mining companies owned and operated most of the county's major coal exploration and mining businesses. The Petroleum Authority of Thailand (PTT), PTT Exploration and Production Public Co. Ltd. (PTTEP) and its joint ventures, and major multinational oil companies owned most of the country's petroleum and natural gas exploration projects and exploitation businesses. Thailand's mineral industry was engaged in mining and processing of metallic and industrial minerals and exploring for crude oil and natural gas (table 2).

Mineral Trade

In 2012, Thailand's exports were valued at about $226 billion compared with $219 billion in 2011, which was an increase of 3%. Imports were valued at about $218 billion compared with $202 billion in 2011, which was an increase of 8%. Thailand exported industrial commodities, including automobiles and automotive parts, computer parts, electrical appliances, electronics, machinery, and equipment, mainly to China (12%), the United States (11.5%), Japan (10%), Hong Kong (6%), and Malaysia (5%). Thailand imported raw materials and fuels for its domestic manufacturing industry mainly from Japan (20%), China (15%), the United Arab Emirates (6%), and Malaysia and the United States (5% each) (U.S. Census Bureau, 2012a, b; Bank of Thailand, 2013b, p. 11–17; U.S. Central Intelligence Agency, 2013).

Historically, Thailand's trading partners were several European countries, Japan, and the United States. Thailand's traditional export markets were the members of the Association of Southeast Asian Nations (ASEAN)—Indonesia, Malaysia, the Philippines, Singapore, and Vietnam. In addition, the countries of Burma, Cambodia, and Laos had gradually become more important to the Thai economy as high-potential export and investment destinations. The numbers of Thai export markets and trade partners were continuing to increase to include Australia, China, India, the Middle East, and South Africa (Bank of Thailand, 2013b, p. 11–17; U.S. Central Intelligence Agency, 2013).

In 2012, Thailand's imports from the United States were valued as about $11 billion, and included semiconductors valued at $1.3 billion; gold, $500 million; petroleum products, $319 million; and steelmaking materials, $148 million. Thailand's exports to the United States were valued at about $26 billion, and included crude oil valued at $813 million; iron and steel products, $195 million; tin, $38 million; and bauxite and aluminum, $11 million. Thailand also exported manufacturing products to the United States, including $4.7 billion worth of computer accessories, peripherals, and parts; $2.4 billion worth of telecommunications equipment; and $1.4 billion worth of metal jewelry, such as watches and rings (U.S. Census Bureau, 2012a, b).

Production

In 2012, Thailand's tin output decreased by 56% to 124 metric tons (t) from 282 t in 2011, crude oil production decreased by 27% to 37.2 million barrels (Mbbl) from 50.9 Mbbl in 2011, and natural gas production decreased by 25% to 21.8 million cubic meters from 29.1 million cubic meters in 2011. Manganese production of 6,300 t was about 34 times the 187 t in 2011, gold production increased by 75% to 4,200 kilograms (kg) from 2,400 kg in 2011, and silver production increased by 60% to 31,000 kg from 19,000 kg in 2011. Manufactured iron (iron sheet) production increased by 20% to 0.6 million metric tons (Mt) from 0.5 Mt in 2011, and steel manufacturing (steel bar and shape steel) production increased by 6.7% to 1.6 Mt from 1.5 Mt in 2011. Cement production increased by 11% to 41.0 Mt from 36.6 Mt in 2011 (Bank of Thailand, 2013c, p. 6–8).

Commodity Review

Metals

Copper.—Thailand-registered Puthep Co. Ltd., which was a joint venture of Padaeng Industry Public Co. Ltd. of Thailand (51%) and PanAust Ltd. of Australia through its wholly owned subsidiary PNA Pty Ltd. (49%), owned and explored the Puthep copper project in Loei, northern Thailand. In mid-2012, PanAust and Padaeng began to divest the project. The joint venture also evaluated the effect of the changes in the copper price and mining costs on the results of previous studies, which were completed 2 years earlier (PanAust Ltd., 2013).

Gold.—Kingsgate Consolidated Ltd. of Australia owned and operated the Chatree gold mine in central Thailand. In fiscal year 2012 (which ran from the beginning of July 2011 to the end of June 2012), the company produced about 4,200 kg of gold and about 31,000 kg of silver. The company mined 5.7 Mt of ore during the year. Because the milled ore had a higher grade than the planned head grade, gold production increased by 75% from the 2,400 kg produced in 2011. Kingsgate's new processing plant had an ore processing capacity of more than 5 million metric tons per year (Mt/yr), but operation of the new plant was delayed by 63 days in 2012 (Kingsgate Consolidated, Ltd., 2013 p. 3).

Zinc.—Padaeng was engaged in mining, milling, and smelting zinc and producing zinc alloys. Padaeng owned the only zinc mine in Thailand—the Mae Sod Mine, which is located in the Mae Sod district of Tak Province. Padaeng's smelter was located in the Muang district of Tak Province, the roaster plant was located in Rayong Province, and the head office was located in Bangkok. In 2012, the Mae Sod Mine produced less zinc silicate ore output than in previous years, and the Tak smelter processed increased volumes of secondary raw material, such as zinc carbonate and sulfide and nonsulfide zinc. to meet the mill's zinc production demand. Therefore, despite the decrease in zinc ore and primary zinc metal output, Padaeng produced 53,000 t of zinc ingots, 44,000 t of zinc alloy, and 101,000 t of zinc cathodes. Padaeng increased its domestic zinc sales to about 91,000 t, or by 13% compared with those of 2011, and increased its zinc exports to about 16,000 t, or by 14% (Padaeng Industry Public Co. Ltd., 2013, p. 19–24).

Industrial Minerals

Gypsum.—Thailand's gypsum deposits occur in both the northern and southern areas of the country; however, mining was concentrated in the southern part of the country because of the south's proximity to the country's seaports. Siam Cement Group and Thai Gypsum Products were the two major suppliers to the Thai gypsum market. Thai Gypsum Products had the capacity to produce 75 million square meters per year of gypsum products, and the company exported about 40% of its output. In recent years, Thai Gypsum Products focused on the development and production of new gypsum products and introduced a thermal insulation board and thermal tile to its line of products. BuilderSmart Public Co. Ltd. (BSM), which received 51% of its revenue from gypsum products, refocused on its export markets, mainly India and Myanmar. BSM's exports aimed to increase the company's revenue to $31.5 million within the next 2 years and to increase its net profit by 50% each year (Global Gypsum, 2013).

Mineral Fuels

Natural Gas and Petroleum.—Chevron Corp. of the United States operated the Platong II natural gas project in the Gulf of Thailand. Platong II is located in shallow waters 193 kilometers (km) from the southern coastline of Thailand. Chevron invested $3.1 billion in this project and expected to reach production of 9.3 million cubic meters per day of natural gas and 18,000 barrels per day of natural gas liquids when the project is completed. Platong II was aimed at increasing the country's domestic natural gas production by more than 10% to meet Thailand's growing energy demand; it would increase Chevron's net natural gas production from the Gulf of Thailand by more than 20%. Chevron operated and held a 69.9% interest in the Platong II project; Mitsui Oil Exploration Co. Ltd. of Japan held a 27.4% interest in the project, and PTTEP held the remaining 2.7% (Chevron Corp., 2011).

Outlook

The joint venture of PanAust and Padaeng was to continue its Puthep project asset sale in 2013 and to shift the investment to other projects that would be more profitable. Kingsgate expects that the gold production at the Chatree Mine in fiscal year 2013 will be in the range of between 3,700 kg and 4,100 kg. Kingsgate planned further exploration for gold in the Chatree Mine leasing area in hopes of substantially extending the mine's life. Chevron acquired production rights in the Pakarang oil and gas field, which covers 118 square kilometers in the Gulf of Thailand, and is expected to invest about $605 million to develop the project. The World Bank upgraded Thailand's income categorization from a lower-middle income economy to an upper-middle income economy in July 2011 owing to the country's progress in economic development. According to the Bank of Thailand, the country's GDP is forecast to grow by 4.9% in 2013, and the country's exports will gradually improve (Dow Jones & Company, Inc., 2012; Bank of Thailand, 2013b; Kingsgate Consolidated Ltd., 2013; Padaeng Industry Public Co. Ltd., 2013; PanAust Ltd., 2013).

References Cited

Bank of Thailand, 2013a, Table 2—Growth rate of domestic production in major sectors: Bank of Thailand, 1 p. (Accessed October 18, 2013, at http://www.bot.or.th/English/Statistics/Indicators/Docs/tab02.pdf.)

Bank of Thailand, 2013b, Thailand's economic conditions in 2012: Bank of Thailand, 66 p. (Accessed October 22, 2013, at http://www.bot.or.th/English/EconomicConditions/Thai/report/AnnualReport_Doc/AnnualReport_2012.pdf.)

Bank of Thailand, 2013c, Table 3—Major non-agricultural products: Bank of Thailand, 3 p. (Accessed October 23, 2013, at http://www.bot.or.th/English/Statistics/Indicators/Docs/tab03.pdf.)

Carlin, J.F., Jr., 2013, Tin: U.S. Geological Survey Mineral Commodity Summaries 2013, p. 170–171.

Chevron Corp., 2011, Chevron announces first gas from Platong II project in Gulf of Thailand: Chevron Corp., October 24. (Accessed October 31, 2013, at http://www.chevron.com/chevron/pressreleases/article/10242011_chevronannouncesfirstgasfromplatongiiprojectingulfofthailand.news.)

Crangle, R.D., Jr., 2013, Gypsum: U.S. Geological Survey Mineral Commodity Summaries 2013, p. 70–71.

Department of Mineral Resources, 2012, About us: Department of Mineral Resources. (Accessed September 27, 2012, at http://www.dmr.go.th/ewtadmin/ewt/dmr_web/main.php?filename=web_En.)

Dow Jones & Company, Inc., 2012, Chevron, Apico get Thailand OK for production rights in 2 fields: Dow Jones & Company, Inc., August 22. (Accessed August 25, 2012, at http://www.rigzone.com/news/oil_gas/a/120179/Chevron_Apico_Get_Thailand_OK_for_Production_Rights_in_2_Fields.)

Global Gypsum, 2013, Thai supplier refocuses on exports: Global Gypsum, September 5. (Accessed October 31, 2013, at http://www.globalgypsum.com/news/item/852-thai-supplier-refocuses-on-exports.)

Kingsgate Consolidated Ltd., 2013, Annual reports 2012: Kingsgate Consolidated Ltd., 95 p. (Accessed October 31, 2013, at http://www.kingsgate.com.au/investors/annuals-quarterlies.htm.)

Padaeng Industry Public Co. Ltd., 2013, Padaeng Industry Plc. annual report of year 2012: Padaeng Industry Public Company Ltd., 102 p. (Accessed October 31, 2013, at http://www.padaeng.com/media/annual-report.php.)

PanAust Ltd., 2013, Puthep copper project: PanAust Ltd. (Accessed October 30, 2013, at http://www.panaust.com.au/thailand-puthep.)

Tanner, A.O., 2013, Feldspar: U.S. Geological Survey Mineral Commodity Summaries 2013, p. 54–55.

U.S. Census Bureau, 2013a, U.S. exports to Thailand by 5-digit end-use code 2003–2012: U.S. Census Bureau. (Accessed October 23, 2013, at http://www.census.gov/foreign-trade/statistics/product/enduse/exports/c5490.html.)

U.S. Census Bureau, 2013b, U.S. imports from Thailand by 5-digit end-use code 2003–2012: U.S. Census Bureau. (Accessed October 23, 2013, at http://www.census.gov/foreign-trade/statistics/product/enduse/imports/c5490.html.)

U.S. Central Intelligence Agency, 2013, Thailand, *in* The world factbook: U.S. Central Intelligence Agency. (Accessed October 17, 2013, at https://www.cia.gov/library/publications/the-world-factbook/geos/th.html.)

van Oss, H.G., 2013, Cement: U.S. Geological Survey Mineral Commodity Summaries 2013, p. 38–39.

TABLE 1
THAILAND: PRODUCTION OF MINERAL COMMODITIES[1]

(Metric tons unless otherwise specified)

Commodity[2]		2008	2009	2010	2011	2012
METALS						
Antimony:						
Ore, gross weight		--	--	--	25 [r]	28
Metal, smelter[e]		422 [3]	555 [3]	500	500	500
Copper, metal, refined, secondary[e]		438	490 [3]	490 [3]	500	500
Gold, mine output, Au content	kilograms	2,721	5,400	4,125	2,372	4,158
Iron and steel:						
Iron ore:						
Gross weight		1,586,250 [r]	528,899 [r]	969,937	285,566 [r]	103,009
Fe content[e]		779,000 [r]	259,000 [r]	475,000 [r]	140,000 [r]	50,000
Crude steel	thousand metric tons	5,211	3,646	4,145	4,000 [e]	4,000 [e]
Lead, metal		73,303	55,504	55,500	55,000 [e]	55,000 [e]
Manganese ore:						
Metallurgical grade, gross weight, 46% to 50% MnO_2		111,000	64,930	50,450	398 [r]	13,435
Mn content		52,700	31,200	24,200 [e]	187 [r]	6,300
Silicon, metal, gross weight		--	--	NA	22,500 [e]	22,500 [e]
Silver, mine output, Ag content	kilograms	5,465	15,300	17,092	19,456 [r]	31,121
Tin:						
Concentrate, Sn content		215	166	291	282	124
Metal, primary[e]		21,860 [3]	19,423 [3]	20,000	20,000	20,000
Tungsten concentrate, W content[e]		-- [r]	-- [r]	-- [r]	40 [r]	83
Zinc:						
Ore:						
Gross weight		118,728 [r]	184,505	146,470	148,391 [r]	95,671
Zn content		17,811	34,000	25,529	25,000 [r,e]	31,000
Metal, primary		107,753	104,695	100,000 [e]	68,203	53,000
Alloy, Zn content		70,000	31,000	30,000 [e]	35,163	44,000

See footnotes at end of table.

TABLE 1—Continued
THAILAND: PRODUCTION OF MINERAL COMMODITIES[1]

(Metric tons unless otherwise specified)

Commodity[2]		2008	2009	2010	2011	2012
INDUSTRIAL MINERALS						
Barite		9,180	51,895	33,465	67,703 r	64,499
Cement, hydraulic	thousand metric tons	31,651	33,562	36,496	36,602 r	41,047
Clays:e						
Ball clay		1,499,993 [3]	1,000,000	1,000,000	1,000,000	1,000,000
Kaolin, marketable:						
Beneficiated, washed		162,215 [3]	160,000	160,000	160,000	160,000
Nonbeneficiated, unwashed		479,443 [3]	500,000	500,000	500,000	500,000
Filler		6,061 [3]	6,000	6,000	6,000	6,000
Diatomite		4,075	5,600 r	7,100 r	38,130 r	8,500
Feldspar		670,618	718,692 r	641,900 r	1,041,152 r	1,100,619
Fluorspar, crude, metallurgical grade		26,118	86,365	2,222	5,093 r	9,602
Gypsum	thousand metric tons	8,500	8,679	9,985	8,955 r	6,259
Perlite		7,000 e	13,500 r	14,700 r	26,500 r	41,400
Phosphate rock, crude		3,675	658 r	35,783 r	3,300 r	1,990
Sand, silica, glasse		495,848 [3]	500,000	500,000	500,000	500,000
Stone:						
Calcitee		823,706 [3]	750,000 [3]	750,000	750,000	750,000
Dolomitee		1,353,763 [3]	1,200,000 [3]	1,200,000	1,200,000	1,200,000
Granite:						
Dimension stone	cubic meters	10,579	6,352 r	6,123 r	5,267 r	4,820
Industrial rock	thousand metric tons	5,190	5,210 r	5,259 r	5,648 r	6,339
Limestone	do.	142,118 r	132,521 r	135,022 r	145,599 r	147,657
Marble, dimension stone and fragment	cubic meters	664,930	735,216 r	779,234 r	509,237 r	311,839
Marl for cement manufacture only		41,720	98,000 r	68,000 r	65,000 r	100,000
Quartz		3,290	12,954 r	49,064 r	152,576 r	401,710
Shale for cement manufacture only	thousand metric tons	3,767 r	4,181 r	4,000	4,593 r	4,792
Travertine		3,640	2,910 r	1,760 r	900 r	900 e
Talc		3,264	504 r	672 r	2,304 r	5,856
MINERAL FUELS AND RELATED MATERIALS						
Coal, lignite	thousand metric tons	18,095	16,360	17,907	21,324 r	12,072
Natural gas, gross production	million cubic meters	27,576	26,362	29,583	29,059 r	21,766
Petroleum:						
Crude	thousand 42-gallon barrels	52,805	56,302	55,906	50,976 r	37,164
Natural gas condensate	do.	31,157	30,625	31,730	30,693 r	21,169
Refinery productse	do.	219,000 r	229,000 r	229,000 r	229,000 r	229,000

eEstimated; estimated data are rounded to no more than three significant digits. rRevised. do. Ditto. NA Not available. -- Zero.

[1]Table includes data available through October 29, 2013.

[2]In addition to the commodities listed, Thailand produced gemstones and pyrophyllite, but available information is inadequate to make reliable estimates of output.

[3]Reported figure.

Sources: Department of Mineral Resources, Mineral Statistics of Thailand; Department of Primary Industries and Mines; Ministry of Energy, Energy Policy and Planning Office; and U.S. Geological Survey Minerals Questionnaires, 2008–2012.

TABLE 2
THAILAND: STRUCTURE OF THE MINERAL INDUSTRY IN 2012

(Thousand metric tons unless otherwise specified)

Commodity		Major operating companies and major equity owners	Location of main facilities	Annual capacity
Antimony	metric tons	Amco Thai Mining Co. (Hibino Metal Industry)	Antimony smelter, Ban Pin, Phrae Province	555
Do.		Other companies	Located in different Thai Provinces	NA
Barite		Asian Mineral Resources Co. Ltd.	Loei, Mae Hong Son, Nakhon Si Thammarat, and Satun Provinces	60
Do.		P&S Barite Mining Co. Ltd.	Loei and Nakhon Si Thammarat Province	60
Cement		Asia Cement Co. Ltd.	Pra Phutthabath, Saraburi Province	4,800
Do.		Jalaprathan Cement Co. Ltd. (Cement Francais S.A., 37%; Veatprapat Holding Co. Ltd., 19%; others, 44%)	Takli and Nakhorn, Sawarn Province, and Cha-Am, Petchburi Province	2,350
Do.		Samukee Cement Ltd.	Pakchong, Nakhon Ratchasima Province	125
Do.		Saraburi Cement Co. Ltd. (CEMEX Asia Holdings Ltd., 99%)	Chalerm Phrakiat, Saraburi Province	700
Do.		Siam Cement Industry Co. Ltd. (Bureau of the Crown Property, 30%; Thai Security Depository Co. Ltd., 6.94%; CPB Equity Co. Ltd., 5.6%; other financial institutions and the general public, 57.46%)	Kaeng Khoi, Phabhudhabat, and Khao Wong, Saraburi Province; Chae Hom, Lampang Province; Thung Song, Thammarat Province; and Ta Luang, Ayutthaya Province	23,200
Do.		Siam City Cement Co. Ltd. (SCCC) (Holcim Ltd., 33.7%; Rattanarak family, 27%; other investors, 39.3%)	Kaeng Khoi, Saraburi Province	14,500
Do.		TPI Polene Co. Ltd.	do.	9,900
Coal, lignite		Electricity Generating Authority of Thailand (EGAT) (Government, 100%)	Mae Moh, Lampang Province	20,000
Do.		Lanna Lignite Public Co. Ltd.	Ban Pakha, Lamphun Province	1,000
Copper		Thai Copper Industries Public Co. Ltd. (TCI)	Rayong Industrial Park, Rayong Province	165
Feldspar, concentrate		Asia Mineral Processing Co. Ltd.	Nakhon Si Thammarat and Trang Provinces	500
Fluorspar, concentrate		Asian Mineral Resources Co. Ltd.	Mae Hong Son Province	14
Gas, natural	million cubic meters per day	Esso Exploration and Production Khorat Inc.	Namphong, Khon Kaen Province	4
Do.	do.	TOTAL Exploration and Production (Thailand)	Bongkot in the Gulf of Thailand	15
Do.	do.	Unocal Thailand Ltd.	Baanpot, Erawan, Funan, Kaphong, Pladang, Satun, Pailin, and Trat, all in the Gulf of Thailand	33
Do.	do.	Chevron Corp.	Platong II project	NA
Gold	kilograms	Akara Mining Ltd. (Kingsgate Consolidated Ltd., 100%)	Chatree, Phichit Province	5,000
Gypsum		Vanich Gypsum Co. Ltd.	Khlong Prab, Mai Riang. Thoong Yai Mai in Nakhon Si Thammarat and Surat Thani Provinces	8,500
Do.		Siam Cement Group	NA	NA
Do.	thousand square meters	Thai Gypsum Products Public Co. Ltd.	NA	75,000
Do.		Lotus Mines Co. Ltd.	Nakornsawan	NA
Do.		General Mining and Trading Co. Ltd.	Talad, Muang	NA
Iron ore, gross weight		P.T.K. Mining Co. Ltd.	Phu Ang, Loei Province	720
Lead, in concentrate		Kanchanaburi Exploration and Mining Co. Ltd.	Song Toh, Nong Phai, and Bo Ngam in Kanchanaburi Province	55
Petroleum, crude, including condensate	thousand 42-gallon barrels per day	Chevron Corp.	Benjamas and Tantawan, offshore in the Gulf of Thailand	35
Do.	do.	PTT Exploration and Production Public Co. Ltd. (PTTEP)	Arthit, Songkhla, Gulf of Thailand	20
Do.	do.	Thai Shell Exploration and Production Co. Ltd.	Sirikit in Kamphaenghet Province	24
Do.	do.	TOTAL Exploration and Production (Thailand)	Bongkot, offshore in the Gulf of Thailand	12
Do.	do.	Unocal Thailand Ltd.	Baanpot, Erawan, Funan, Gomin, Jakrawan, Kaphong, Pailin, Platon, Satun, Surat, and Trat Plamuk, offshore in the Gulf of Thailand	38

See footnotes at end of table.

TABLE 2—Continued
THAILAND: STRUCTURE OF THE MINERAL INDUSTRY IN 2012

(Thousand metric tons unless otherwise specified)

Commodity		Major operating companies and major equity owners	Location of main facilities	Annual capacity
Silicon, metal (gross weight)	metric tons	G.S. Energy Co., Ltd.	Ratchaburi Silicon Plant	25,000
Steel, rolled		The Bangkok Iron and Steel Works Co. Ltd.	Phrapradaeng, Samutprakarn Province	120
Do.		Bangkok Steel Industry Public Co. Ltd.	do.	300
Do.		Tata Steel (Thailand) Plc (Tata Steel Ltd., 67.11%; McDonald Investment, 6.5%; other investors, 26.39%)	Map Ta Phut, Rayong Province; Sriracha, Chonburi Province; and Ban Mon, Saraburi Province	1,700
Do.		Namheng Steel Co. Ltd.	Lopburi Province	300
Do.		Sahaviriya Group Corp. Ltd.	Bang Saphan, Prachuap Khiri Khan Province	2,400
Do.		Siam United Steel Co. Ltd.	Rayong Province	1,000
Do.		G-Steel Plc (formerly Siam Ystrip Mill Plc)	Map Ta Phut, Rayong Province	600
Tantalum, metal powder and oxides	metric tons	H.C. Starck (Thailand) Co. Ltd. (H.C. Starck GmbH, 94.98%, and others, 5.02%)	do.	250
Tin:				
Concentrate		Numerous small companies	Nakhon Si Thammarat, Phangnga, Phuket, and Rayong Provinces	3
Refined		Thailand Smelting and Refining Co. Ltd. (Thaisarco) (Amalgamated Metal Corp., 75.25%, and other, 24.75%)	Phuket, Phuket Province	30
Tungsten, in concentrate	metric tons	SC Mining Co. Ltd. (Som Chai family, 100%)	Ban Pin, Phrae Province	650
Zinc:				
In concentrate		Padaeng Industry Public Co. Ltd. (Bali Ventures Ltd, 21.7%; Thai Ministry of Finance, 13.81%; RAK Minerals & Metals Investments, 12.5%; and others, 52%)	Mae Sod district, Tak Province	65
Refined		do.	Smelter in Muang district, Tak Province; roaster plant in Rayong Province	115

Do., do. Ditto. NA Not available

THE MINERAL INDUSTRY OF VIETNAM

By Yolanda Fong-Sam

In 2012, Vietnam ranked seventh in the production of crude petroleum in the Asia and the Pacific region. Vietnam also produced about 2.3%, 1.8%, and 1% of the world's tin, cement, and barite, respectively (Carlin, 2013; Miller, 2013; U.S. Energy Information Administration, 2013; van Oss, 2013). Other minerals produced in the country included chromium ore, coal, natural gas, lead, crude petroleum, phosphate rock, salt, and zirconium. As for major processed minerals, Vietnam produced refined copper, rolled steel, refined tin, and zinc (table 1).

Minerals in the National Economy

According to the General Statistics Office of Vietnam (2012b), the output value of the mining and quarrying sector (which included mineral fuels and nonfuel minerals) in 2012 increased by about 3.5% to an estimated $11.2 billion[1] (in 2010 constant dollars[2]) from $10.8 billion in 2011. The mining and quarrying sector made up 9.57% of the country's total estimated gross domestic product of $116.6 billion (in 2010 constant dollars) compared with about 9.62% in 2011.

Government Policies and Programs

In 2012, the Government of Vietnam created and approved several decisions and decrees that supported the implementation of the 2010 Mineral Law and hence attracted international interest in the country's potential for mineral mining. In January 9, the Government released Directive 02/CT–TTg (Directive 02), which contains stricter provisions with respect to the exploration, mining, processing, usage, and exportation of mineral resources. Directive 02 outlines a specific licensing process for such mineral commodities as apatite, bauxite, cement, chromite, coal, copper, gold, lead and zinc, manganese, and rare earths. The new licensing provisions are compatible with the new Mineral Law and the country's master plan for the mineral industry. The master plan calls for the use and mining of minerals to be accomplished in a reasonably economical and efficient manner while maintaining technological progress and protecting cultural sites and the environment. Under Directive 02, licenses for new exploration for and mining of bauxite and alumina-related products will not be granted until the Tan Rai and Nhan Co Mines, which are located in Lam Dong and Dak Nong, are commissioned and operational. Licenses for the exploration of coal and minerals related to the manufacturing of cement will continue to be granted provided they are compliant with the country's mineral law. Directive 02 also prohibits the issuance of new licenses for the mining of placer gold, bans the exportation of unprocessed titanium ore starting on July 1, 2012, and establishes that all mining projects must be appraised by the Appraisal Council of the Ministry of Industry and Trade, the Ministry of Construction, and the Provincial or municipal committee (Mayer Brown JSM, 2012c).

In March, Decree No. 15/2012/ND–CP (Decree 15) was issued, and it went into effect on April 25. The decree designates the ministries and Government offices that are to oversee matters related to the country's mining industry. Decree 15 designates the Ministry of Industry and Trade (MIT), along with the Ministry of Natural Resources and the Environment (MONRE), as the responsible entities to coordinate provisions for the export of minerals. The MIT sets the conditions and standards for the export of most minerals, with the exception of construction materials. The export of construction materials is coordinated by the Ministry of Construction. Under the MONRE, the General Department of Geology and Minerals processes applications for mining licenses as well as applications for mine closures; the Office of the Council for Assessment of National Mineral Resources certifies the validity of mineral resource assessments (after analyzing the results of feasibility studies). In addition, the decree stipulates that the costs associated with geologic surveys performed with Government funds be reimbursed by the end user of the data. The decree regulates the selection and licensing of companies to explore for minerals in areas that are not subject to the auctioning of mining rights. The decree requests license holders to apply for an extension of the mining license, if interested, at no later than 45 days prior to the expiration of the license and to provide a report on the results of the preliminary work done on the property. In areas where the mining license is obtained through Government auction, the Government issued Decree No. 22/2012/ND–CP (Decree 22) on March 26, which regulates the conditions, procedures, and requirements governing such licenses. Decree 22 went into effect on May 15 (Mayer Brown JSM, 2012a, b).

On December 21, the MIT released Circular No. 41/2012/TT–BCT (Circular 41), which went into effect on February 4, 2013, and provides the guidelines for the export of minerals. The circular includes a list of minerals allowed to be exported, the specific quality of the material (percentage of contained mineral), and the conditions under which the minerals may be exported; the goods include metallic minerals and industrial minerals. The circular does not regulate the export of coal, natural gas, petroleum, minerals used as construction materials, and minerals used as raw materials for the production of cement. Minerals that are allowed to be exported included barite powder, bismuth (≥70%), products processed from bauxite ores [alumina ≥ 98.5% and (or) aluminum hydroxide ≥64%] and titanium ores (titanium slag grade 1 with titanium oxide content ≥85%; grade 2 with titanium oxide

---

[1]Where necessary, values have been converted from Vietnam dong (VND) to U.S. dollars (US$) at the rate of VND20,693=US$1.00 for 2012 and VND20,454=US$1.00 for 2011.

[2]In 2012, the General Statistics Office of Vietnam started reporting its data at constant 2010 dollars, whereas in the past it was reported in constant 1994 dollars.

content from 70% to 85%). Other minerals allowed for export include copper (≥20%) and wolframite (≥55%) from the Nui Phao mining and mineral processing company, fluoride (calcium fluoride ≥90%), white marble (powder and lump), nickel ore from Ban Phuc Nickel Mines LLC, and total rare-earth-oxide powder (≥99%). Circular 41 specifies that minerals may be exported only by enterprises operating under the Enterprise Law and that are in compliance with the Commercial Law established by the country; both laws require enterprises to meet certain conditions related to the export, processing, and trade of goods with foreign partners. Parties interested in exporting minerals must comply with the conditions set out in the circular, including that the minerals must be in a processed state (to encourage domestic processing of ore) and listed as being allowed for export; that they must have a minimum percentage of mineral content in the ore; and that they must have been exploited legally in mines under a valid license or imported lawfully. All material processed for export must be accompanied by supporting documentation certifying its lawful origin and quality. The Heavy Industry Department under the MIT is the designated entity responsible for the coordination and logistics of mineral exports and is responsible for overseeing compliance with Circular 41 (General Department of Vietnam Customs, 2013).

The main law that regulates the mining industry in Vietnam is the 2010 Mineral Law, which was passed by the National Assembly of Vietnam in November 2010 and became fully effective on July 1, 2011. The new Mineral Law replaces the 1996 Mineral Law, as amended in 2005. The new law regulates geologic surveys for mineral resources, the protection of unexploited minerals, mineral exploration, mineral mining, and the management of minerals located within all the territory under the control of Vietnam. To promote sustainable development and ensure that mined products are used in a cost-effective and efficient way, the Government had begun developing mineral master plans (mineral strategies) every 10 years that include outlooks for 20 years. The plans cover such topics as the future of the country's mineral resources, as well as the promotion of sustainable socioeconomic development, national defense, and security. The master plans also ensure the protection of minerals and that the mining of minerals is executed in an environmentally friendly manner and in the most cost-effective way. The MONRE oversees and coordinates with other Ministries to prepare the mineral strategy and then submits it to the Prime Minister for approval. The Government assigned the MONRE as the authority to regulate all minerals throughout the country (Mayer Brown JSM, 2011).

Production

In 2012, the greatest increases in mineral production were for salt (by about 36.7%), titanium (36.1%), zinc slab (12.5%), zirconium (an estimated 11.4%), natural gas (10.9%), and petroleum (10.2%). Production decreased for manganese (by 75.5%), iron ore and tungsten (35.8% each), zinc ore (an estimated 26.5%), building stone (12.7%), and sand and gravel (10.8%). Data on mineral production are in table 1.

Structure of the Mineral Industry

According to the General Statistics Office of Vietnam (2011, 2012a), the number of employees working in the mining and quarrying sector in 2011 (the latest year for which data were available) was approximately 279,100, which accounted for less than 1% of the total number of employed people in the country. Investments in the mining and quarrying sector for 2012 accounted for approximately $2.6 billion (in 2010 constant dollars), which represented about 6.9% of the total investment in the country. Table 2 is a list of major mineral industry facilities and their ownership. Many of the mineral industry facilities are Government owned.

Mineral Trade

In 2012, total trade in Vietnam increased by about 12% to $228.3 billion (preliminary) from $203.7 billion in 2011. The total value of exports for 2012 was about $114.5 billion compared with about $97 billion in 2011 (an increase of 18%). Exports of coal decreased by 11% to about 15.2 million metric tons (Mt) from about 17 Mt in 2011; exports of crude oil increased by 12.3% to 68 million barrels (Mbbl) from 60.6 Mbbl in 2011. In 2012, the total value of imports increased by about 6.6% to $113.8 billion from $106.7 billion in 2011. Imports of steel increased by 3% to about 7.6 Mt from about 7.4 Mt in 2011; imports of copper decreased by about 5% to about 1.10 Mt from about 1.16 Mt in 2011 (General Statistics Office of Vietnam, 2012c, f, g).

Vietnam's main trading partners in 2012 were Australia, Burma, China, Germany, Hong Kong, India, Japan, the Republic of Korea, Malaysia, Taiwan, Thailand, and the United States. The United States was Vietnam's leading export partner; the United States imported $19.7 billion in Vietnamese goods (which was equivalent to 17.2% of Vietnam's total exports), followed by Japan, which imported $13.1 billion (11.4% of Vietnam's total exports), and China, which imported $12.4 billion (10.8% of Vietnam's total exports). Vietnam's imports came mainly from India (which supplied 25.3% of Vietnam's total imports valued at an estimated $28.8 billion), Hong Kong (13.7% of total imports valued at an estimated $15.5 billion), and China (10.2% of total imports valued at about $11.6 billion) (General Statistics Office of Vietnam, 2012d, e).

Commodity Review

Metals

Antimony.—Total antimony resources in Vietnam were estimated to be approximately 845,000 metric tons (t), which would make them some of the largest in the world. In recent years, the Government of Vietnam had been investing in prospecting for, mining, and processing antimony ore, specifically in the Chiem Hoa zone in northern Vietnam in Tuyen Quang Province; the Na Bac zone on the northeastern coast of Vietnam (includes the Provinces of Hoa Binh, Ninh Binh, and Thanh Hoa); the Quang Ninh ore zone located in the northeastern part of the country in Quang Ninh Province; and the Yen Minh ore zone of northern Vietnam in Ha Giang Province. Exploratory studies have revealed that the Chiem Hoa zone had an estimated resource of 400,000 t of antimony. The Lang Vai Mine, which is located within the Chiem Hoa zone, also was estimated to host 139,000 t of arsenic, 103 t of silver, and 10 t of gold. The Quang Ninh ore zone had an estimated resource of about 338,000 t containing 35,000 t of antimony concentrate distributed in two deposits—the Khe Chim and the Tan Mai deposits. The Vietnam Institute of Geosciences and Mineral Resources (VIGMR) announced the start of the construction of an antimony oxide preparation plant in Quang Ninh Province. The plant, which would have a processing capacity of 200,000 to 300,000 metric tons per year (t/yr) of ore, would produce battery components, chemicals, and flame retardants (Industrial Minerals, 2013).

Bauxite and Alumina and Aluminum.—According to the Ministry of Industry and Trade, as of 2012, Vietnam imported approximately 500,000 t/yr of aluminum at a cost of about $1 billion. Domestic consumption of aluminum was projected to increase by between 750,000 t/yr and 1 million metric tons per year (Mt/yr) by 2020, and by 1.6 to 2 Mt/yr by 2030. Domestically, aluminum is used mainly in the country's manufacturing industry. Government studies have estimated Vietnam's bauxite reserves to be between 10 and 11 billion metric tons (Gt). Most of the reserves are located in the Central Highlands of the country, of which about 4.6 Gt is located in Dak Nong Province and 2 Gt is located in Lam Dong Province (Vietnam National Coal-Mineral Industries Holding Corp. Ltd., 2013).

By yearend 2012, the Vietnam National Coal-Mineral Industries Holding Corp. Ltd. (Vinacomin) announced that the Tan Rai alumina and bauxite complex had successfully carried out the trial production of aluminum products. In addition, the Government had been investing in infrastructural upgrades, such as roads, to connect the Tan Rai plant to highways and ports to facilitate the transportation of the alumina for export. In 2012, the construction of the Tan Rai complex, which is located in the Bao Lam District in Lam Dong Province in the country's Central Highlands, was 90% completed. The project began construction in 2008 with a capital investment of $700 million; the estimated production capacity of the plant would be 600,000 t/yr of alumina when completed. The estimated initial output for the first year after commissioning was 300,000 t of alumina, and the following year, 520,000 t. The estimated life of the project was 30 years (Vietnam National Coal-Mineral Industries Holding Corp. Ltd., 2012a–c; 2013).

On November 15, Vinacomin announced that it had signed a loan agreement with a consortium of banks led by Citi Vietnam to finance projects to mine bauxite and produce alumina in Lam Dong and Dak Nong Provinces. The consortium included the Bank of Tokyo-Mitsubishi UFJ Ltd., Mizuho Corp. Bank Ltd., and Sumitomo Mitsui Trust Bank Ltd., all of Japan. The agreement was valued at $300 million and consisted of a 13-year loan that was guaranteed by the Ministry of Finance and the Agency for Export and Investment Insurance of Japan (Thomson Reuters, 2012; Vietnam National Coal-Mineral Industries Holding Corp. Ltd., 2012c).

In Dak Nong Province, Vinacomin was also building the Nhan Co bauxite mining and refinery complex, which was expected to start operations in 2014 at a cost of $665 million. Initial output was projected to be 300,000 t/yr of alumina; the complex was projected to reach full capacity of 650,000 t/yr of alumina by 2016. The project would include an alumina refinery and a bauxite ore sorting plant. With the commissioning of the Tan Rai and the Nhan Co projects, the country was expected to produce about 1,250,000 t/yr of alumina at full capacity, of which between 600,000 and 900,000 t/yr would be sold to the Yunnan Metallurgical Group of China in a 30-year sales agreement (Thomson Reuters, 2012; Vietnam National Coal-Mineral Industries Holding Corp. Ltd., 2012a, c).

Copper.—In July 2012, Vinacomin officially updated the Sin Quyen copper mine resource estimate to 106 Mt from the previous estimate of 56 Mt. The Sin Quyen Mine, which is located in the Bat Xat District, was operated by Vinacomin's subsidiary Sin Quyen Copper Co. The Sin Quyen Mine processed 1.2 Mt/yr of ore at an average grade of 1.03% copper. During 2012, the company was waiting for Government approval to increase the Lao Cai copper complex's mining capacity to 3 Mt/yr from 1.2 Mt/yr, and also to extend the mine life to between 40 and 50 years from between 20 and 30 years. The mine originally started production in 2006. According to Vinacomin, all the copper processed in Lao Cai's copper refinery was consumed by the domestic market (Vietnam National Coal-Mineral Industries Holding Corp. Ltd., 2012d).

Gold.—Besra Gold Inc. of Canada, formerly Olympus Pacific Minerals Ltd., owns interest in several gold projects in Quang Nam Province in the Central Highlands of Vietnam. During 2012, the company discovered a high-grade gold vein in the Nui Kem deposit within the company's Bong Mieu property. After the discovery, the company refocused an ongoing 4-year underground drilling program on the Nui Kem deposit and the Ho Gan South deposit, which is part of the Bong Mieu property. The new objectives of the drilling program were to verify if the new discovery shows continuity in the mineralization between the two deposits. Gold resources in the Bong Mieu property were estimated to be about 22,000 kilograms (kg) (reported as 700,000 troy ounces). Preliminary studies revealed that the average grade in the Ho Gan South deposit was between 6 and 8 grams per metric ton (g/t) gold. Another project in the Bong Mieu premises was the Ho Ray open pit, which was under a feasibility study. The company was expected to produce about 3,000 kg of gold (reported as 100,000 troy ounces) in 2013 at the Bong Mieu property owing mainly to the company's plans to upgrade the facilities (Gold Mining Journal, 2012).

Iron and Steel.—In October 2012, ABB Group (an automation and power technology group of Switzerland) announced that Formosa Plastic Group of Taiwan had granted the group a contract worth about $50 million to produce a series of substations to supply power and enhance the power transmission capacity to a new steel complex. The substations would consist of four gas-insulated switchgears that would supply power to a steel complex built by Hung Nghiep Formosa Ha Tinh Steel Corp. (Formosa Steel Complex) located in the Vung Ang Economic Zone in Ha Tinh Province, 400 km southeast of the capital city of Hanoi. The steel complex would have the capacity to house four steel blast furnaces with a combined production capacity of 15 Mt/yr of steel, a 1,600-megawatt thermal power station, and the Son Duong deep sea port, which would be designed with an annual handling capacity of 300 Mt. The complex was also expected to produce hot-rolled steel sheets and steel bars. The project was expected to be completed by 2014 (ABB Group, 2012).

Tungsten.—On November 30, Hazelwood Resources Ltd. of Australia announced that the commissioning of the ATC Ferrotungsten project in Vietnam was programmed for February 2013. The company claimed that the ATC plant was the largest ferrotungsten plant, in terms of capacity, to be built outside of China. The plant had a designed capacity of 4,000 t/yr of ferrotungsten alloy. The company expected to start ferrotungsten production in the second quarter of 2013 (Yao, 2012).

Industrial Minerals

Cement.—In 2012, cement production decreased by 4.7% to 55.53 Mt from 58.27 Mt (revised) in 2011. The Vietnam Cement Association (VNCA) reported that in 2012, the sale of cement for domestic use was 45.5 Mt, and the sale of cement and clinker for export was 8.5 Mt. For 2013, the VNCA expected an increase in cement domestic consumption to about 49 Mt. According to the Vietnam Cement Industry Corp. (VICEM), the country had excess cement production capacity; the designed production capacity was 70 Mt/yr of cement, and in 2012, the country produced at about 80% of total capacity. The Vietnam Building Material Association (VBMA) estimated that producers incurred losses of approximately $80 million in 2012 because of the reduction in cement prices, as producers increased exports in order to decrease the amount of surplus inventory. The Ministry of Construction encouraged local cement producers to seek export markets because cement production in the country exceeded domestic demand. Cement producers, however, have encountered some challenges in the export process, owing mainly to poor local infrastructure (which impeded the transportation of goods), the high cost of transportation, and the lack of a port facility with enough capacity to handle large shipments (table 1; World Cement, 2012; GlobalCement.com, 2013; International Cement Review, 2013).

In August, PT Semen Gresik of Indonesia announced its interest in acquiring a cement company in Vietnam to supply cement to Indonesia. No further details were released (International Cement Review, 2012).

Fluorspar.—In 2012, several fluorspar mining projects were under development, many of which were expected to be commissioned in 2013. Phu Yen Mineral Joint-Stock Co. (PYMICO), which was one of the leading fluorspar suppliers in Vietnam, invested about $14.2 million in the construction of a fluorspar processing plant in Phu Yen Province in eastern Vietnam. The new processing plant was expected to have an estimated production capacity of 20,000 t/yr of fluorspar. PYMICO also owned the Xuan Lanh Mine, which is also located in Phu Yen Province. Another project expected to be commissioned by the end of 2013 was Masan Group's Nui Phao, which is located in Thai Nguyen Province in northeastern Vietnam. The estimated resource for this project was 8.5 Mt of fluorspar, and the project had an estimated production capacity of 214,000 t/yr. Another fluorspar deposit in Vietnam is the Binh Duong fluorspar deposit, which is located in Nguyen Binh District, Cao Bang Province, and had proven reserves of 5,341 t. The fluorspar in the Dong Pao deposit, which is located in Lai Chau Province, is associated with such minerals as barite and rare earths; the fluorspar reserve at Dong Pao is estimated to be about 1 Mt. The Xuan Lanh fluorspar deposit is located in Phu Province in central Vietnam; the proven reserve of fluorspar at this deposit is about 277,500 t, and the probable reserve is about 105,500 t (Salwan, 2013).

Rare Earths.—During 2012, the Dong Pao Rare Earth Mine, which is located in Ban Hon Commune, Tam Duong District, Lai Chau Province, was mined by artisanal miners. The miners dug for rare-earth minerals, which they then sold to traders. The Dong Pao Mine was considered one of the largest rare-earth mines in the country; it covered an area of 11 square kilometers and had a reserve estimate of 5 Mt. The mine was operated by Lai Chau-Vietnam National Minerals Corp. (VIMICO) Rare Earth Joint Stock Co. and the Japanese Dong Pao Rare Earth Development Co. During 2012, the mine was still under development (the commissioning date was yet to be determined), when commissioned, the mine was expected to produce about 10,000 t/yr of ore. In October 2012, according to Vinacomin, the operators of the mine signed a memorandum of understanding for the exploitation and processing of rare-earth oxide from the Dong Pao Mine. The agreement could potentially lead to ore output of 720,000 t/yr (TalkVietnam.com, 2012; Vietnam National Coal-Mineral Industries Holding Corp. Ltd., 2012e).

Mineral Fuels

Coal.—In May, Vinacomin announced the country's coal production plan for 2012–15. During that period, the company was planning to upgrade and expand the production capacity of about 60 coal mines located in the northeastern part of the country. Vinacomin was also planning to invest in starting 28 new coal mines, which would have a collective capacity of about 2 Mt/yr of coal. By 2015, the country was expected to have the capacity to produce 55 Mt/yr of coal (Vietnam National Coal-Mineral Industries Holding Corp. Ltd., 2012f).

Crude Petroleum.—In June 2011, the operator of the Dung Quat oil refinery, Binh Son Refining and Petrochemical Co. of Vietnam, announced the company's plans to expand the capacity of the refinery at an estimated cost of up to $2 billion. The expansion, which was expected to be commissioned by 2017, would increase the processing capacity to 200,800 barrels per day (bbl/d) from 130,500 bbl/d. The facility was designed to process crude petroleum from the Middle East and Venezuela (Tran, 2011).

In 2012, Vietnam's second refinery, the Nghi Son refinery, was still under construction. The refinery, which was located in the Nghi Son Economic Zone in northeastern Vietnam in Thanh Hoa Province, would be managed by Nghi Son Refinery & Petrochemical LLC. The 200,000-bbl/d facility had an estimated construction cost of $9 billion; it was expected to be commissioned by yearend 2016 and to begin commercial operations by mid-2017. The Nghi Son refinery's ownership was as follows: Idemitsu Kosan Co. Ltd. of Japan and Kuwait Petroleum International of Kuwait (35.1% each), PetroVietnam (25.1%), and Mitsui Chemicals Inc. of Japan (4.7%) (Oil & Gas Journal, 2013).

In November 2012, Vietnam Oil and Gas Group (PetroVietnam) announced that the Su Tu Trang offshore field had started gas production. The field, identified as Block 15.1, was located in the Cuu Long Basin and was operated by the Cuu Long Joint Operating Co. Data on total capacity were not available (Vu, 2012).

Natural Gas.—In April, Russian-Vietnam Joint Operating Co. (Vietgazprom) announced the start of the third round of exploratory drilling in the Bao Vang deposit, which is located on the country's continental shelf. The same deposit was drilled previously in 2009 and 2010. During 2012, in addition to the exploratory drillings, the company planned to complete a feasibility study of the Bao Vang deposit, carry out 3-dimensional seismic exploration, and determine the reserves of the deposit (OAO Gazprom International, 2012; undated).

Outlook

Since Vietnam's 2010 Mineral Law became fully effective in July 2011, the Government has been diligent in the creation and approval of decisions and decrees to support the implementation of the Mineral Law to attract international interest in the country's mining of metals and industrial minerals. As a result, the country has seen an increase in foreign investment in the mineral industry, which has resulted in new exploration drillings, resource discoveries, and expansions and commissioning of many projects in the bauxite, cement, copper, gold, and other sectors. In the near future, an increase in the production of fluorspar, and tungsten and its byproducts, is expected as projects that were under development in 2012 come online in 2013 and 2014. VNCA forecasted a steady increase in domestic consumption of cement for 2013 as a result of an increase in demand to support infrastructure projects in the country.

In the next 5 years and beyond, the production of metals and industrial minerals is expected to increase as the development of the mineral projects that started in 2011 and 2012 progress and mines and plants start being commissioned. Among these projects are the Tan Rai alumina and bauxite complex, which is expected to start production in early 2013. Other mining projects expected to be commissioned in 2013 include the Dong Pao rare-earths project, the Nhan Co bauxite mining and refinery complex, and the Nui Phao polymetallic mining project. Exports of cement are likely to increase over time, mainly because cement output is expected to continue to exceed local demand.

Vietnam's trading of minerals with neighboring countries is highly dependent on the fluctuation of the demand for such commodities in the region, although the demand will also be dependent on the world market and the economic conditions of neighboring countries.

References Cited

ABB Group, 2012, ABB wins $50 million orders to power new steel complex in Vietnam: Zurich, Switzerland, ABB Group press release, October 23, 1 p.

Carlin, J.F., Jr., 2013, Tin: U.S. Geological Survey Mineral Commodity Summaries 2013, p. 170–171.

General Department of Vietnam Customs, 2013, Circular No. 41/2012/TT–BCT of December 21, 2012, providing the export of minerals: General Department of Vietnam Customs. (Accessed November 22, 2013, at http://www.customs.gov.vn/Lists/EnglishDocuments/ViewDetails.aspx?ID=1142&language=en-US.)

General Statistics Office of Vietnam, 2011, Population and employment—Employed population at 15 years of age and above as of annual 1 July by kinds of economic activity, in Statistical yearbook of Vietnam: General Statistics Office of Vietnam. (Accessed September 24, 2012, at http://www.gso.gov.vn/default_en.aspx?tabid=467&idmid=3&ItemID=12892.)

General Statistics Office of Vietnam, 2012a, Investments—Investment at constant 2010 prices by kinds of economic activity, in Statistical yearbook of Vietnam: General Statistics Office of Vietnam. (Accessed November 18, 2013, at http://www.gso.gov.vn/default_en.aspx?tabid=471&idmid=3&ItemID=14373.)

General Statistics Office of Vietnam, 2012b, National accounts—Gross domestic product—Constant 2010 prices by types of ownership and by kinds of economic activity, in Monthly statistical information: General Statistics Office of Vietnam. (Accessed November 18, 2013, at http://www.gso.gov.vn/default_en.aspx?tabid=468&idmid=3&ItemID=14494.)

General Statistics Office of Vietnam, 2012c, Trade, price and tourism—Export and import of goods in 2012, in Statistical yearbook of Vietnam: General Statistics Office of Vietnam. (Accessed November 18, 2013, at http://www.gso.gov.vn/default_en.aspx?tabid=472&idmid=3&ItemID=14616.)

General Statistics Office of Vietnam, 2012d, Trade, price and tourism—Exports of goods by country group, by country and territory, in Statistical yearbook of Vietnam: General Statistics Office of Vietnam. (Accessed November 18, 2013, at http://www.gso.gov.vn/default_en.aspx?tabid=472&idmid=3&ItemID=14611.)

General Statistics Office of Vietnam, 2012e, Trade, price and tourism—Imports of goods by country group, by country and territory, in Statistical yearbook of Vietnam: General Statistics Office of Vietnam. (Accessed November 18, 2013, at http://www.gso.gov.vn/default_en.aspx?tabid=472&idmid=3&ItemID=14606.)

General Statistics Office of Vietnam, 2012f, Trade, price and tourism—Some main goods for exportation, in Statistical yearbook of Vietnam: General Statistics Office of Vietnam. (Accessed November 18, 2013, at http://www.gso.gov.vn/default_en.aspx?tabid=472&idmid=3&ItemID=14610.)

General Statistics Office of Vietnam, 2012g, Trade, price and tourism—Some main goods for importation, in Statistical yearbook of Vietnam: General Statistics Office of Vietnam. (Accessed November 18, 2013, at http://www.gso.gov.vn/default_en.aspx?tabid=472&idmid=3&ItemID=14605.)

GlobalCement.com, 2013, Vietnam cement producers lost US$80m in 2012: GlobalCement.com, April 12. (Accessed October 29, 2013, at http://www.globalcement.com/news/item/1588-vietnam-cement-producers-lost-us$80m-in-2012.)

Gold Mining Journal, 2012, Vietnam holds more for Olympus: Gold Mining Journal, v. 1, no. 108, p. 56.

Industrial Minerals, 2013, Vietnam antimony—Exploration accelerates: Industrial Minerals, no. 549, June, p. 28–33.

International Cement Review, 2012, Vietnam—Gresik investments: International Cement Review, August, p. 9.

International Cement Review, 2013, Vietnam—Full year sales down: International Cement Review, February, p. 10.

Mayer Brown JSM, 2011, Vietnam's 2010 mineral law: Mayer Brown JSM. (Accessed January 19, 2012, at http://www.mayerbrown.com/publications/article.asp?id=10599&nid=6.)

Mayer Brown JSM, 2012a, Decree on auctioning mining rights: Mayer Brown JSM. (Accessed November 21, 2013, at http://www.mayerbrown.com/Decree-on-Auctioning-Mining-Rights-05-30-2012/.)

Mayer Brown JSM, 2012b, Vietnam issues decree to implement the Mineral Law: Mayer Brown JSM. (Accessed November 21, 2013, at http://www.mayerbrown.com/Vietnam-Issues-Decree-to-Implement-the-Mineral-Law-05-31-2012/.)

Mayer Brown JSM, 2012c, Vietnam's long-term strategy for exploitation of mineral resources: Mayer Brown JSM, June 1. (Accessed November 21, 2013, at http://www.mayerbrown.com/Vietnams-Long-Term-Strategy-for-Exploitation-of-Mineral-Resources-06-01-2012/.)

Miller, M.M., 2013, Barite: U.S. Geological Survey Mineral Commodity Summaries 2013, p. 24–25.

OAO Gazprom International, 2012, Vietgazprom started drilling the third exploratory well in the Bao Vang deposit: OAO Gazprom International. (Accessed November 22, 2013, at http://gazprom-international.com/en/news-media/articles/vietgazprom-started-drilling-third-exploratory-well-bao-vang-deposit.)

OAO Gazprom International, [undated], Vietnam—Current state: OAO Gazprom International. (Accessed November 22, 2013, at http://www.gazprom.com/about/production/projects/deposits/vietnam/.)

Oil & Gas Journal, 2013, Plans advance for Vietnam's second refinery: Oil & Gas Journal, January 1. (Accessed on January 31, 2013, at http://www.ogj.com/articles/2013/01/plans-advance-for-vietnams-second-refinery.html.)

Salwan, Shruti, 2013, Uncertainty rules over Vietnam fluorspar: Industrial Minerals, April 30, p. 1, 2. (Accessed September 29, 2014, at http://www.indmin.com/Print.aspx?ArticleId=3198795.)

TalkVietnam.com, 2012, Rare earth minerals illegally stripped in Lai Chau: TalkVietnam.com, June 10. (Accessed November 18, 2013, at http://talkvietnam.com/2012/06/rare-earth-minerals-illegally-stripped-in-lai-chau/#.UqZRdz-ma9u.)

Thomson Reuters, 2012, Vietnam receives $300 mln Citi loan for alumina project: Thomson Reuters, November 15. (Accessed November 15, 2012, at http://www.reuters.com/article/2012/11/15/vietnam-alumina-loan-idUSL3E8MF0WR20121115.)

Tran, Le Thuy, 2011, Vietnam Dung Quat refinery to hit 200,800 bpd 2017: Thomson Reuters, June 13, 2 p. (Accessed September 29, 2014, at http://www.reuters.com/article/2011/06/13/us-climate-summit-vietnam-refinery-idUSTRE75C0OE20110613.)

U.S. Energy Information Administration, 2013, International energy statistics from 2008–2012—Production of crude oil, NGPL, and other liquids: U.S. Energy Information Administration. (Accessed November 18, 2013, at http://www.eia.gov/cfapps/ipdbproject/iedindex3.cfm?tid=5&pid=55&aid=1&cid=r1,&syid=2008&eyid=2012&unit=TBPD.)

van Oss, H.G., 2013, Cement: U.S. Geological Survey Mineral Commodity Summaries 2013, p. 38–39.

Vietnam National Coal-Mineral Industries Holding Corp. Ltd., 2012a, Many countries interested in Vietnam alumina: Vietnam National Coal-Mineral Industries Holding Corp. Ltd., March 6. (Accessed November 27, 2013, at http://www.vinacomin.vn/en/news/News-of-Vinacomin/Many-countries-interested-in-Vietnam-alumina-95.html.)

Vietnam National Coal-Mineral Industries Holding Corp. Ltd., 2012b, Ore-bauxite factory to open in April: Vietnam National Coal-Mineral Industries Holding Corp. Ltd., February 28. (Accessed November 27, 2013, at http://www.vinacomin.vn/en/news/News-of-Vinacomin/Ore-bauxite-factory-to-open-in-April-93.html.)

Vietnam National Coal-Mineral Industries Holding Corp. Ltd., 2012c, Tan Rai bauxite plant to release first product in mid-December: Vietnam National Coal-Mineral Industries Holding Corp. Ltd., November 27. (Accessed November 27, 2013, at http://www.vinacomin.vn/en/news/News-of-Vinacomin/Tan-Rai-bauxite-plant-to-release-first-product-in-mid-December-404.html.)

Vietnam National Coal-Mineral Industries Holding Corp. Ltd., 2012d, Vinacomin doubles copper resource estimate to 106m/t: Vietnam National Coal-Mineral Industries Holding Corp. Ltd., July 30. (Accessed December 6, 2013, at http://www.vinacomin.vn/en/news/News-of-Vinacomin/Vinacomin-doubles-copper-resource-estimate-to-106m-t-269.html.)

Vietnam National Coal-Mineral Industries Holding Corp. Ltd., 2012e, Vinacomin & Japanese firm to exploit and process rare earth in Lai Chau: Vietnam National Coal-Mineral Industries Holding Corp. Ltd., October 8. (Accessed December 9, 2013, at http://www.vinacomin.vn/en/news/News-of-Vinacomin/Vinacomin-Japanese-firm-to-exploit-and-process-rare-earth-in-Lai-Chau-278.html.)

Vietnam National Coal-Mineral Industries Holding Corp. Ltd., 2012f, Vinacomin plans to produce 55m tonnes of clean coal by 2015: Vietnam National Coal-Mineral Industries Holding Corp. Ltd., May 16. (Accessed November 27, 2013, at http://www.vinacomin.vn/en/news/News-of-Vinacomin/Vietnam-plans-to-produce-55m-tonnes-of-clean-coal-by-2015-205.html.)

Vietnam National Coal-Mineral Industries Holding Corp. Ltd., 2013, Developing bauxite industry is sound policy: Vietnam National Coal-Mineral Industries Holding Corp. Ltd., March 21. (Accessed November 27, 2013, at http://www.vinacomin.vn/en/news/Home-News/Developing-bauxite-industry-is-sound-policy-515.html.)

Vu, Trong Khanh, 2012, PetroVietnam—Su Tu Trang field starts gas production: Rigzone.com, November 16. (Accessed November 19, 2012, at http://www.rigzone.com/news/oil_gas/a/122188/PetroVietnam_Su_Tu_Trang_Field_Starts_Gas_Production.)

Yao, Bevis, 2012, Hazelwood Resources is closer to commissioning Vietnam ferrotungsten plant: Proactive Investors Australia Pty Ltd. press release. (Accessed December 9, 2013, at http://www.proactiveinvestors.com.au/companies/news/36659/hazelwood-resources-is-closer-to-commissioning-vietnam-ferrotungsten-plant-36659.html.)

World Cement, 2012, Cement industry trends—Asia: World Cement, September 7. (Accessed October 29, 2013, at http://www.worldcement.com/news/cement/articles/Cement_market_export_expansion_Asia_1160.aspx#.UqiP1z-ma9t.)

TABLE 1
VIETNAM: PRODUCTION OF MINERAL COMMODITIES[1]

(Metric tons unless otherwise specified)

Commodity[2]		2008	2009	2010	2011	2012
METALS						
Antimony ore		540	664	608	714 [r]	755
Bauxite[e]		80,000	80,000	80,000	100,000	100,000
Chromium ore, gross weight[e]		55,880 [3]	37,105 [3]	40,000	40,000	40,000
Copper:						
Concentrate, gross weight		46,079	51,741	49,038	47,552	45,065
Concentrate, Cu content[e]		11,000 [3]	11,300 [3]	11,300	11,300 [r]	11,300
Metal, smelter[e]		2,200	6,000	8,000	8,000	8,000
Gold[e]	kilograms	3,000	3,000	3,500	3,500	3,500
Iron and steel:						
Iron ore, Fe content		1,371,600	1,904,500	1,972,100	2,371,300 [r]	1,523,100
Metal:						
Steel, crude	thousand metric tons	937 [r]	1,702 [r, 3]	2,906 [r]	2,931 [r]	2,992 [r]
Steel, rolled	do.	5,001	7,498 [r]	8,415 [r]	8,085 [r]	7,640
Lead, mine output, Pb content[e]		14,200	7,700	6,500 [r]	6,400 [r]	6,300
Manganese:[e]						
Gross weight		62,300 [3]	92,200 [3]	82,700	64,600	15,800
Mn content (43%)		26,800 [3]	39,600 [3]	35,600	27,800	6,800
Tin:						
Mine output, Sn content[e]		5,400	5,400	5,400	5,400	5,400
Metal, smelter		3,583	2,747	3,042	3,900 [r]	4,000 [e]
Titanium concentrate, gross weight[4]		709,500	698,700	912,000	840,600 [r]	1,143,800
Tungsten, mine output, W content		--	725	1,150	1,635	1,050
Zinc:[e]						
Mine output, Zn content		42,000	38,000	36,000	34,000 [r]	25,000
Slab		16,000 [r]	17,000	16,000	16,000 [r]	18,000
Zirconium, gross weight[e, 5]		22,000	6,800	6,900	14,000	15,600
INDUSTRIAL MINERALS						
Barite[e]		90,000 [3]	75,000 [3]	85,000	85,000	85,000
Cement, hydraulic	thousand metric tons	40,009	48,810	55,801	58,271 [r]	55,531
Lime	do.	1,619	1,584	1,454	1,500 [e]	1,500 [e]
Phosphate rock:						
Gross weight	do.	2,101	2,047	2,325	2,395 [r]	2,365
P_2O_5 content[e]	do.	630	614	680	720	710
Salt	do.	717	679	975	862 [r]	1,178
Sand and gravel, and silica sand	do.	112,000	123,000	110,300	101,295 [r]	90,354
Stone, building stone	do.	317,429	355,932	381,828	404,421 [r]	352,823
MINERAL FUELS AND RELATED MATERIALS						
Coal, anthracite	thousand metric tons	39,777	44,078	44,835	46,611 [r]	42,383
Gas, natural, gross	million cubic meters	7,499	8,010	9,402	8,480	9,403
Petroleum, crude	thousand 42-gallon barrels	109,291	119,968	110,098	111,351 [r]	122,747

[e]Estimated; estimated data are rounded to no more than three significant digits. [r]Revised. do. Ditto. -- Zero.

[1]Table includes data available through November 19, 2013.

[2]In addition to the commodities listed, bentonite, construction aggregates, gemstones, granite, graphite, gypsum, kaolin clay, lignite, marble, nitrogen, pig iron, pyrite, pyrophyllite, rare earths, refractory clay, silver, and sulfur were mined, but not reported, and available information is inadequate to make reliable estimates of output.

[3]Reported figure.

[4]Figures based on Vietnam's inferred exports of titanium ores to China, Japan, the Republic of Korea, Malaysia, and the United States.

[5]Estimated figures based on Vietnam's inferred exports of zirconium ore to China.

Sources: General Statistics Office of Vietnam, Statistical Yearbook, 2009–2012; World Steel Association, Steel Statistical Yearbook, 2008; World Metal Statistics, December 2009; South East Asia Iron and Steel Institute, Crude Steel Production, Annual Statistics, 2009–2011; The Barytes Association, World Barytes Production, 2008–2010; Copper Bulletin of the International Copper Study Group, 2012; International Chromium Development Association, Statistical Bulletin, 2010–2011; International Tungsten Industry Association, 2008–2012; U.S. Geological Survey Minerals Questionnaire 2011–2012.

TABLE 2
VIETNAM: STRUCTURE OF THE MINERAL INDUSTRY IN 2012

(Thousand metric tons unless otherwise specified)

Commodity	Major operating companies and major equity owners	Location of main facilities	Annual capacity
Alumina	Vietnam National Coal-Mineral Industries Holding Corp. Ltd. (Vinacomin)	Tan Rai alumina complex in Lam Dong Province	600.
Barite	NA	Ao Sen deposit located in Son Duong District, Tuyen Quang Province	80.
Bauxite	Vietnam National Coal-Mineral Industries Holding Corp. Ltd. (Vinacomin)	Tan Rai plant, located in Bao Lam District, Lam Dong Province	600.
Cement	An Giang Cement Co.	An Giang cement plant, An Giang Province	400.
Do.	Binh Phuoc Cement Co.	Binh Phuoc cement plant, Binh Phuoc Province	2,000.
Do.	Building Materials Corp. No. 1	Fico Tay Ninh cement plant, in Tan Chau District, Tay Ninh Province	2,000.
Do.	Cement X18 Factory Co.	Cement X18 plant, Lang Son Province	100.
Do.	Chin Fon Cement Co.	Chin Fon cement plant, Ha Giang Province	1,400.
Do.	Chinfong Hai Phong Cement Corp. [Chingfong Group of Taiwan, 70%; Hai Phong Municipal Government, 15.56%; Vietnam National Cement Corp. (VICEM), 14.44%]	Min Duc cement near Hai Phong City	1,400.
Do.	Cong Thanh Cement Joint Stock Co.	Cong Thanh cement plant, Thanh Hoa Province	1,000.
Do.	Cao Ngan Cement Co.	Cao Ngan cement plant, Thai Nguyen Province	600.
Do.	Dong Banh Cement Co.	Dong Banh cement plant, Lang Son Province	1,000.
Do.	Dong Son Cement Co.	Dong Son cement plant, Thai Nguyen Province	1,500.
Do.	Dong Thanh Cement Co.	Dong Thanh cement plant, Dong Nai Province	1,000.
Do.	Ha Long Cement Co.	Ha Long cement plant, Ho Chi Minh City	2,000.
Do.	Ha Tien Kien Giang Cement Co.	Ha Tien Kien Giang cement plant, Binh Duong Province	200.
Do.	Lafarge (Vietnam) Cement	Cement grinding station in Dong Nai Province	500.
Do.	La Hien Cement Co.	La Hien cement plant, Thai Nguyen Province	600.
Do.	Langbang Cement Co.	Langbang cement plant, Quang Ninh Province	1,500.
Do.	Luckvaxi Cement Co.	Luckvaxi cement plant, Thua Thien-Hue Province	1,200.
Do.	Luck's Group (Vietnam Holdings) Co. Ltd.	Kim Dinh cement plant and Ninh Thuan grinding plant, Thua Thien-Hue Province	2,800.
Do.	Lucky Group Ltd. and Phuc Son Cement Corp.	Phuc Son cement plant, Hai Duong Province	4,000.
Do.	Mai Son Cement Co.	Mai Son cement plant, Son La Province	1,200.
Do.	Midland Construction Corp. (COSEVCO)	Song Gianh cement plant, Quang Binh Province	1,400.

See footnotes at end of table.

TABLE 2—Continued
VIETNAM: STRUCTURE OF THE MINERAL INDUSTRY IN 2012

(Thousand metric tons unless otherwise specified)

Commodity	Major operating companies and major equity owners	Location of main facilities	Annual capacity
Cement—Continued	Morning Star Cement Ltd. [Holcim Group, 65%, and Vietnam National Cement Corp. (VICEM), 35%]	Cat Lai grinding plant, Hiep Phuoc grinding plant, Thi Vai grinding plant Hon Chong, Kien Giang Province	4,700.
Do.	Nghi Son Cement Corp. [Taiheiyo Cement Corp., 45.5%; Mitsubishi Materials Corp. of Japan, 19.5%; Vietnam National Cement Corp. (VICEM), 35%]	Nghi Son cement plant, Thanh Hoa Province	4,300.
Do.	Quang Ninh Cement and Construction Joint Stock Co.	Quang Ninh cement plant, Ha Long, Quang Ninh Province	1,200.
Do.	Quan Trieu Cement Joint Stock Co. [Viet Bac Mining Industry Corp. and Vietnam National Coal-Mineral Industries Holding Corp. Ltd. (Vinacomin)]	Quan Trieu cement plant, Thai Nguyen Province	820.
Do.	ROLI-Quang Tri Cement Co.	ROLI-Quang cement plant, Quang Tri Province	600.
Do.	Song Thao Cement Co.	Song Thao cement plant, Phu Tho Province	1,000.
Do.	Thai Nguyen Cement Co.	Thai Nguyen cement plant, Thai Nguyen Province	1,400.
Do.	Thang Long Cement	Thang Long cement grinding plant in Hiep Phuoc Industrial Zone	2,300.
Do.	Tuyen Quang Cement Group 1	Tuyen Quang cement plant, Tuyen Quang Province	600.
Do.	Vietnam Construction and Import-Export Joint Stock Corp. (VINACONEX)	Cam Pha cement grinding plant in Phu Tho Province	2,300.
Do.	do.	Luong Son cement plant, Hoa Binh Province	1,200.
Do.	do.	Yen Bai cement plant, Yen Bai Province	200.
Do.	Vietnam National Cement Corp. (VICEM) (100% state owned)	Bim Son cement, Thanh Hoa Province	3,800.
Do.	do.	But Son cement, Ha Nam Province	1,600.
Do.	do.	Hai Phong cement, Ha Giang Province	1,700.
Do.	do.	Ha Tien I, Ho Chi Minh City	1,500.
Do.	do.	Ha Tien II, Kien Giang Province	1,200.
Do.	do.	Hai Van cement, Danang City	600.
Do.	do.	Hoang Mai cement, Nghe An Province	1,400.
Do.	do.	Hoang Thach cement, Hai Duong Province	2,300.
Do.	do.	Tam Diep cement, Ninh Binh Province	1,400.
Do.	Vietnam Industrial Construction Corp. (VINAINCON)	Quang Son cement factory, in Quang Son Commune, Dong Hy District, Thai Nguyen Province	1,500.
Chromite, gross weight	Thai Nguyen Nonferrous Metal Co. [wholly owned subsidiary of state-owned Vietnam National Minerals Corp. (VIMICO)]	Nui Nua, Thanh Hoa Province	10.
Coal, anthracite	Vietnam National Coal Corp. (VINACOAL) (100% state owned)	Cam Pha, Cao Son, Coc Sau, Vang Danh, Dong Trieu, Ha Lam, Ha Tu, Hong Gai, Khe Cham, Mao Khe, Mong Duong, Deo Nai, Cua Ong, Uong Bi in Quang Ninh Province	42,000.
Copper:			
Concentrate, Cu content	Lao Cai Copper Complex [wholly owned subsidiary of Vietnam National Minerals Corp. (VIMICO)]	Sin Quyen, Lao Cai Province	11.
Ore	Sin Quyen Copper Co. [a subsidiary of Vietnam National Coal-Mineral Industries Holding Corp. Ltd. (Vinacomin)]	Sin Quyen Mine, Bat Xat District, Lao Cai Province	1,200.

See footnotes at end of table.

TABLE 2—Continued
VIETNAM: STRUCTURE OF THE MINERAL INDUSTRY IN 2012

(Thousand metric tons unless otherwise specified)

Commodity	Major operating companies and major equity owners	Location of main facilities	Annual capacity
Copper—Continued			
Refined	Tang Loong Lao Cai Copper Smelting Enterprise [wholly owned subsidiary of Vietnam National Coal-Mineral Industries Holding Corp. Ltd. (Vinacomin)]	Tang Loong Long Commune, Bao Tang District, Lao Cai Province	10.
Fertilizer:			
Nitrogen, ammonia	Vietnam National Chemical Corp. (VNCC) (100% state owned), and Phy My Nitrogenous Fertilizer and Chemical Joint Stock Corp.	Ha Bac, northern Vietnam; Phu My, Ba Ria-Vung Tau Province	375.
Superphosphate	do.	Lam Thao, Phu Tho Province	800.
Gas, natural million cubic meters per day	VietSovPetro (a joint venture of Vietnam Oil and Gas Corp. and Zarubeznheft), and the joint venture of PetroVietnam, BP p.l.c., Oil and Natural Gas Co., and ConocoPhilips Co.	Offshore Bach Ho oilfield, Rang Dong oilfield, and Lan Tay and Lan Do gasfields	20.
Do.	Vietnam Oil and Gas Group (PetroVietnam) and operated by Cuu Long Joint Operating Co.	Su Tu Trang offshore field (Block 15.1) located in Cuu Long Basin	NA.
Gold, gold content kilograms of mine output	Bong Mieu Gold Mining Company Ltd. (Bong Mieu Holdings Ltd. [a wholly owned subsidiary of Olympus Pacific Minerals Inc.], 80%; Mineral Development Co., 10%; Quang Nam Mineral Joint Stock Co., 10%)	Ho Gan open pit and Nui Kem underground mines, Quang Nam Province	400.
Do.	Besra Gold Inc.	Bai Dat and Bai Go deposit within the Phuoc Son gold property, Quang Nam Province	NA.
Iron ore, gross weight	Thai Nguyen Iron and Steel Corp. [wholly owned subsidiary of Vietnam National Steel Corp. (VNSTEEL)]	Trai Cau and Tein Bo in Thai Nguyen Province; Thach Khe in Ha Tinh Province	850.
Petroleum, crude thousand 42-gallon barrels per day	VietSovPetro (a joint venture of Vietnam Oil and Gas Corp. and Zarubeznheft)	Offshore Bach Ho, Rong, Rang Dong, Ruby, Bunga Kekwa, Dai Hung, and Su Tu Trang oilfields	320.
Do. thousand 42-gallon barrels	Vietnam Oil and Gas Group (PetroVietnam), 50%; ConocoPhillips Co., 23.25%; Korea National Oil Corp., 14.25%; SK Innovation, 9%; Geopetrol SA, 3.5%. Operated by Binh Son Refining and Petrochemical Co.	Dung Quat refinery, in Quang Ngai Province	47,600.
Phosphate rock, gross weight	Vietnam Apatite Limited Co. [Vietnam National Chemical Corp. (VNCC), 100%]	Cam Duong and Tang Loong, Lao Cai Province	1,250.
Rare earths	Lai Chau-Vietnam National Minerals Corp. (VIMICO) Rare Earth Joint Stock Co. and the Japanese Dong Pao Rare Earth Development Co.	Dong Pao Rare Earth Mine, located in Tam Duong District, Lai Chau Province	NA.
Salt	Vietnam National Salt Corp.	Nam Dinh, Nghe An, and Ha Tinh Provinces	12,000.
Steel:			
Crude	Vietnam National Steel Corp. (VNSTEEL)	Cai Lan, Thai Nguyen Province, and Phu My, Ba Ria-Vung Tau Province	2,000.
Products	Shengli (Vietnam) Special Steel Co. Ltd., established by Shengli Group Corp., and Guangdong Metals and Minerals Import & Export Corp.	Cau Nghin Industry billets plant, in Quynh Phu, Thai Binh Province	500.
Do.	do.	Bar & wire rod plant in Quynh Phu, Thai Binh Province	600.
Rolled	Lotus Group	Cold-rolled steel plant in Phu My Industrial Park in Ba Ria-Vung Tau Province	1,000.

See footnotes at end of table.

TABLE 2—Continued
VIETNAM: STRUCTURE OF THE MINERAL INDUSTRY IN 2012

(Thousand metric tons unless otherwise specified)

Commodity	Major operating companies and major equity owners	Location of main facilities	Annual capacity
Steel—Continued			
Rolled—Continued	POSCO-Vietnam, 100% owned by POSCO Group	POSCO Special Steel in Phu My Industrial Park in Ba Ria-Vung Tau Province	700 cold-rolled steel; 3,000 hot-rolled steel.
Do.	Vietnam Shipbuilding Industry Group (VINASHIN)	Cai Lan steel plate hot-rolling plant, Ha Long City, Quang Ninh Province	1,000.
Do.	Viet Steel Corp.	Bar mill in Ba Ria-Vung Tau Province	450.
Tin:			
Concentrate, Sn content	Cao Bang Nonferrous Metal Co. and Nghe Tinh Nonferrous Metal Co. [wholly owned subsidiaries of state-owned Vietnam National Minerals Corp. (VIMICO)]	Pia Oac, Cao Bang Province; Quy Hop, Nghe An Province; and Tam Dao, Tuyen Quang Province	4.
Refined	Thai Nguyen Nonferrous Metals Co.	Thai Nguyen city, Thai Nguyen Province	2.
Titanium, ilmenite	Bimal Minerals Co. Ltd. (Malaysia Mining Corp. and Syarikat Pendorong Sdn. Bhd., 60%, and Binh Dinh Minerals Co., 40%)	Cat Khanh Mine, Qui Nhon, Binh Dinh Province	70.
Do.	Ha Tinh Minerals and Trading Co.	Cam Hoa, Ky Annh-Cam, Xuyen, Ky Khan, and Ky Ninh, Ha Tinh Province	450.
Do.	Mineral Development Co. No. 4 and No. 5 [wholly owned subsidiaries of Vietnam National Minerals Corp. (VIMICO)]	Vinh City, Nghe An Province; Tuy Hoa, Dong Xuan in Phu Yen Province; and Quang Ngan, Vinh My in Thua Thien-Hue Province	50.
Tungsten, concentrates	Vietnam Youngsun Tungsten Industry Co.	Thienke tungsten mine in Tuyen Quang Province	2.
Do.	Do.	Philieng tungsten mine in Lam Dong Province	1.
Tungsten, ferrotungsten, W content	Do.	Quang Ninh plant in Halong, Quang Ninh Province	3.
Zinc:			
Concentrate, Zn content	Thai Nguyen Nonferrous Metal Co. [wholly owned subsidiary of state-owned Vietnam National Minerals Corp. (VIMICO)]	Cho Dien, Backan Province	50.
Refined	Ta Pan Zinc-Lead Plant (a Chinese private firm, 70.2%, and Ha Giang Mineral Exploiting and Engineering Co., 29.8%)	Lung Vay, Bac Me District, Ha Giang Province	6.
Do.	Thai Nguyen Zinc Refinery [wholly owned subsidiary of state-owned Vietnam National Minerals Corp. (VIMICO)]	Thai Nguyen City, Thai Nguyen Province	10.

Do., do. Ditto. NA Not available.